Springer INdAM Series

Volume 2

For further volumes:
www.springer.com/series/10283

Rolando Magnanini · Shigeru Sakaguchi ·
Angelo Alvino
Editors

Geometric Properties for Parabolic and Elliptic PDE's

Springer

Editors
Rolando Magnanini
Dipartimento di Matematica "U. Dini"
Università di Firenze
Firenze, Italy

Angelo Alvino
Dipartimento di Matematica e Applicazioni
"R. Caccioppoli"
Università di Napoli "Federico II"
Napoli, Italy

Shigeru Sakaguchi
Graduate School of Information Sciences
Division of Mathematics
Tohoku University
Sendai, Japan

ISSN 2281-518X ISSN 2281-5198 (electronic)
Springer INdAM Series
ISBN 978-88-470-2840-1 ISBN 978-88-470-2841-8 (eBook)
DOI 10.1007/978-88-470-2841-8
Springer Milan Heidelberg New York Dordrecht London

Library of Congress Control Number: 2012948272

Printed on acid-free paper

Springer is part of Springer Science+Business Media (www.springer.com)

Preface

The study of geometric properties of partial differential equations has always attracted the interest of researchers and is now a broad and well-established research area, with contributions that often come from experts from disparate areas of mathematics, such as differential and convex geometry, functional analysis, calculus of variations, mathematical physics, to name a few.

The interplay between partial differential equations and geometry has two main aspects: on the one hand, the former is classically a powerful tool for the investigation of important problems coming from differential geometry and, on the other hand, the latter gives useful and often decisive insights in the study of PDE's. Now that basic questions about PDE's, such as existence, uniqueness, stability and regularity of solutions for initial/boundary value problems, have been fairly understood, research on topological and/or geometric properties of their solutions have become more vigorous.

Research on geometric aspects for parabolic and elliptic PDE's provides a vast variety of possibilities. Issues currently and actively studied comprehend among others: positivity of solutions; critical points: their structure, possible occurrence and evolution; spike-shaped solutions; symmetry and non-symmetry for ground states and overdetermined boundary value problems; stability of symmetric configurations; convexity, quasi-convexity or starshape of level sets; estimates on geometrically or physically relevant quantities such as surface area and curvature of level surfaces or torsional creep, eigenvalues and eigenfunctions; impact of curvature of the domain on the relevant solutions and their possible behavior for large or short times, and so on. Similarly wide is the assortment of mathematical tools and techniques, analytic and geometric, employed to analyze such issues: functional inequalities such as isoperimetric, Hardy or Brunn-Minkowski inequalities; Pohozaev-type identities; maximum principles; Harnack inequalities; asymptotics for solutions; moving-planes or sliding methods; Bernstein and Liouville-type theorems, viscosity-solutions techniques, et cetera.

This volume aims to promote scientific collaboration in this very active area of research, by presenting recent results and informative surveys and by exploring new trends and techniques. It contains original papers and a few survey articles.

As it appears from the table of contents, apart for one or two exceptions, all authors are either Italian or Japanese. The Italian and Japanese mathematical schools have a long tradition of research in PDE's and count various research groups active and steadily collaborating in the study of geometric properties of their solutions. For this reason, the successful idea of E. Yanagida and K. Ishige, at the time (2008) in Sendai at Tohoku University, to have a (regularly meeting) joint conference on these topics was enthusiastically welcome by the first Editor of this book. Consequently, a first workshop met in Sendai in 2009. Contributors to this volume are some of the participants to the *Second Italian-Japanese Workshop on Geometric Properties for Parabolic and Elliptic PDE's* that met in Cortona (Italy) on June 20–24, 2011, that was kindly sponsored by the Istituto Nazionale di Alta Matematica "F. Severi" (INdAM) and, besides the Editors of this book, was organized by A. Cianchi (Università di Firenze), F. Gazzola (Politecnico di Milano), K. Ishige (Tohoku University) and E. Yanagida (Tokyo Institute of Technology). This meeting was a great occasion to blend common and different experiences in the field both at senior and junior level.

The Editors wish to thank INdAM and its President Vincenzo Ancona who made possible both the Cortona workshop and the publication of this book of articles.

The meeting took place only a few months after the catastrophic earthquake and tsunami that hit Japan at the beginning of 2011, particularly in the Sendai and Fukushima area. The option of a cancellation was seriously taken into account by the organizers. Thanks to the serene courage of the Japanese part, the conference finally took place and was hailed as a promising sign in the way to normality.

This book is certainly the best confirmation of that sign.

Firenze, Italy — Rolando Magnanini
Sendai, Japan — Shigeru Sakaguchi
Napoli, Italy — Angelo Alvino

Contents

Stability and Instability of Group Invariant Asymptotic Profiles for Fast Diffusion Equations

Goro Akagi

Abstract This paper is concerned with group invariant solutions for fast diffusion equations in symmetric domains. First, it is proved that the group invariance of weak solutions is inherited from initial data. After briefly reviewing previous results on asymptotic profiles of vanishing solutions and their stability, the notions of stability and instability of group invariant profiles are introduced under a similarly invariant class of perturbations, and moreover, some stability criteria are exhibited and applied to symmetric domain (e.g., annulus) cases.

Keywords Fast diffusion equation · Asymptotic profile · Group invariance · Stability

1 Introduction

In this paper, we are concerned with the Cauchy-Dirichlet problem for the fast diffusion equation,

$$\partial_t\big(|u|^{m-2}u\big) = \Delta u \quad \text{in } \Omega \times (0,\infty), \tag{1}$$

$$u = 0 \quad \text{on } \partial\Omega \times (0,\infty), \tag{2}$$

$$u(\cdot,0) = u_0 \quad \text{in } \Omega, \tag{3}$$

where Ω is a bounded domain of $\mathbb{R}^N$ with smooth boundary $\partial\Omega$, $m > 2$, $\partial_t = \partial/\partial t$, $u_0 \in H^1_0(\Omega)$ and Δ stands for the N-dimensional Laplacian. By putting $w = |u|^{m-2}u$, Eq. (1) can be rewritten in a usual form of *fast diffusion equation*,

$$\partial_t w = \Delta\big(|w|^{r-2}w\big) \quad \text{in } \Omega \times (0,\infty) \tag{4}$$

with the exponent $r = m/(m-1) < 2$. Fast diffusion equations arise in the studies of plasma physics (see [6]), kinetic theory of gases, solid state physics and so on.

G. Akagi (✉)
Graduate School of System Informatics, Kobe University, 1-1 Rokkodai-cho, Nada-ku, Kobe 657-8501, Japan
e-mail: akagi@port.kobe-u.ac.jp

R. Magnanini et al. (eds.), *Geometric Properties for Parabolic and Elliptic PDE's*, Springer INdAM Series 2, DOI 10.1007/978-88-470-2841-8_1,

It is well known that every solution $u = u(x,t)$ of (1)–(3) vanishes at a finite time $t_* = t_*(u_0) \geq 0$ such that

$$c(t_* - t)_+^{1/(m-2)} \leq \|u(\cdot,t)\|_{H_0^1(\Omega)} \leq c^{-1}(t_* - t)_+^{1/(m-2)} \quad \text{for all } t \geq 0$$

with some constant $c > 0$, provided that $2 < m \leq 2^* := 2N/(N-2)_+$ (see [5, 7, 18]). Moreover, for the case that $2 < m < 2^*$, Berryman and Holland [7] studied asymptotic profiles

$$\phi(x) := \lim_{t \nearrow t_*} (t_* - t)^{-1/(m-2)} u(x,t)$$

of solutions $u = u(x,t)$ for (1)–(3).

Now, let us address ourselves to the stability and instability of asymptotic profiles. Namely, our question is the following: For any initial data $u_0 \in H_0^1(\Omega)$ sufficiently close to an asymptotic profile ϕ, does the asymptotic profile of the unique solution $u = u(x,t)$ for (1)–(3) also coincide with ϕ or not? In [7] and [15], the stability of the unique positive asymptotic profile is discussed for nonnegative initial data in some special cases (e.g., $N = 1$). Recently, in [8], further detailed behaviors of nonnegative solutions near the extinction time are investigated. Moreover, in [3], the notions of stability and instability of asymptotic profiles are precisely defined for (possibly) sign-changing initial data, and furthermore, some criteria for the stability and instability are presented under $2 < m < 2^*$. Furthermore, they are applied to several concrete cases of the domain Ω (e.g., ball domains) and the exponent m. However, there are still cases (e.g., annular domain case) which do not fall within the scope of the criteria.

In this paper, we treat symmetric domain cases and discuss the stability and instability of group invariant asymptotic profiles. More precisely, for a subgroup G of $O(N)$ and a G-invariant domain Ω, we only deal with G-invariant (e.g., radial) initial data and solutions of (1)–(3). Furthermore, the stability and instability of profiles are also discussed only under G-invariant perturbations.

In the next section, we prove the G-invariance of weak solutions for parabolic problems such as (1)–(3) with G-invariant initial data and domains. This issue would be obvious in classical formulations, where one can directly calculate the change of variables. However, one should pay careful attention in weak formulations of parabolic problems such as nonlinear diffusion equations because of the lack of pointwise representation of the time-derivative of solution in a dual space $H^{-1}(\Omega) = (H_0^1(\Omega))^*$. In Sect. 3, we first briefly review previous studies, particularly [3], on asymptotic profiles of vanishing solutions for fast diffusion equations and their stability. We next define the notions of stability and instability of G-invariant asymptotic profiles under G-invariant perturbations, and then, stability criteria will be presented for them under $2 < m < 2^*$. Finally, we discuss applications of the stability criteria to some cases (e.g., the annular domain case) which do not fall within the scope of the criteria presented in [3].

Notation Let $H_0^1(\Omega)$ be the closure of $C_0^\infty(\Omega)$ in the usual Sobolev space $H^1(\Omega) = W^{1,2}(\Omega)$. Let us denote by $\|\cdot\|_m$ the usual norm of $L^m(\Omega)$-space, and moreover, $\|\cdot\|_{1,2} := \|\nabla\cdot\|_2$ stands for the norm of $H_0^1(\Omega)$. For a function $u = u(x,t) : \Omega \times (0,\infty) \to \mathbb{R}$, we often write $u(t) := u(\cdot,t)$, which is a function from Ω into $\mathbb{R}$, for a fixed time $t > 0$.

2 Group Invariance of Weak Solutions for Parabolic Problems

In this section, we shall prove that the group invariance of weak solutions for parabolic problems such as (1)–(3) is inherited from initial data and domains. More precisely, let G be a subgroup of $O(N)$ and let Ω be a G-*invariant domain* of $\mathbb{R}^N$, i.e., $g(\Omega) = \Omega$ for any $g \in G$, with smooth boundary $\partial\Omega$. Here let us treat

$$\partial_t\left(|u|^{m-2}u\right) = \Delta u + \lambda|u|^{m-2}u \quad \text{in } \Omega \times (0,\infty), \tag{5}$$

$$u = 0 \quad \text{on } \partial\Omega \times (0,\infty), \tag{6}$$

$$u(\cdot,0) = u_0 \quad \text{in } \Omega \tag{7}$$

with

$$\lambda \in \mathbb{R}, \quad 1 < m < \infty \quad \text{and} \quad u_0 \in H_0^1(\Omega) \cap L^m(\Omega)$$

(as an independent interest, we also treat $1 < m < 2$ and $\lambda \in \mathbb{R}$). We shall prove that u is G-*invariant*, i.e., $u(g^{-1}x,t) = u(x,t)$ for all $g \in G$, provided that u_0 is G-invariant. This fact can be easily checked for classical solutions by directly calculating the change of variables. As for weak formulations of differential equations, one should more carefully treat this issue. There are many papers on this topic for elliptic problems. However, there seems to be very few works on weak formulations for parabolic problems (see [4]).

We start with the definition of weak solutions for (5)–(7) by setting

$$X := H_0^1(\Omega) \cap L^m(\Omega)$$

with the norm $\|\cdot\|_X := \|\cdot\|_{1,2} + \|\cdot\|_m$. Then the dual space X^* of X is equivalent to $H^{-1}(\Omega) + L^{m'}(\Omega)$. In particular, X and X^* coincide with $H_0^1(\Omega)$ and $H^{-1}(\Omega)$, respectively, provided that $1 < m \le 2^*$.

Definition 1 (Weak solution of (5)–(7)) A function $u : \Omega \times (0,\infty) \to \mathbb{R}$ is said to be a (weak) solution of (5)–(7), if the following conditions are all satisfied:

- $u \in C([0,\infty); X)$ and $|u|^{m-2}u \in C^1([0,\infty); X^*)$, where X^* is the dual space of X.
- For all $t \in (0,\infty)$ and $\psi \in C_0^\infty(\Omega)$,

$$\left\langle \frac{\mathrm{d}}{\mathrm{d}t}\left(|u|^{m-2}u\right)(t), \psi \right\rangle_X + \int_\Omega \nabla u(x,t)\cdot\nabla\psi(x)\,\mathrm{d}x = \lambda \int_\Omega \left(|u|^{m-2}u\right)(x,t)\psi(x)\,\mathrm{d}x,$$

  where $\langle\cdot,\cdot\rangle_X$ denotes a duality pairing between X and its dual space X^*.
- $u(\cdot,t) \to u_0$ strongly in X as $t \to +0$.

Hence the weak formulation of (5)–(7) stated above can be written as an evolution equation for $u(t) := u(\cdot, t)$ in X^*,

$$\frac{\mathrm{d}}{\mathrm{d}t}\big(|u|^{m-2}u\big)(t) - \Delta u(t) = \lambda\big(|u|^{m-2}u\big)(t) \quad \text{in } X^*,\ t > 0, \qquad u(0) = u_0.$$

Then for any $u_0 \in X$, the problem (5)–(7) admits a unique solution (see, e.g., [9, 20, 21] and also [2]).

Remark 1 In case $m > 2$, where (5) is the fast diffusion equation, every sign-definite solution becomes a classical solution (see [12]). However, sign-changing solutions should be treated in the weak formulation, because the transformed equation from (5) in a similar way to (4) has a singularity when $u(x,t) = 0$. In case $m < 2$, where (5) is the porous medium equation, the weak formulation is essentially required for sign-definite solutions as well as for sign-changing solutions because of the lack of regularity of solution.

Let G be a subgroup of $O(N)$ whose elements leave Ω invariant. For $g \in G$ and a function $u : \Omega \to \mathbb{R}$, we define a function $gu : \Omega \to \mathbb{R}$ by

$$(gu)(x) := u\big(g^{-1}x\big) \quad \text{for } x \in \Omega.$$

Then X and X^* become Banach G-spaces. More precisely, we have a representation π_X of G over X given by

$$\pi_X(g)u := gu \quad \text{for } u \in X \text{ and } g \in G,$$

where $\pi_X(g)$ is a bounded linear operator in X. Moreover, define a representation π_{X^*} of G over X^* by

$$\big\langle \pi_{X^*}(g)f, u\big\rangle_X := \big\langle f, \pi_X\big(g^{-1}\big)u\big\rangle_X \quad \text{for } u \in X,\ f \in X^* \text{ and } g \in G.$$

The following facts are well known in the variational analysis of elliptic problems. For the convenience of the reader, we briefly give a proof.

Proposition 1 (G-equivariance of $-\Delta u$ and $|u|^{m-2}u$) *Define operators $A, B : X \to X^*$ by*

$$A(u) := -\Delta u, \qquad B(u) := |u|^{m-2}u \quad \textit{for } u \in X.$$

Then A and B are G-equivariant, i.e.,

$$\pi_{X^*}(g)\big(A(u)\big) = A\big(\pi_X(g)u\big), \qquad \pi_{X^*}(g)\big(B(u)\big) = B\big(\pi_X(g)u\big)$$

for all $u \in X$ and $g \in G$.

Proof It is well known that $A = \phi_A'$ and $B = \phi_B'$ with $\phi_A, \phi_B : X \to [0, \infty)$ given by

$$\phi_A(u) := \frac{1}{2}\int_\Omega \big|\nabla u(x)\big|^2\,\mathrm{d}x, \qquad \phi_B(u) := \frac{1}{m}\int_\Omega \big|u(x)\big|^m\,\mathrm{d}x \quad \text{for } u \in X.$$

Moreover, ϕ_A and ϕ_B are G-invariant, i.e., $\phi_A(gu) = \phi_A(u)$ for all $u \in X$ and $g \in G$. Hence ϕ'_A and ϕ'_B are G-equivariant from the general fact that the derivative of G-invariant functional is G-equivariant. Indeed, for an G-invariant Gâteaux differentiable functional $\phi : X \to \mathbb{R}$, the Gâteaux differential ϕ' of ϕ satisfies

$$\begin{aligned}
\langle \phi'(\pi_X(g)u), e\rangle_X &= \lim_{h\to 0} \frac{\phi(\pi_X(g)u + he) - \phi(\pi_X(g)u)}{h} \\
&= \lim_{h\to 0} \frac{\phi(u + h\pi_X(g^{-1})e) - \phi(u)}{h} \\
&= \langle \phi'(u), \pi_X(g^{-1})e\rangle_X \\
&= \langle \pi_{X^*}(g)\phi'(u), e\rangle_X \quad \text{for any } e, u \in X \text{ and } g \in G,
\end{aligned}$$

which implies $\phi'(\pi_X(g)u) = \pi_{X^*}(g)\phi'(u)$ for all $u \in X$ and $g \in G$. One can obtain a similar conclusion for Fréchet differentials as well. □

A tiny novelty of this section is the following proposition, where the G-equivariance of the time-differential operator is shown in a space of vector functions with values in X^*. A similar attempt has been done for a Gel'fand triplet setting in [4], where a parabolic version of the so-called "principle of symmetric criticality" is established.

To this end, we work on a large space,

$$\mathscr{H} := L^2(0, T; X^*).$$

Then the representation $\pi_{\mathscr{H}}$ of G over $\mathscr{H}$ is given by

$$(\pi_{\mathscr{H}}(g)u)(t) = \pi_{X^*}(g)u(t) \quad \text{for } t \in (0, T),\ u \in \mathscr{H} \text{ and } g \in G.$$

Moreover, we define the time-differential operator,

$$\frac{\mathrm{d}}{\mathrm{d}t} : \mathscr{H} \to \mathscr{H}$$

with the domain

$$D(\mathrm{d}/\mathrm{d}t) := \{u \in \mathscr{H}\colon \mathrm{d}u/\mathrm{d}t \in \mathscr{H}\} = W^{1,2}(0, T; X^*).$$

Proposition 2 (G-equivariance of $\mathrm{d}/\mathrm{d}t$) *The differential operator* $\mathrm{d}/\mathrm{d}t$ *is* G-*equivariant, i.e.,*

$$\pi_{\mathscr{H}}(g)\frac{\mathrm{d}u}{\mathrm{d}t} = \frac{\mathrm{d}}{\mathrm{d}t}(\pi_{\mathscr{H}}(g)u) \quad \textit{for all } u \in D(\mathrm{d}/\mathrm{d}t) \textit{ and } g \in G,$$

which is equivalently rewritten as

$$\pi_{X^*}(g)\frac{\mathrm{d}u}{\mathrm{d}t}(t) = \frac{\mathrm{d}}{\mathrm{d}t}(\pi_{X^*}(g)u(t)) \quad \textit{for all } u \in D(\mathrm{d}/\mathrm{d}t) \textit{ and } g \in G$$

for a.e. $t \in (0, T)$.

Proof For $u \in D(\mathrm{d}/\mathrm{d}t)$, $g \in G$, $\eta \in C_0^\infty(0,T)$ and $e \in X$, it follows that

$$\begin{aligned}
\left\langle \int_0^T \left(\pi_{\mathscr{H}}(g)\frac{\mathrm{d}u}{\mathrm{d}t}\right)(t)\eta(t)\,\mathrm{d}t, e\right\rangle_X &= \int_0^T \left\langle \pi_{X^*}(g)\frac{\mathrm{d}u}{\mathrm{d}t}(t), e\right\rangle_X \eta(t)\,\mathrm{d}t \\
&= \int_0^T \left\langle \frac{\mathrm{d}u}{\mathrm{d}t}(t), \pi_X\left(g^{-1}\right)e\right\rangle_X \eta(t)\,\mathrm{d}t \\
&= \left\langle \int_0^T \frac{\mathrm{d}u}{\mathrm{d}t}(t)\eta(t)\,\mathrm{d}t, \pi_X\left(g^{-1}\right)e\right\rangle_X \\
&= \left\langle -\int_0^T u(t)\frac{\mathrm{d}\eta}{\mathrm{d}t}(t)\,\mathrm{d}t, \pi_X\left(g^{-1}\right)e\right\rangle_X \\
&= -\int_0^T \left\langle u(t), \pi_X\left(g^{-1}\right)e\right\rangle_X \frac{\mathrm{d}\eta}{\mathrm{d}t}(t)\,\mathrm{d}t \\
&= -\int_0^T \left\langle \pi_{X^*}(g)u(t), e\right\rangle_X \frac{\mathrm{d}\eta}{\mathrm{d}t}(t)\,\mathrm{d}t \\
&= \left\langle -\int_0^T \left(\pi_{\mathscr{H}}(g)u\right)(t)\frac{\mathrm{d}\eta}{\mathrm{d}t}(t)\,\mathrm{d}t, e\right\rangle_X.
\end{aligned}$$

Thus we have

$$\int_0^T \left(\pi_{\mathscr{H}}(g)\frac{\mathrm{d}u}{\mathrm{d}t}\right)(t)\eta(t)\,\mathrm{d}t = -\int_0^T \left(\pi_{\mathscr{H}}(g)u\right)(t)\frac{\mathrm{d}\eta}{\mathrm{d}t}(t)\,\mathrm{d}t \quad \text{in } X^*,$$

which implies

$$\pi_{\mathscr{H}}(g)\frac{\mathrm{d}u}{\mathrm{d}t} = \frac{\mathrm{d}}{\mathrm{d}t}\left(\pi_{\mathscr{H}}(g)u\right)$$

in the sense of distribution. Hence $\mathrm{d}/\mathrm{d}t$ is G-equivariant in $\mathscr{H}$. □

Remark 2 In [4], the G-equivariance of the time-differential operator is also proved in a different setting. More precisely, let V and V^* be a reflexive Banach space and its dual space, respectively, and suppose that there exists a Hilbert space H satisfying the following Gel'fand triplet:

$$V \hookrightarrow H \equiv H^* \hookrightarrow V^*,$$

where H^* stands for the dual space of H, with densely defined continuous canonical injections. Then the G-equivariance of the time-differential operator $\mathrm{d}/\mathrm{d}t$ defined from $\mathscr{V} := L^2(0,T;V)$ to $\mathscr{V}^* := L^2(0,T;V^*)$ is proved.

Combining all these facts, we are now in position to prove the following theorem.

Theorem 1 (G-invariance of weak solutions) *Let G be a subgroup of $O(N)$ and let Ω be a G-invariant bounded domain of $\mathbb{R}^N$ with smooth boundary. Let $u = u(x,t)$ be a weak solution of* (5), (6). *Then so is $gu := u(g^{-1}x, t)$ for any $g \in G$.*

In addition, if the initial data u_0 is G-invariant, then so is the unique weak solution of (5)–(7).

Proof Let $u = u(x,t)$ be a weak solution of (5), (6) and put $w(x,t) = u(g^{-1}x,t)$. Then, $w(t) = \pi_X(g)u(t)$. Then for each $\phi \in X$, it follows that

$$\left\langle \frac{\mathrm{d}}{\mathrm{d}t} B\big(u(t)\big) + A\big(u(t)\big) - \lambda B\big(u(t)\big), \pi_X\big(g^{-1}\big)\phi \right\rangle_X = 0.$$

Then by Propositions 1 and 2, we have

$$\begin{aligned}
&\left\langle \frac{\mathrm{d}}{\mathrm{d}t} B\big(u(t)\big) + A\big(u(t)\big) - \lambda B\big(u(t)\big), \pi_X\big(g^{-1}\big)\phi \right\rangle_X \\
&= \left\langle \pi_{X^*}(g)\left(\frac{\mathrm{d}}{\mathrm{d}t} B\big(u(t)\big)\right) + \pi_{X^*}(g)A\big(u(t)\big) - \lambda\pi_{X^*}(g)B\big(u(t)\big), \phi \right\rangle_X \\
&= \left\langle \frac{\mathrm{d}}{\mathrm{d}t}\big(\pi_{X^*}(g)B\big(u(t)\big)\big) + A\big(w(t)\big) - \lambda B\big(w(t)\big), \phi \right\rangle_X \\
&= \left\langle \frac{\mathrm{d}}{\mathrm{d}t} B\big(w(t)\big) + A\big(w(t)\big) - \lambda B\big(w(t)\big), \phi \right\rangle_X .
\end{aligned}$$

Therefore w also solves (5), (6).

In addition, if the initial data u_0 is G-invariant, all the solutions $w(x,t) = u(g^{-1}x,t)$ for any $g \in G$ solve (5)–(7) with the same data u_0. Therefore from the uniqueness of weak solution, w coincides with u for all $g \in G$. Consequently, the unique solution u is G-invariant. □

3 Stability Analysis of Group Invariant Asymptotic Profiles

This section is devoted to a stability analysis of asymptotic profiles invariant under a symmetry group for vanishing solutions of (1)–(3). Throughout this section, we assume that

$$2 < m < 2^* := \begin{cases} 2N/(N-2) & \text{if } N \geq 3, \\ \infty & \text{if } N = 1,2 \end{cases} \quad \text{and} \quad u_0 \in H_0^1(\Omega) \tag{8}$$

(then $H_0^1(\Omega)$ is compactly embedded in $L^m(\Omega)$). In Sect. 3.1, we overview preliminary facts on the stability analysis of asymptotic profiles for fast diffusion equations. In Sect. 3.2, we define the notions of stability and instability of asymptotic profiles under group invariant perturbations and present some stability criteria. Moreover, these stability criteria will be proved in the following two subsections. The contents in these subsections would be similar to those in [3], even though the setting under consideration here is not covered by [3]. However, results in Sect. 3.5 to be obtained by applying these criteria would be noteworthy, because they enable us to discuss the asymptotic stability of radial profiles under radial perturbations as well as to investigate further information on the instability of some symmetric sign-changing profiles.

3.1 Asymptotic Profiles for Fast Diffusion Equations

In this subsection, we briefly review previous results on asymptotic profiles for fast diffusion equations. The finite-time extinction of solutions for fast diffusion equations is first proved by Sabinina [18] (for $N=1$), and then, generalized by Bénilan and Crandall [5]. We denote by $t_*(u_0)$ the *extinction time* of the unique solution u of (1)–(3) for the initial data u_0. Berryman and Holland [7] obtained an optimal rate of the finite-time extinction for each solution u of (1)–(3),

$$c(t_* - t)_+^{1/(m-2)} \le \|u(t)\|_{1,2} \le c^{-1}(t_* - t)_+^{1/(m-2)}$$

with the extinction time t_* of u and a positive constant $c > 0$. Moreover, they showed the existence of *asymptotic profiles*

$$\phi(x) := \lim_{t_n \nearrow t_*} (t_* - t_n)_+^{-1/(m-2)} u(x, t_n) \quad \text{in } H_0^1(\Omega)$$

with some sequence $t_n \nearrow t_*$ for positive classical solutions.

In order to characterize ϕ, let us apply the following transformation:

$$v(x,s) := (t_* - t)^{-1/(m-2)} u(x,t) \quad \text{and} \quad s := \log\big(t_*/(t_* - t)\big) \ge 0. \tag{9}$$

Then s tends to infinity as $t \nearrow t_*$. Moreover, the asymptotic profile $\phi = \phi(x)$ of $u = u(x,t)$ is reformulated as

$$\phi(x) := \lim_{s_n \nearrow \infty} v(x, s_n) \quad \text{in } H_0^1(\Omega) \text{ with } s_n := \log\big(t_*/(t_* - t_n)\big) \to \infty.$$

Furthermore, the Cauchy-Dirichlet problem (1)–(3) for $u = u(x,t)$ is rewritten as the following rescaled problem:

$$\partial_s\big(|v|^{m-2}v\big) = \Delta v + \lambda_m |v|^{m-2} v \quad \text{in } \Omega \times (0,\infty), \tag{10}$$

$$v = 0 \quad \text{on } \partial\Omega \times (0,\infty), \tag{11}$$

$$v(\cdot, 0) = v_0 \quad \text{in } \Omega, \tag{12}$$

where the initial data v_0 and the constant λ_m are given by

$$v_0 = t_*(u_0)^{-1/(m-2)} u_0 \quad \text{and} \quad \lambda_m = (m-1)/(m-2) > 0. \tag{13}$$

Then (10)–(12) can be regarded as a generalized gradient system,

$$\frac{\mathrm{d}}{\mathrm{d}s} |v|^{m-2} v(s) = -J'\big(v(s)\big) \quad \text{for } s > 0, \tag{14}$$

where $J : H_0^1(\Omega) \to \mathbb{R}$ is given by

$$J(w) := \frac{1}{2} \int_\Omega |\nabla w(x)|^2 \,\mathrm{d}x - \frac{\lambda_m}{m} \int_\Omega |w(x)|^m \,\mathrm{d}x \quad \text{for } w \in H_0^1(\Omega),$$

and moreover, the function $s \mapsto J(v(s))$ is nonincreasing. One can prove the following theorem (see [3, 7, 8, 15, 19]):

Theorem 2 (Existence of asymptotic profiles and their characterization) *For any sequence $s_n \to \infty$, there exist a subsequence (n') of (n) and $\phi \in H_0^1(\Omega) \setminus \{0\}$ such that $v(s_{n'}) \to \phi$ strongly in $H_0^1(\Omega)$. Moreover, ϕ is a nontrivial stationary solution of* (10)–(12), *that is, ϕ solves the Dirichlet problem,*

$$-\Delta\phi = \lambda_m |\phi|^{m-2}\phi \quad \text{in } \Omega, \qquad \phi = 0 \quad \text{on } \partial\Omega. \tag{15}$$

The Dirichlet problem (15) is an Euler-Lagrange equation for J.

Remark 3

(i) If ϕ is a nontrivial solution of (15), then the function $U(x,t) = (1-t)_+^{1/(m-2)} \times \phi(x)$ solves (1)–(3) with $U(x,0) = \phi(x)$. Hence $t_*(\phi) = 1$ and the profile of $U(x,t)$ coincides with $\phi(x)$.
(ii) Hence, by Theorem 2, the set of all asymptotic profiles of solutions for (1)–(3) coincides with the set of all nontrivial solutions of (15). Obviously, they also coincide with the set of all nontrivial critical points of J. We shall denote these sets by $\mathscr{S}$.
(iii) Due to [13], the asymptotic profile is uniquely determined for each nonnegative data $u_0 \geq 0$.

In [7, 8, 15, 19], the stability of positive profiles is discussed for nonnegative solutions. However, until the work in [3], sign-changing profiles had not been treated, and moreover, the stability of positive profiles had not been discussed under a wider class of perturbations which allow sign-changing initial data.

In [3], the notions of stability and instability of asymptotic profiles of solutions for (1)–(3) were first precisely defined for possibly sign-changing solutions by introducing a set,

$$\mathscr{X} := \left\{ t_*(u_0)^{-1/(m-2)} u_0 \colon u_0 \in H_0^1(\Omega) \setminus \{0\} \right\},$$

which coincides with the level set $\{v_0 \in H_0^1(\Omega) \colon t_*(v_0) = 1\}$ of the functional $t_* : H_0^1(\Omega) \to [0, \infty)$. Here we note

Lemma 1 (Property of $\mathscr{X}$, [3]) *Let v be a solution of* (10)–(12) *for an initial data v_0.*

(i) *If $v_0 \in \mathscr{X}$, then $v(s) \in \mathscr{X}$ for all $s \geq 0$.*
(ii) *If $v_0 \in \mathscr{X}$, then for any $s_n \to \infty$, up to a subsequence, $v(s_n) \to \phi$ for some $\phi \in \mathscr{S}$ (by Theorem* 2).
(iii) *It follows that $\mathscr{S} \subset \mathscr{X}$.*

Moreover, the following criteria for the stability and instability of profiles were presented:

- Each least energy solution ϕ of (15) is (resp., asymptotically) stable in the sense of asymptotic profiles, if ϕ is isolated from the other least energy (resp., sign-definite) solutions.

- All the sign-changing solutions ψ are not asymptotically stable profiles. Moreover, ψ is unstable, if it is isolated from all the other profiles with lower energies.

As a by-product of [3], the whole of the energy space $H_0^1(\Omega)$ of initial data is completely classified in terms of large-time behaviors of solutions for (10)–(12). In particular, the set $\mathscr{X}$ turns out to be a separatrix between stable and unstable sets (cf. see [14] for a semilinear heat equation).

The criteria stated above do not cover all the situations. Indeed, in case Ω is a thin annulus, there exists a positive radial profile ϕ_1 which may not take the least energy among $\mathscr{S}$.

In the following subsections, we introduce the notions of stability and instability of G-invariant profiles under similarly invariant perturbations for a subgroup G of $O(N)$ and slightly modify the argument of [3] to obtain stability criteria for G-invariant profiles. As a typical application of the criteria, we shall discuss the stability of the unique positive radial profile in the annulus case under $O(N)$-invariant perturbations.

3.2 *Stability and Instability of G-Invariant Profiles*

Let G be a subgroup of $O(N)$ and let Ω be a G-invariant domain of $\mathbb{R}^N$ with smooth boundary. Assume (8) and denote the space of G-invariant functions of class $H_0^1(\Omega)$ by

$$H_{0,G}^1(\Omega) := \{u \in H_0^1(\Omega) \colon\ gu = u \text{ for all } g \in G\}.$$

Each asymptotic profile lying on $H_{0,G}^1(\Omega)$ is called a *G-invariant asymptotic profile*. Let us introduce the notions of stability and instability of G-invariant asymptotic profiles of solutions for (1)–(3) under G-invariant perturbations. To this end, we first introduce the set,

$$\mathscr{X}_G := \{t_*(u_0)^{-1/(m-2)}u_0 \colon\ u_0 \in H_{0,G}^1(\Omega) \setminus \{0\}\} = \mathscr{X} \cap H_{0,G}^1(\Omega).$$

Then we define:

Definition 2 (Stability and instability of profiles under G-invariant perturbations) Let $\phi \in H_{0,G}^1(\Omega)$ be an asymptotic profile of vanishing solutions for (1)–(3).

(i) ϕ is said to be *stable under G-invariant perturbations*, if for any $\varepsilon > 0$ there exists $\delta = \delta(\varepsilon) > 0$ such that any solution v of (10), (11) satisfies

$$v(0) \in \mathscr{X}_G \cap B_{H_0^1}(\phi;\delta) \quad \Rightarrow \quad \sup_{s\in[0,\infty)} \|v(s) - \phi\|_{1,2} < \varepsilon,$$

where $B_{H_0^1}(\phi;\delta) := \{w \in H_0^1(\Omega) \colon\ \|\phi - w\|_{1,2} < \delta\}$.

(ii) ϕ is said to be *unstable under G-invariant perturbations*, if ϕ is not stable under G-invariant perturbations.

(iii) ϕ is said to be *asymptotically stable under G-invariant perturbations*, if ϕ is stable under G-invariant perturbations, and moreover, there exists $\delta_0 > 0$ such that any solution v of (10), (11) satisfies

$$v(0) \in \mathscr{X}_G \cap B_{H_0^1}(\phi; \delta_0) \quad \Rightarrow \quad \lim_{s \nearrow \infty} \left\| v(s) - \phi \right\|_{1,2} = 0.$$

Remark 4 Apparently, $\mathscr{X}_G$ is a subset of $\mathscr{X}$. Hence if an asymptotic profile ϕ is (asymptotically) stable in the sense of [3], then so is it under G-invariant perturbations. On the other hand, if ϕ is unstable or not asymptotically stable under G-invariant perturbations, then so is ϕ without restriction of perturbation.

To state our stability criteria under G-invariant perturbations, we set up notation. Let $\mathscr{S}_G = \mathscr{S} \cap H^1_{0,G}(\Omega)$, which is the set of all G-invariant nontrivial solutions. A function $\phi \in \mathscr{S}_G$ is called a *least energy G-invariant solution* if ϕ attains the infimum of J over $\mathscr{S}_G$. Then our criteria read as follows.

Theorem 3 (Stability of G-invariant profiles) *Assume* (8). *Let $\phi \in H^1_{0,G}(\Omega)$ be a least energy G-invariant solution of* (15). *Then it follows that*

(i) *ϕ is a stable profile under G-invariant perturbations, if ϕ is isolated in $H^1_0(\Omega)$ from the other least energy G-invariant solutions.*
(ii) *ϕ is an asymptotically stable profile under G-invariant perturbations, if ϕ is isolated in $H^1_0(\Omega)$ from the other sign-definite G-invariant solutions.*

Theorem 4 (Instability of G-invariant profiles) *Assume* (8). *Let $\phi \in H^1_{0,G}(\Omega)$ be a sign-changing G-invariant solution of* (15). *Then it follows that*

(i) *ϕ is not an asymptotically stable profile under G-invariant perturbations.*
(ii) *ϕ is an unstable profile under G-invariant perturbations, if ϕ is isolated in $H^1_0(\Omega)$ from any $\psi \in \mathscr{S}_G$ satisfying $J(\psi) < J(\phi)$.*

3.3 *Proof of Theorem 3*

We first prepare a couple of lemmas.

Lemma 2 (Properties of $\mathscr{X}_G$) *Let v be a solution of* (10)–(12) *for an initial data v_0.*

(i) *If $v_0 \in \mathscr{X}_G$, then $v(s) \in \mathscr{X}_G$ for all $s \geq 0$.*
(ii) *If $v_0 \in \mathscr{X}_G$, then for any $s_n \to \infty$, up to a subsequence, $v(s_n) \to \phi$ for some $\phi \in \mathscr{S}_G$.*
(iii) *It holds that $\mathscr{S}_G \subset \mathscr{X}_G$.*

Proof Combining Theorem 1 with (i) of Lemma 1, we have (i). Let $v_0 \in \mathscr{X}_G$. By Theorem 2, there exist $s_n \to \infty$ and $\phi \in \mathscr{S}$ such that $v(s_n) \to \phi$ strongly in $H^1_0(\Omega)$.

Since $v(s_n) \in H^1_{0,G}(\Omega)$ by (i) and $H^1_{0,G}(\Omega)$ is closed, ϕ is G-invariant. Thus (ii) holds. Recall $\mathscr{X}_G = \mathscr{X} \cap H^1_{0,G}(\Omega)$, $\mathscr{S}_G = \mathscr{S} \cap H^1_{0,G}(\Omega)$ and (iii) of Lemma 1 to obtain (iii). □

Lemma 3 (Weak closedness of $\mathscr{X}_G$) *If $u_n \in \mathscr{X}_G$ and $u_n \to u$ weakly in $H^1_0(\Omega)$, then $u \in \mathscr{X}_G$.*

Proof The (sequentially) weak closedness of $\mathscr{X}$ is proved in [3]. Moreover, $H^1_{0,G}(\Omega)$ is also weakly closed, and hence, so is $\mathscr{X}_G = \mathscr{X} \cap H^1_{0,G}(\Omega)$. □

Lemma 4 (Variational feature of $\mathscr{X}_G$) *Let $d_1 = \inf_{\mathscr{S}_G} J$. Then*

$$\mathscr{X}_G \subset [d_1 \le J] := \big\{v_0 \in H^1_0(\Omega)\colon\ d_1 \le J(v_0)\big\}.$$

Moreover, if $v_0 \in \mathscr{X}_G$ and $J(v_0) = d_1$, then $J'(v_0) = 0$.

Proof Let $v_0 \in \mathscr{X}_G$ and let $v(s)$ be a solution of (10)–(12) with $v(0) = v_0$. Then by (ii) of Lemma 2 there exist $s_n \to \infty$ and $\phi \in \mathscr{S}_G$ such that $v(s_n) \to \phi$ strongly in $H^1_0(\Omega)$. From the nonincrease of $J(v(\cdot))$, we deduce that

$$J(v_0) \ge J\big(v(s)\big) \ge J(\phi) \ge d_1 = \inf_{\mathscr{X}_G} J.$$

Hence $d_1 \le J(v_0)$.

If $v_0 \in \mathscr{X}_G$ and $J(v_0) = d_1$, then $J(v_0) = \min_{\mathscr{X}_G} J$. Hence $v(s) \equiv v_0$ by (14). □

Denote by $\mathscr{L}\mathscr{E}\mathscr{S}_G$ the set of all least energy G-invariant solutions of (15). Let us assume that

$$B_{H^1_0}(\phi; r) \cap \mathscr{L}\mathscr{E}\mathscr{S}_G = \{\phi\} \tag{16}$$

with some $r > 0$. Here we write $B_{H^1_0}(\phi; r) := \{w \in H^1_0(\Omega)\colon\ \|w - \phi\|_{1,2} < r\}$.

Claim *For any $\varepsilon \in (0, r)$, it holds that*

$$c := \inf\big\{J(v)\colon\ v \in \mathscr{X}_G,\ \|v - \phi\|_{1,2} = \varepsilon\big\} > d_1.$$

Proof Assume on the contrary that $c = d_1$, i.e., there exists $v_n \in \mathscr{X}_G$ such that

$$\|v_n - \phi\|_{1,2} = \varepsilon \quad \text{and} \quad J(v_n) \to d_1.$$

Since $m < 2^*$, it entails that, up to a subsequence,

$$v_n \to v_\infty \quad \text{weakly in } H^1_0(\Omega) \text{ and strongly in } L^m(\Omega).$$

By Lemmas 3 and 4, we obtain

$$v_\infty \in \mathscr{X}_G, \quad \text{and hence,} \quad d_1 \le J(v_\infty).$$

Therefore it follows that

$$\begin{aligned}\frac{1}{2}\|v_n\|_{1,2}^2 &= J(v_n) + \frac{\lambda_m}{m}\|v_n\|_m^m \\ &\to d_1 + \frac{\lambda_m}{m}\|v_\infty\|_m^m \le J(v_\infty) + \frac{\lambda_m}{m}\|v_\infty\|_m^m = \frac{1}{2}\|v_\infty\|_{1,2}^2.\end{aligned}$$

By using the weak lower semicontinuity,

$$\liminf_{n\to\infty}\|v_n\|_{1,2} \ge \|v_\infty\|_{1,2},$$

and the uniform convexity of $\|\cdot\|_{1,2}$, we deduce that $v_n \to v_\infty$ strongly in $H_0^1(\Omega)$. Hence $\|v_\infty - \phi\|_{1,2} = \varepsilon$ and $J(v_\infty) = d_1$. Thus $v_\infty \in \mathscr{LES}_G$ by Lemma 4. However, the fact that $\|v_\infty - \phi\|_{1,2} = \varepsilon < r$ contradicts (16). □

Let $\varepsilon \in (0, r)$ be arbitrarily given. Choose $\delta \in (0, \varepsilon)$ so small that

$$J(v) < c \quad \text{for all } v \in B_{H_0^1}(\phi; \delta).$$

Here it is possible, because $c > d_1 = J(\phi)$ by the last claim, and J is continuous in $H_0^1(\Omega)$. For any $v_0 \in \mathscr{X}_G \cap B_{H_0^1}(\phi; \delta)$, let $v(s)$ be a solution of (10)–(12). Then $v(s) \in \mathscr{X}_G$ for $s \ge 0$ by (i) of Lemma 2.

Claim *For any $s \ge 0$, $v(s) \in B_{H_0^1}(\phi; \varepsilon)$, and hence ϕ is stable.*

Proof Assume on the contrary that $v(s_0) \in \partial B_{H_0^1}(\phi; \varepsilon)$ at some $s_0 > 0$. By the definition of c, it holds that $c \le J(v(s_0))$. However, it contradicts the fact that $J(v(s_0)) \le J(v_0) < c$. Thus $v(s) \in B_{H_0^1}(\phi; \varepsilon)$ for all $s \ge 0$. □

Moreover, if ϕ is isolated in $H_0^1(\Omega)$ from all the other sign-definite G-invariant solutions of (15), then so is it from all sign-changing solutions by the Hopf maximum principle for (15). One can prove that $v(s_n)$ converges strongly in $H_0^1(\Omega)$ to ϕ along any sequence $s_n \to \infty$ whenever $v(0) \in \mathscr{X}_G \cap B_{H_0^1(\Omega)}(\phi; \delta_0)$ with some $\delta_0 > 0$ (see [3] for more details).

3.4 *Proof of Theorem 4*

Let ϕ be a sign-changing G-invariant solution of (15) (hence ϕ admits more than two nodal domains).

We first prove (i). Let D be a nodal domain of ϕ and define

$$\phi_\mu(x) := \begin{cases} \mu\phi(x) & \text{if } x \in D, \\ \phi(x) & \text{if } x \in \Omega \setminus D \end{cases} \qquad \text{for } \mu \ge 0.$$

(Note: ϕ_μ might not belong to $\mathscr{X}_G$.) Then one can observe that

- ϕ_μ is G-invariant;
- $\phi_\mu \to \phi$ strongly in $H_0^1(\Omega)$ as $\mu \to 1$;
- if $\mu \ne 1$, then $J(c\phi_\mu) < J(\phi)$ for any $c \ge 0$.

Moreover, we set

$$u_{0,\mu} := \phi_\mu, \qquad \tau_\mu := t_*(u_{0,\mu}), \qquad v_{0,\mu} := \tau_\mu^{-1/(m-2)} u_{0,\mu} \in \mathscr{X}_G.$$

As in [3], it then follows that

- $\tau_\mu \to t_*(\phi) = 1$ and $v_{0,\mu} \to \phi$ strongly in $H_0^1(\Omega)$ as $\mu \to 1$;
- if $\mu \neq 1$, then $J(v_{0,\mu}) < J(\phi)$.

Hence the solution $v_\mu(s)$ of (10)–(12) with $v_\mu(0) = v_{0,\mu}$ never converges to ϕ as $s \to \infty$. Therefore ϕ is not an asymptotically stable profile under G-invariant perturbations.

Let us move on to (ii). Here we further assume that

$$\overline{B_{H_0^1}(\phi; R)} \cap \{\psi \in \mathscr{S}_G \colon J(\psi) < J(\phi)\} = \emptyset \tag{17}$$

with some $R > 0$.

Claim *If $\mu \neq 1$, then $v_\mu(s) \notin \overline{B_{H_0^1}(\phi; R)}$ for any $s \gg 1$.*

Proof Assume on the contrary that $v_\mu(s_n) \in \overline{B_{H_0^1}(\phi; R)}$ with some sequence $s_n \to \infty$. Then by (ii) of Lemma 2, we deduce that, up to a subsequence,

$$v_\mu(s_n) \to \psi \quad \text{strongly in } H_0^1(\Omega)$$

with some $\psi \in \overline{B_{H_0^1}(\phi; R)} \cap \mathscr{S}_G$. Moreover, we have

$$J(\psi) \leq J(v_{0,\mu}) < J(\phi),$$

which contradicts (17). Thus ϕ is an unstable profile. □

3.5 Applications of Stability Criteria

We first apply the preceding stability criteria to the case that Ω is an annular domain given by

$$\Omega := \{x \in \mathbb{R}^N \colon a < |x| < b\}$$

with constants $0 < a < b$. Then it is known that (15) admits a unique positive radial solution ϕ_1 (see [17]) and an arbitrary number of positive nonradial solutions by properly choosing a, b. Particularly, least energy solutions of (15) are nonradial, provided that $(b-a)/a \ll 1$ (see [10, 11, 16]). Hence the unique positive radial solution ϕ_1 is out of the scope of the stability criteria proposed in [3]. It is also known that (15) admits infinitely many radial sign-changing solutions under (8) (see also [17]) and all the sign-changing solutions turn out to be not asymptotically stable under a wider class of perturbations by [3].

Corollary 1 *Let Ω be an annular domain in $\mathbb{R}^N$ and assume* (8). *Then the sign-definite radial solutions $\pm\phi_1$ of* (15) *are asymptotically stable profiles under $O(N)$-invariant perturbations. Furthermore, the other radial solutions of* (15) *are not asymptotically stable profiles under $O(N)$-invariant perturbations.*

Proof Let $G = O(N)$. Since ϕ_1 is the unique positive G-invariant solution, we conclude by Theorem 3 that $\pm\phi_1$ are asymptotically stable profiles under G-invariant perturbations. The other radial solutions are sign-changing, so by Theorem 4 they are not asymptotically stable under $O(N)$-invariant perturbations. □

Remark 5 The frame of stability analysis for positive radial profiles in annular domains without restriction of perturbation will be discussed in a forthcoming joint paper with Ryuji Kajikiya.

We next give a corollary for general G-invariant domains. Here we call ϕ a *least energy sign-changing G-invariant solution* if ϕ is a sign-changing G-invariant solution of (15) and takes the least energy among sign-changing G-invariant solutions. Such a least energy sign-changing G-invariant solution always exists for any subgroup $G \subset O(N)$ under (8).

Corollary 2 *Let Ω be a G-invariant domain of $\mathbb{R}^N$ with smooth boundary and assume* (8). *Then least energy sign-changing G-invariant solutions of* (15) *are unstable asymptotic profiles under G-invariant perturbations.*

Proof As in [3], one can prove that every least energy sign-changing G-invariant solution ϕ is distinct from all the G-invariant nontrivial solutions taking lower energies (i.e., sign-definite G-invariant solutions) by the Hopf maximum principle. Thus ϕ is unstable under G-invariant perturbations by Theorem 4. □

This fact is new also in view of the stability analysis as in [3]. When Ω is an annulus, least energy sign-changing solutions are nonradial by [1] and unstable in the sense of asymptotic profiles for (1)–(3) by [3]. By Remark 4, this corollary further assures that least energy sign-changing G-invariant solutions of (15) are also unstable profiles for any subgroup G of $O(N)$.

Acknowledgements This work is partially supported by KAKENHI #22740093 and Hyogo Science and Technology Association.

References

1. Aftalion, A., Pacella, F.: Qualitative properties of nodal solutions of semilinear elliptic equations in radially symmetric domains. C. R. Math. Acad. Sci. Paris **339**, 339–344 (2004)
2. Akagi, G.: Energy solutions of the Cauchy-Neumann problem for porous medium equations. Discrete Contin. Dyn. Syst. **suppl.**, 1–10 (2009)

3. Akagi, G., Kajikiya, R.: Stability analysis of asymptotic profiles for sign-changing solutions to fast diffusion equations. Manuscr. Math. doi:10.1007/s00229-012-0583-9
4. Akagi, G., Kobayashi, J., Ôtani, M.: Principle of symmetric criticality and evolution equations. Discrete Contin. Dyn. Syst. **suppl.**, 1–10 (2003)
5. Bénilan, P., Crandall, M.G.: The continuous dependence on φ of solutions of $u_t - \Delta\varphi(u) = 0$. Indiana Univ. Math. J. **30**, 161–177 (1981)
6. Berryman, J.G., Holland, C.J.: Nonlinear diffusion problem arising in plasma physics. Phys. Rev. Lett. **40**, 1720–1722 (1978)
7. Berryman, J.G., Holland, C.J.: Stability of the separable solution for fast diffusion. Arch. Ration. Mech. Anal. **74**, 379–388 (1980)
8. Bonforte, M., Grillo, G., Vazquez, J.L.: Behaviour near extinction for the fast diffusion equation on bounded domains. J. Math. Pures Appl. **97**, 1–38 (2012)
9. Brézis, H.: Monotonicity methods in Hilbert spaces and some applications to non-linear partial differential equations. In: Zarantonello, E. (ed.) Contributions to Nonlinear Functional Analysis, pp. 101–156. Academic Press, New York (1971)
10. Byeon, J.: Existence of many nonequivalent nonradial positive solutions of semilinear elliptic equations on three-dimensional annuli. J. Differ. Equ. **136**, 136–165 (1997)
11. Coffman, C.V.: A nonlinear boundary value problem with many positive solutions. J. Differ. Equ. **54**, 429–437 (1984)
12. DiBenedetto, E., Kwong, Y., Vespri, V.: Local space-analyticity of solutions of certain singular parabolic equations. Indiana Univ. Math. J. **40**, 741–765 (1991)
13. Feiresl, E., Simondon, F.: Convergence for semilinear degenerate parabolic equations in several space dimension. Dyn. Partial Differ. Equ. **12**, 647–673 (2000)
14. Gazzola, F., Weth, T.: Finite time blow-up and global solutions for semilinear parabolic equations with initial data at high energy level. Differ. Integral Equ. **18**, 961–990 (2005)
15. Kwong, Y.C.: Asymptotic behavior of a plasma type equation with finite extinction. Arch. Ration. Mech. Anal. **104**, 277–294 (1988)
16. Li, Y.Y.: Existence of many positive solutions of semilinear elliptic equations in annulus. J. Differ. Equ. **83**, 348–367 (1990)
17. Ni, W.-M.: Uniqueness of solutions of nonlinear Dirichlet problems. J. Differ. Equ. **50**, 289–304 (1983)
18. Sabinina, E.S.: On a class of non-linear degenerate parabolic equations. Dokl. Akad. Nauk SSSR **143**, 794–797 (1962)
19. Savaré, G., Vespri, V.: The asymptotic profile of solutions of a class of doubly nonlinear equations. Nonlinear Anal. **22**, 1553–1565 (1994)
20. Vázquez, J.L.: Smoothing and Decay Estimates for Nonlinear Diffusion Equations. Equations of Porous Medium Type. Oxford Lecture Series in Mathematics and Its Applications, vol. 33. Oxford University Press, Oxford (2006)
21. Vázquez, J.L.: The Porous Medium Equation. Mathematical Theory. Oxford Mathematical Monographs. The Clarendon Press, Oxford University Press, Oxford (2007)

A Family of Hardy-Rellich Type Inequalities Involving the L^2-Norm of the Hessian Matrices

Elvise Berchio

Abstract We derive a family of Hardy-Rellich type inequalities in $H^2(\Omega) \cap H_0^1(\Omega)$ involving the scalar product between Hessian matrices. The constants found are optimal and the existence of a boundary remainder term is discussed.

Keywords Hardy-Rellich inequality · Optimal constants · Biharmonic equation

1 Introduction

Let $\Omega \subset \mathbf{R}^N (N \geq 2)$ be a bounded domain (open and connected) with Lipschitz boundary. By combining interpolation inequalities (see [1, Corollary 4.16]) with the classical Poincaré inequality, the Sobolev space $H^2(\Omega) \cap H_0^1(\Omega)$ becomes a Hilbert space when endowed with the scalar product

$$(u, v) := \int_\Omega D^2 u \cdot D^2 v \, dx = \sum_{i,j=1}^N \int_\Omega \partial_{ij}^2 u \, \partial_{ij}^2 v \, dx \quad \text{for all } u, v \in H^2(\Omega) \cap H_0^1(\Omega), \tag{1}$$

which induces the norm $\|D^2 u\|_2 := (\int_\Omega D^2 u \cdot D^2 u \, dx)^{1/2} = (\int_\Omega |D^2 u|^2 \, dx)^{1/2}$.

If, furthermore, Ω satisfies a uniform outer ball condition, see [3, Definition 1.2], some of the derivatives in (1) may be dropped. Then, the bilinear form

$$\langle u, v \rangle := \int_\Omega \Delta u \, \Delta v \, dx \quad \text{for all } u, v \in H^2(\Omega) \cap H_0^1(\Omega) \tag{2}$$

defines a scalar product on $H^2(\Omega) \cap H_0^1(\Omega)$ with corresponding norm $\|\Delta u\|_2 := (\int_\Omega |\Delta u|^2 \, dx)^{1/2}$. Easily, $\|D^2 u\|_2^2 \geq 1/N \|\Delta u\|_2^2$, for every $u \in H^2(\Omega) \cap H_0^1(\Omega)$. The converse inequality follows from [3, Theorem 2.2].

E. Berchio (✉)
Dipartimento di Matematica, Politecnico di Milano, Piazza L. da Vinci 32, 20133 Milano, Italy
e-mail: elvise.berchio@polimi.it

R. Magnanini et al. (eds.), *Geometric Properties for Parabolic and Elliptic PDE's*,
Springer INdAM Series 2, DOI 10.1007/978-88-470-2841-8_2,

A well-known generalization of the first order Hardy inequality [15, 16] to the second order is the so-called Hardy-Rellich inequality [19] which reads

$$\int_\Omega |\Delta u|^2\,dx \geq \frac{N^2(N-4)^2}{16}\int_\Omega \frac{u^2}{|x|^4}\,dx \quad \text{for all } u \in H_0^2(\Omega). \tag{3}$$

Here $\Omega \subset \mathbf{R}^N$ $(N \geq 5)$ is a bounded domain such that $0 \in \Omega$ and the constant $\frac{N^2(N-4)^2}{16}$ is optimal, in the sense that it is the largest possible. Further generalizations to (3) have appeared in [9] and in [17]. In [11] the validity of (3) was extended to the space $H^2 \cap H_0^1(\Omega)$, see also [12]. One may wonder what happens in (3), if we replace the L^2-norm of the Laplacian with $\|D^2u\|_2^2$. In $H_0^2(\Omega)$, a density argument and two integrations by parts yield that $\frac{N^2(N-4)^2}{16}$ is still the "best" constant. In $H^2 \cap H_0^1(\Omega)$ the answer is less obvious and, to our knowledge, the corresponding inequality is not known, not even when Ω is smooth. This regard motivates the present paper.

Let ν be the exterior unit normal at $\partial\Omega$, we set

$$c_0 = c_0(\Omega) := \inf_{H^2\cap H_0^1(\Omega)\setminus H_0^2(\Omega)} \frac{\int_\Omega |D^2u|^2\,dx}{\int_{\partial\Omega} u_\nu^2\,d\sigma}. \tag{4}$$

The above definition makes sense as soon as Ω has Lipschitz boundary. Indeed, the normal derivative to a Lipschitz domain is defined almost everywhere on $\partial\Omega$ so that $u_\nu \in L^2(\partial\Omega)$ for any $u \in H^2 \cap H_0^1(\Omega)$. By the compactness of the embedding $H^2(\Omega) \subset H^1(\partial\Omega)$ (see [18, Chap. 2, Theorem 6.2]), the infimum in (4) is attained and $c_0 > 0$.

For $c > -c_0$, we aim to determine the largest $h(c) > 0$ such that

$$\int_\Omega |D^2u|^2\,dx + c\int_{\partial\Omega} u_\nu^2\,d\sigma \geq h(c)\int_\Omega \frac{u^2}{|x|^4}\,dx \quad \text{for all } u \in H^2 \cap H_0^1(\Omega). \tag{5}$$

In Sect. 3, for $\partial\Omega \in C^2$, we prove that there exists $C_N = C_N(\Omega) \in (-c_0, +\infty)$ such that:

- $h(c) < \frac{N^2(N-4)^2}{16}$, for $c \in (-c_0, C_N)$ and the equality is achieved in (5);
- $h(c) = \frac{N^2(N-4)^2}{16}$, for $c \in [C_N, +\infty)$ and, if $c > C_N$ $(u \not\equiv 0)$, the inequality is strict in (5).

When Ω satisfies a suitable geometrical condition (see (25) in the following) and $C = C_N$, we show that the equality cannot be achieved in (5). At last, we derive lower and upper bounds for C_N and we discuss its sign, see Theorem 1 and Remark 3.

If $\Omega = B$, the unit ball in $\mathbf{R}^N$ $(N \geq 5)$, several computations can be done explicitly. In Sect. 5, we show that $c_0(B) = 1$, $C_N(B) = N - 3 - \frac{\sqrt{2(N^2-4N+8)}}{2}$ and we

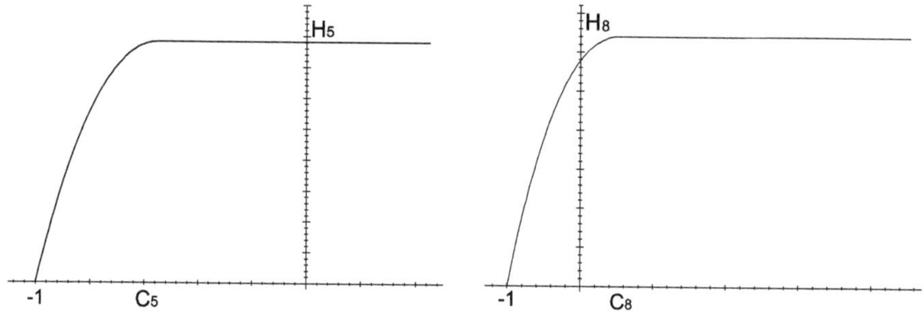

Fig. 1 The plot of the map $(-c_0, +\infty) \in c \mapsto h(c)$ when $\Omega = B$, $N = 5$ or $N = 8$ (*right*). H_5 and H_8 denote the Hardy-Rellich constants, $c_0(B) = 1$

determine the (radial) functions for which the equality holds in (5) (when $c < C_N$). In particular, for all $u \in H^2 \cap H_0^1(B) \setminus \{0\}$, we show that

$$\int_B |D^2 u|^2 \, dx + \left(N - 3 - \frac{\sqrt{2(N^2 - 4N + 8)}}{2}\right) \int_{\partial B} u_\nu^2 \, d\sigma > \frac{N^2(N-4)^2}{16} \int_B \frac{u^2}{|x|^4} \, dx \tag{6}$$

and the constants are optimal.

It's worth noting that $C_N(B)$ is positive when $N \geq 7$, negative when $N = 5, 6$, see Fig. 1. Hence, in lower dimensions, the following Hardy-Rellich inequality (with a boundary remainder term) holds

$$\int_B |D^2 u|^2 \, dx > \frac{N^2(N-4)^2}{16} \int_B \frac{u^2}{|x|^4} \, dx \left(+ |C_N| \int_{\partial B} u_\nu^2 \, d\sigma \right),$$

for all $u \in H^2 \cap H_0^1(B) \setminus \{0\}$, where $|C_5| = \sqrt{13/2} - 2$ and $|C_6| = \sqrt{10} - 3$. While, if $N \geq 7$, the "best" constant $h(0)$ is no longer the classical Hardy-Rellich one and we prove

$$\int_B |D^2 u|^2 \, dx \geq \frac{(N-1)(N-5)(2N-5)}{4} \int_B \frac{u^2}{|x|^4} \, dx \quad \text{for all } u \in H^2 \cap H_0^1(B). \tag{7}$$

Here, $\frac{(N-1)(N-5)(2N-5)}{4} < \frac{N^2(N-4)^2}{16}$ and the equality in (7) is achieved by a unique positive radial function, see Theorem 2 in Sect. 5.

The plan of the paper is the following: in Sect. 2 we prove existence and positivity of solutions to a suitable biharmonic linear problem. The boundary conditions considered arise from (5). In Sect. 3 we state our statement about the family of inequalities (5) while, in Sect. 4, we put its proof. At last, in Sect. 5, we focus on the case $\Omega = B$ and we prove (6) and (7). The Appendix contains the proof of some estimates we need in Sect. 3.

2 Preliminaries

Let $\Omega \subset \mathbf{R}^N$ $(N \geq 2)$ be a Lipschitz bounded domain which satisfies a uniform outer ball condition. We recall the definition of the first *Steklov* eigenvalue

$$d_0 = d_0(\Omega) := \inf_{H^2 \cap H_0^1(\Omega) \setminus H_0^2(\Omega)} \frac{\int_\Omega |\Delta u|^2 \, dx}{\int_{\partial\Omega} u_\nu^2 \, d\sigma}. \tag{8}$$

From the compactness of the embedding $H^2(\Omega) \subset H^1(\partial\Omega)$, the infimum in (8) is attained. Furthermore, due to [6], we know that the corresponding minimizer is unique, positive in Ω and solves the equation $\Delta^2 u = 0$ in Ω, subject the conditions $u = 0 = \Delta u - d_0 u_\nu$ on $\partial\Omega$.

Next, we assume that $\partial\Omega \in C^2$ and we denote with $|\Omega|$ and $|\partial\Omega|$ the Lebesgue measures of Ω and $\partial\Omega$. There holds

$$d_0(\Omega) \leq \frac{|\partial\Omega|}{|\Omega|},$$

see, for instance, [10, Theorem 1.8]. Let $K(x)$ denote the mean curvature of $\partial\Omega$ at x,

$$\underline{K} := \min_{\partial\Omega} K(x) \quad \text{and} \quad \overline{K} := \max_{\partial\Omega} K(x). \tag{9}$$

If Ω is convex, it was proved in [10, Theorem 1.7] that

$$d_0(\Omega) \geq N\underline{K}. \tag{10}$$

Notice that we adopt the convention that K is positive where the domain is convex.

Finally, from [14, Theorem 3.1.1.1] we recall

$$\int_\Omega |\Delta u|^2 \, dx = \int_\Omega \left|D^2 u\right|^2 dx + (N-1) \int_{\partial\Omega} K(x) u_\nu^2 \, d\sigma \quad \text{for all } u \in H^2 \cap H_0^1(\Omega). \tag{11}$$

Identity (11) is the basic ingredient to prove

Proposition 1 *Let Ω be a bounded domain with C^2 boundary, let c_0 and d_0 be as in* (4) *and* (8), *$\overline{K}$ and $\underline{K}$ as in* (9). *There holds*

$$\max\left\{ d_0(\Omega) - (N-1)\overline{K}; \frac{d_0(\Omega)}{N} \right\} \leq c_0(\Omega) \leq d_0(\Omega) - (N-1)\underline{K}. \tag{12}$$

Furthermore, if Ω is convex, then

(i) *$c_0 \geq \underline{K}$ and the equality holds if and only if Ω is a ball;*
(ii) *the minimizer u_0 of* (4) *is unique (up to a multiplicative constant) and, if $u_0(x_0) > 0$ for some $x_0 \in \Omega$, then $u_0 > 0$, $-\Delta u_0 \geq 0$ in Ω and $(u_0)_\nu < 0$ on $\partial\Omega$.*

If $\Omega = B$, the unit ball in $\mathbf{R}^N$, since $K(x) \equiv 1$, Proposition 1-(i) yields $c_0(B) = 1$.

Proof The estimates in (12) follow by combining (11) with (4) and (8). For the lower bound $d_0(\Omega)/N$, we exploit the fact that $\|D^2u\|_2^2 \geq 1/N\|\Delta u\|_2^2$, for every $u \in H^2(\Omega) \cap H_0^1(\Omega)$.

Let Ω be convex, by (10) and (12), $c_0 \geq \underline{K}$. If $c_0 = \underline{K}$, by (10) and (12), we deduce that $d_0 = N\underline{K}$ and, by [10, Theorem 1.7], Ω must be a ball. On the other hand, if Ω is a ball, then $\underline{K} = \overline{K}$ and, by (12), we get $c_0 = d_0 - (N-1)\underline{K}$. Since, from [10], $d_0 = N\underline{K}$, statement (i) follows at once.

To prove statement (ii), by (11), we write (12) as

$$c_0 = \inf_{H^2 \cap H_0^1(\Omega) \setminus H_0^2(\Omega)} \frac{\int_\Omega |\Delta u|^2\,dx - (N-1)\int_{\partial\Omega} K(x) u_\nu^2\,d\sigma}{\int_{\partial\Omega} u_\nu^2\,d\sigma}. \tag{13}$$

Let u_0 be a minimizer to c_0. As in [6], we define $\bar{u}_0 \in H^2 \cap H_0^1(\Omega)$ as the unique (weak) solution to

$$\begin{cases} -\Delta \bar{u}_0 = |\Delta u_0| & \text{in } \Omega \\ \bar{u}_0 = 0 & \text{on } \partial\Omega. \end{cases}$$

By the maximum principle for superharmonic functions,

$$|u_0| \leq \bar{u}_0 \quad \text{in } \Omega \quad \text{and} \quad \big|(u_0)_\nu\big| \leq \big|(\bar{u}_0)_\nu\big| \quad \text{on } \partial\Omega.$$

If Δu_0 changes sign, then the above inequalities are strict and, since K is positive, by (13), we infer

$$\begin{aligned} c_0 &= \frac{\int_\Omega |\Delta u_0|^2\,dx - (N-1)\int_{\partial\Omega} K(x)(u_0)_\nu^2\,d\sigma}{\int_{\partial\Omega} (u_0)_\nu^2\,d\sigma} \\ &> \frac{\int_\Omega |\Delta \bar{u}_0|^2\,dx - (N-1)\int_{\partial\Omega} K(x)(\bar{u}_0)_\nu^2\,d\sigma}{\int_{\partial\Omega} (\bar{u}_0)_\nu^2\,d\sigma}, \end{aligned}$$

a contradiction. This noticed, a further application of the maximum principle yields the positivity issue. Uniqueness follows by standard arguments. That is, by exploiting the fact that a (positive) minimizer to (4) solves the linear problem (15), here below, for $f \equiv 0$ and $c = -c_0$. □

Remark 1 The problem of dealing with domains having a nonsmooth boundary goes beyond the purposes of the present paper. We limit ourselves to make a couple of remarks on the topic.

If we drop the regularity assumption on $\partial\Omega$, identity (11) is, in general, no longer true. Hence, the previous proof cannot be carried out. Assume that $\Omega \subset \mathbf{R}^N$ $(N \geq 2)$ is a bounded domain with Lipschitz boundary which satisfies an outer ball condition. Due to [3], we know that there exist a sequence of smooth domains $\Omega_m \nearrow \Omega$, with $\partial\Omega_m \in C^\infty$, and a real constant C such that the mean curvatures satisfy $K_m(x) \geq C$, for every $x \in \partial\Omega$ and $m \geq 1$. Next, for $u \in H^2 \cap H_0^1(\Omega)$ fixed, define the sequence of functions $\{u_m\}_{m\geq 1}$ such that $u_m \in H^2 \cap H_0^1(\Omega)$ solves

$$\begin{cases} -\Delta u_m = -\Delta u & \text{in } \Omega_m \\ u_m = 0 & \text{on } \partial\Omega_m. \end{cases}$$

When $C \geq 0$, from (11), it is readily deduced that

$$\int_{\Omega_m} |D^2 u_m|^2 \, dx \leq \int_\Omega |\Delta u|^2 \, dx$$

while, if $C < 0$, we get

$$\begin{aligned} \int_{\Omega_m} |D^2 u_m|^2 \, dx &\leq \int_\Omega |\Delta u|^2 \, dx - (N-1)C \int_{\partial\Omega_m} (u_m)_\nu^2 \, d\sigma \\ &\leq \left(1 + \frac{(N-1)|C|}{d_0(\Omega)}\right) \int_\Omega |\Delta u|^2 \, dx, \end{aligned}$$

where d_0 is as in (8). Then, by a standard weak convergence argument, see [14, Theorem 3.2.1.2], one concludes that

$$\int_\Omega |D^2 u|^2 \, dx \leq (1 + \gamma(\Omega)) \int_\Omega |\Delta u|^2 \, dx \quad \text{for all } u \in H^2 \cap H_0^1(\Omega), \tag{14}$$

where $\gamma(\Omega) = 0$, if $C \geq 0$, and $\gamma(\Omega) = ((N-1)|C|)/d_0(\Omega)$, otherwise.

Obviously, (14) does not replace (11). However, it can be exploited to obtain the first part of Proposition 1 for domains satisfying the above mentioned (weaker) regularity assumptions.

For every $c > -c_0$ and for $f \in L^2(\Omega)$, we will consider the linear problem

$$\begin{cases} \Delta^2 u = f(x) & \text{in } \Omega \\ u = 0 & \text{on } \partial\Omega \\ u_{\nu\nu} + c u_\nu = 0 & \text{on } \partial\Omega. \end{cases} \tag{15}$$

This choice of boundary conditions will be convenient in the next section.

By solutions to (15) we mean weak solutions, that is functions $u \in H^2 \cap H_0^1(\Omega)$ such that

$$\int_\Omega D^2 u \cdot D^2 v \, dx + c \int_{\partial\Omega} u_\nu v_\nu \, d\sigma = \int_\Omega f v \, dx \quad \text{for all } v \in H^2 \cap H_0^1(\Omega). \tag{16}$$

Indeed, formally, two integrations by parts give

$$\int_\Omega D^2 u \cdot D^2 v \, dx = \int_\Omega \Delta^2 u v \, dx + \int_{\partial\Omega} u_{\nu\nu} v_\nu \, d\sigma \quad \text{for all } v \in H^2 \cap H_0^1(\Omega), \tag{17}$$

see [5, formula (36)]. Then, plugging (17) into (16), by standard density arguments, we infer that u solves (15) pointwise. Since the boundary conditions in (15) have the same principal part of Navier boundary conditions ($u = 0 = \Delta u$ on $\partial\Omega$), they must satisfy the so-called *complementing conditions* [4]. See also [13, formula (2.22)]. Hence, standard elliptic regularity theory applies. Therefore, if $\partial\Omega \in C^4$ and $f \in L^2(\Omega)$, then $u \in H^4(\Omega)$ and (17) makes sense.

Solutions to (16) correspond to critical points of the functional

$$I_c(u) := \frac{1}{2}\left(\int_\Omega |D^2 u|^2 \, dx + c \int_{\partial\Omega} u_\nu^2 \, d\sigma\right) - \int_\Omega f u \, dx \quad \text{for } u \in H^2 \cap H_0^1(\Omega).$$

For $c > -c_0$, I_c turns to be coercive. Since it is also strictly convex, there exists a unique critical point u_c which is the global minimum of I_c. When $\partial\Omega \in C^2$, thanks to (11), I_c writes

$$I_c(u) = \frac{1}{2}\left(\int_\Omega |\Delta u|^2\,dx - \int_{\partial\Omega} \alpha_c(x) u_\nu^2\,d\sigma\right) - \int_\Omega f u\,dx \quad \text{for } u \in H^2 \cap H_0^1(\Omega),$$

where $\alpha_c(x) := (N-1)K(x) - c$, for every $x \in \partial\Omega$. Then, the minimizer u_c to I_c also satisfies

$$\int_\Omega \Delta u_c\,\Delta v\,dx - \int_{\partial\Omega} \alpha_c(x)(u_c)_\nu v_\nu\,d\sigma = \int_\Omega f v\,dx \quad \text{for all } v \in H^2 \cap H_0^1(\Omega). \tag{18}$$

From [13, Definition 5.21], we know that (18) is the definition of *weak* solutions to the equation $\Delta^2 u = f$ in Ω, subject to *Steklov* boundary conditions (with nonconstant parameter α_c). Namely, $u = 0 = \Delta u - \alpha_c(x)u_\nu$ on $\partial\Omega$. Arguing as in the proof of [13, Theorem 5.22], if $\alpha_c \geq 0$ and $0 \not\equiv f \geq 0$, we infer that the minimizer u_c to I_c is positive. Furthermore, $-\Delta u_c \geq 0$ in Ω and $(u_c)_\nu < 0$ on $\partial\Omega$. We conclude that Δ^2, subject to the boundary conditions in (15), satisfies the *positivity preserving property* (p.p.p. in the following) if

$$-c_0 < c \leq (N-1)K(x) \quad \text{for every } x \in \partial\Omega.$$

Notice that, if only the positivity of u is concerned, the lower bound for p.p.p. ($\alpha_c \geq 0$) can be weakened, see [13, Theorem 5.22].

We collect the conclusions so far drawn in the following

Proposition 2 *Let $\Omega \subset \mathbf{R}^N (N \geq 2)$ be a Lipschitz bounded domain and c_0 be as in* (4). *For every $c > -c_0$, we have*

(i) *for every $f \in L^2(\Omega)$, problem* (15) *admits a unique solution $u \in H^2 \cap H_0^1(\Omega)$. Moreover, if $f \in H^k(\Omega)$ and $\partial\Omega \in C^{k+4}$ for some $k \geq 0$, then $u \in H^{k+4}(\Omega)$.*
(ii) *Assume, furthermore, that Ω is convex, $\partial\Omega \in C^2$ and $\underline{K}$ is as in* (9). *Then, for every $c \in (-c_0, (N-1)\underline{K}]$, if $f \geq 0$ $(f \not\equiv 0)$ in Ω, the solution u of* (15) *satisfies $u > 0$, $-\Delta u \geq 0$ in Ω and $u_\nu < 0$ on $\partial\Omega$.*

Remark 2 The convexity assumption in Proposition 2-(ii) is only needed to assure the non-emptiness of the interval $(-c_0, (N-1)\underline{K}]$ in which p.p.p. holds. If Ω is not convex, by (12), the same goal can be achieved by assuming that Ω satisfies one of the following inequalities

$$N(N-1)|\,\underline{K}\,| < d_0(\Omega) \quad \text{or} \quad (N-1)\big(\overline{K} + |\underline{K}|\big) < d_0(\Omega). \tag{19}$$

Compare with Proposition 3 in the Appendix.

3 Hardy-Rellich Type Inequalities with a Boundary Term

Before stating our results, we recall some facts from [7]. Set $H_N := \frac{N^2(N-4)^2}{16}$. For every bounded domain Ω such that $0 \in \Omega$ and for every $h \in [0, H_N]$, we know that

$$\int_\Omega |\Delta u|^2\, dx \geq h \int_\Omega \frac{u^2}{|x|^4}\, dx + d_1(h) \int_{\partial\Omega} u_\nu^2\, d\sigma \quad \text{for all } u \in H^2 \cap H_0^1(\Omega). \tag{20}$$

The optimal constant $d_1(h)$ is achieved, if and only if $h < H_N$, by a unique positive function $u_h \in H^2 \cap H_0^1(\Omega)$. Furthermore, $0 \leq d_1(h) < d_1(0) = d_0$, with d_0 as in (8). When $d_1(H_N) > 0$ (this was established only for strictly starshaped domains, namely such that $\min_{\partial\Omega}(x \cdot \nu) > 0$), (20) readily gives the Hardy-Rellich inequality (3) (for $u \in H^2 \cap H_0^1(\Omega)$) plus a boundary remainder term. See also the Appendix.

Let c_0 be as in (4). To obtain (5), for $c > -c_0$, we consider the minimization problem

$$h(c) := \inf_{H^2 \cap H_0^1(\Omega) \setminus \{0\}} \frac{\int_\Omega |D^2 u|^2\, dx + c \int_{\partial\Omega} (u_\nu)^2\, d\sigma}{\int_\Omega \frac{u^2}{|x|^4}\, dx}. \tag{21}$$

Clearly, $h(c) \geq 0$ and $h(-c_0) = 0$. On the other hand, since $\int_\Omega |D^2 u|^2\, dx = \int_\Omega |\Delta u|^2\, dx$, for all $u \in H_0^2(\Omega)$, (3) yields $h(c) \leq H_N$.

Formally, for every $c > -c_0$ fixed, the Euler equation corresponding to (21) is the eigenvalue problem

$$\begin{cases} \Delta^2 u = h \frac{u}{|x|^4} & \text{in } \Omega \\ u = 0 & \text{on } \partial\Omega \\ u_{\nu\nu} + c u_\nu = 0 & \text{on } \partial\Omega. \end{cases} \tag{22}$$

Indeed, by solutions to (22) we mean functions $u \in H^2 \cap H_0^1(\Omega)$ such that

$$\int_\Omega D^2 u\, D^2 v\, dx + c \int_{\partial\Omega} u_\nu v_\nu\, d\sigma = h \int_\Omega \frac{uv}{|x|^4}\, dx \quad \text{for all } v \in H^2 \cap H_0^1(\Omega), \tag{23}$$

see Sect. 2. By elliptic regularity, any solution to (22) belongs to $C^\infty(\Omega \setminus \{0\})$, whereas, up to the boundary, the solution is smooth as the boundary, see again Sect. 2. We prove

Theorem 1 *Let $\Omega \subset \mathbf{R}^N$ $(N \geq 5)$ be a bounded domain such that $0 \in \Omega$ and $\partial\Omega \in C^2$. Let c_0 be as in* (4) *and $h(c)$ be as in* (21). *If $c > -c_0$, then $h(c) > 0$ and*

$$\int_\Omega \left|D^2 u\right|^2 dx + c \int_{\partial\Omega} u_\nu^2\, dS \geq h(c) \int_\Omega \frac{u^2}{|x|^4}\, dx \quad \text{for all } u \in H^2 \cap H_0^1(\Omega). \tag{24}$$

Furthermore, there exists $C_N = C_N(\Omega) \in (-c_0, (N-1)\overline{K} - d_1(H_N)]$, where $\overline{K}$ is as in (9) *and $d_1(h)$ is as in* (20), *such that*

(i) *$h(c)$ is increasing, concave and continuous with respect to $c \in (-c_0, C_N]$;*
(ii) *$h(c) = H_N$ for every $c \geq C_N$.*

Moreover, the infimum in (21) *is not achieved if* $c > C_N$*, achieved if* $-c_0 < c < C_N$ *and the minimizer* $u_c \in H^2 \cap H_0^1(\Omega)$ *solves* (22) *with* $h = h(c)$.

Let now Ω *be such that the following inequality is satisfied*

$$(N-1)(\overline{K} - \underline{K}) \leq d_1(H_N) \quad \textit{for every } N \geq 5, \tag{25}$$

where $\underline{K}$ *is as in* (9)*. Then,* $h(C_N)(= H_N)$ *is not achieved. Furthermore, for every* $-c_0 < c < C_N$*, the minimizer* u_c *of* $h(c)$ *is unique, strictly positive, superharmonic in* Ω *and* $(u_c)_\nu < 0$ *on* $\partial\Omega$.

Condition (25) excludes domains for which the curvature of the boundary has wide oscillations. This requirement is trivially satisfied if Ω is a ball ($\overline{K} = \underline{K}$). On the other hand, if Ω is not a ball, (25) yields $d_1(H_N) > 0$. To our knowledge, this issue has only been proved for strictly starshaped domains, see [7]. In the Appendix, by slightly modifying the proof of [7, Theorem 1], we provide an explicit constant $D_N = D_N(\Omega) > 0$ such that $d_1(H_N) \geq m\, D_N$, where $m := \min_{\partial\Omega}(x \cdot \nu) > 0$. Hence, when Ω is strictly starshaped, in stead of (25), one may check that

$$(N-1)(\overline{K} - \underline{K}) \leq m D_N \quad \text{for every } N \geq 5,$$

where D_N comes from (42) with $h = H_N$.

Theorem 1 as the following

Corollary 1 *Let* $\Omega \subset \mathbf{R}^N$ $(N \geq 5)$ *be a bounded domain such that* $0 \in \Omega$ *and* $\partial\Omega \in C^2$*. There exists an optimal constant* $C_N \in (-c_0, (N-1)\overline{K} - d_1(H_N)]$ *such that*

$$\int_\Omega |D^2 u|^2\, dx + C_N \int_{\partial\Omega} u_\nu^2\, dS \geq \frac{N^2(N-4)^2}{16} \int_\Omega \frac{u^2}{|x|^4}\, dx \quad \forall u \in H^2 \cap H_0^1(\Omega). \tag{26}$$

Furthermore, if Ω *satisfies* (25)*, the inequality in* (26) *is strict (for* $u \not\equiv 0$*).*

Remark 3 When $\Omega = B$, the unit ball in $\mathbf{R}^N (N \geq 5)$, C_N can be computed explicitly and we get

$$C_N(B) = N - 1 - d_1(H_N) = N - 3 - \frac{\sqrt{2(N^2 - 4N + 8)}}{2},$$

see Sect. 5 for the details. Hence, in this case, the upper bound for C_N (given in Corollary 1) is sharp. As already remarked in the Introduction, $C_N(B) > 0$ if and only if $N \geq 7$. In the next section (see, Lemma 2) we show that, if Ω is such that the following inequality is satisfied

$$(N-1)(\overline{K} - \underline{K}) < d_0 - d_1(H_N - \delta) \quad \text{for every } N \geq 5 \text{ and for some } \delta > 0, \tag{27}$$

then $C_N \geq (N-1)\underline{K} - d_1(H_N)$. When Ω is convex, this estimate supports the conjecture

$$\text{there exists } \overline{N} = \overline{N}(\Omega) \geq 5: \quad C_N(\Omega) > 0, \quad \text{for } N \geq \overline{N}.$$

This issue could be proved by providing a suitable upper bound for $d_1(H_N)$. Notice that, in view of (10), the estimate $d_1(H_N) < d_0(\Omega)$ does not suffices to deduce the sign of C_N.

On the other hand, if (25) holds and $\underline{K} < 0$ (Ω is not convex), the upper bound for C_N in Corollary 1 yields $C_N < 0$, for every $N \geq 5$.

4 Proof of Theorem 1 and Corollary 1

We use the same notations of the previous section. First we prove

Lemma 1 *Let $\Omega \subset \mathbf{R}^N$ $(N \geq 5)$ be a Lipschitz bounded domain which satisfies a uniform outer ball condition and such that $0 \in \Omega$. If $h(c) < H_N$ for some $c > -c_0$, then the infimum in* (21) *is attained. Moreover, a minimizer weakly solves problem* (22) *for $h = h(c)$.*

Proof Let $\{u_m\} \subset H^2 \cap H_0^1(\Omega)$ be a minimizing sequence for $h(c)$ such that

$$\int_\Omega \frac{u_m^2}{|x|^4}\, dx = 1. \tag{28}$$

Then,

$$\int_\Omega |D^2 u_m|^2\, dx + c \int_{\partial\Omega} (u_m)_\nu^2\, d\sigma = h(c) + o(1) \quad \text{as } m \to +\infty. \tag{29}$$

For $c > -c_0$, this shows that $\{u_m\}$ is bounded in $H^2 \cap H_0^1(\Omega)$. Exploiting the compactness of the trace map $H^2(\Omega) \to H^1(\partial\Omega)$, we conclude that there exists $u \in H^2 \cap H_0^1(\Omega)$ such that

$$\begin{aligned} &u_m \rightharpoonup u \quad \text{in } H^2 \cap H_0^1(\Omega), \qquad (u_m)_\nu \to u_\nu \quad \text{in } L^2(\partial\Omega), \\ &\frac{u_m}{|x|^2} \to \frac{u}{|x|^2} \quad \text{in } L^2(\Omega) \end{aligned} \tag{30}$$

up to a subsequence.

Now, from [10] we know that the space $H^2 \cap H_0^1(\Omega)$, endowed with (2), admits the following orthogonal decomposition

$$H^2 \cap H_0^1(\Omega) = W \oplus H_0^2(\Omega), \tag{31}$$

where W is the completion of

$$V = \left\{ v \in C^\infty(\overline{\Omega}) : \Delta^2 v = 0,\ v = 0 \text{ on } \partial\Omega \right\}$$

with respect to the norm induced by (2). Furthermore, if $u \in H^2 \cap H_0^1(\Omega)$ and if $u = w + z$ is the corresponding orthogonal decomposition with $w \in W$ and $z \in H_0^2(\Omega)$, then w and z are weak solutions to

$$\begin{cases} \Delta^2 w = 0 & \text{in } \Omega \\ w = 0 & \text{on } \partial\Omega \\ (w)_\nu = u_\nu & \text{on } \partial\Omega \end{cases} \quad \text{and} \quad \begin{cases} \Delta^2 z = \Delta^2 u & \text{in } \Omega \\ z = 0 & \text{on } \partial\Omega \\ (z)_\nu = 0 & \text{on } \partial\Omega. \end{cases}$$

By this, the functions u_m, as given at the beginning, may be written as $u_m = w_m + z_m$, where $w_m \in W$ and $z_m \in H_0^2(\Omega)$. Assume now that (30) holds with $u \equiv 0$. By the first of the above Dirichlet problems, we deduce that $w_m \to 0$ in $H^2 \cap H_0^1(\Omega)$ and, in particular, that $\frac{w_m}{|x|^2} \to 0$ in $L^2(\Omega)$. This yields

$$\int_\Omega \left|D^2 u_m\right|^2 dx = \int_\Omega \left|D^2 z_m\right|^2 dx + o(1) = \int_\Omega |\Delta z_m|^2 dx + o(1) \qquad (32)$$

and

$$\int_\Omega \frac{u_m^2}{|x|^4} dx = \int_\Omega \frac{z_m^2}{|x|^4} dx + o(1).$$

Then, by (3), (28)–(29)–(30) and the fact that $h(c) < H_N$, we infer that

$$H_N > h(c) + o(1) = \int_\Omega \left|D^2 u_m\right|^2 dx + o(1) = \int_\Omega |\Delta z_m|^2 dx + o(1) \geq H_N + o(1),$$

a contradiction. Hence, $u \neq 0$. If we set $v_m := u_m - u$, from (30) we obtain

$$\begin{aligned} &v_m \rightharpoonup 0 \quad \text{in } H^2 \cap H_0^1(\Omega), \qquad (v_m)_\nu \to 0 \quad \text{in } L^2(\partial\Omega), \\ &\frac{v_m}{|x|^2} \to 0 \quad \text{in } L^2(\Omega). \end{aligned} \qquad (33)$$

In view of (33), we may rewrite (29) as

$$\int_\Omega \left|D^2 u\right|^2 dx + \int_\Omega \left|D^2 v_m\right|^2 dx + c \int_{\partial\Omega} u_\nu^2 \, d\sigma = h(c) + o(1). \qquad (34)$$

Moreover, by (28), (33) and the Brezis-Lieb Lemma [8], we have

$$\begin{aligned} 1 = \int_\Omega \frac{u_m^2}{|x|^4} dx &= \int_\Omega \frac{u^2}{|x|^4} dx + \int_\Omega \frac{v_m^2}{|x|^4} dx + o(1) \\ &\leq \int_\Omega \frac{u^2}{|x|^4} dx + \frac{1}{H_N} \int_\Omega |\Delta v_m|^2 dx + o(1) \\ &= \int_\Omega \frac{u^2}{|x|^4} dx + \frac{1}{H_N} \int_\Omega \left|D^2 v_m\right|^2 dx + o(1) \end{aligned}$$

where the last equality is achieved by exploiting the decomposition (31), as explained above. Since $h(c) \geq 0$, the just proved inequality gives

$$h(c) \leq h(c) \int_\Omega \frac{u^2}{|x|^4} dx + \frac{h(c)}{H_N} \int_\Omega \left|D^2 v_m\right|^2 dx + o(1).$$

By combining this with (34), we obtain

$$\begin{aligned} &\int_\Omega \left|D^2 u\right|^2 dx + c \int_{\partial\Omega} u_\nu^2 \, d\sigma \\ &\quad \leq h(c) \int_\Omega \frac{u^2}{|x|^4} dx + \left(\frac{h(c)}{H_N} - 1\right) \int_\Omega \left|D^2 v_m\right|^2 dx + o(1) \\ &\quad \leq h(c) \int_\Omega \frac{u^2}{|x|^4} dx + o(1) \end{aligned}$$

which shows that $u \neq 0$ is a minimizer. □

Remark 4 If $\partial\Omega \in C^2$, to deduce (32), one may exploit (11) instead of the decomposition (31). We leave here this (longer) proof since it highlights that the regularity assumption on $\partial\Omega$ (in the statement of Theorem 1) is not due to the existence issue.

Next, we show

Lemma 2 *Let $\Omega \subset \mathbf{R}^N$ $(N \geq 5)$ be a bounded domain, with $\partial\Omega \in C^2$ and such that $0 \in \Omega$. The map $(-c_0, +\infty) \ni c \mapsto h(c)$ is nondecreasing (increasing when achieved), concave, hence, continuous and*

$$h(c) = H_N \quad \textit{for every } c \geq (N-1)\overline{K} - d_1(H_N).$$

Moreover, if Ω satisfies (27) and $H_N - \delta < h < H_N$, then

$$h(c) \leq h \quad \textit{for every } -c_0 < c \leq (N-1)\underline{K} - d_1(h).$$

Proof The properties of $h(c)$ follow from its definition, we only need to prove the estimates. By (11), the infimum in (21) may be rewritten as

$$h(c) = \inf_{u \in H^2 \cap H_0^1(\Omega)\setminus\{0\}} \frac{\int_\Omega |\Delta u|^2\, dx - \int_{\partial\Omega} \alpha_c(x)(u_\nu)^2\, d\sigma}{\int_\Omega \frac{u^2}{|x|^4}\, dx}, \tag{35}$$

where $\alpha_c(x) = (N-1)K(x) - c$, as defined in Sect. 2. Then, if $\alpha_c(x) \leq d_1(H_N)$ for every $x \in \partial\Omega$, by (20), $h(c) \equiv H_N$ and the first estimate follows. Similarly, if $\alpha_c(x) \geq d_1(h)$ for every $x \in \partial\Omega$, by (20), we get the second estimate. Notice that assumption (27), suitably combined with (12), ensures that $(N-1)\underline{K} - d_1(h) > -c_0$, for every $H_N - \delta < h < H_N$. □

By Lemma 2, the number

$$C_N := \inf\{c > -c_0 : h(c) = H_N\} \tag{36}$$

is well-defined. Furthermore, we have

$$(N-1)\underline{K} - d_1(H_N) \leq C_N \leq (N-1)\overline{K} - d_1(H_N), \tag{37}$$

where the lower bound has been proved for Ω satisfying (27). Then, we show

Lemma 3 *Let $\Omega \subset \mathbf{R}^N$ $(N \geq 5)$ be a bounded domain such that $0 \in \Omega$ and $\partial\Omega \in C^2$. Let C_N be as in (36), then the infimum in (21) is not achieved if $c > C_N$, achieved if $-c_0 < c < C_N$ and the minimizer (weakly) solves problem (22) for $h = h(c)$.*

Assume, furthermore, that Ω satisfies (25). Then, for every $-c_0 < c < C_N$, $h(c)$ is achieved by a unique positive function u_c which satisfies $-\Delta u_c \geq 0$ in Ω and $(u_c)_\nu < 0$ on $\partial\Omega$ while, $h(C_N)$ is not achieved.

Proof The first part of the statement comes from the definition of C_N combined with the previous lemmata. To prove the second part, we write (21) as in (35). From (25), combined with (37), we have that $C_N \leq (N-1)\underline{K}$. Then, $\alpha_c(x) \geq 0$

for every $x \in \partial\Omega$ and for every $-c_0 < c < C_N$. Hence, we may argue as in the proof of Proposition 1-(ii), to deduce the positivity of a minimizer u_c, together with the fact that $-\Delta u_c \geq 0$ in Ω and $(u_c)_\nu < 0$ on $\partial\Omega$. Since problem (22) is linear, once the positivity of a minimizer is known, the proof of its uniqueness is standard.

It remains to show that $h(c)$ is not achieved for $c = C_N$. If a minimizer of $h(C_N)$ exists, it would be a positive and superharmonic solution, vanishing on $\partial\Omega$, to the equation in (22) with $h = H_N$. Then, the same argument of [2, Theorem 2.2-(ii)] gives a contradiction. □

The proofs of Theorem 1 and Corollary 1 follow by combining the statements of the above lemmata.

5 Radial Setting

When $\Omega = B$, the unit ball in $\mathbf{R}^N (N \geq 5)$, the mean curvature $K \equiv 1$. Then, for what remarked in Sect. 2, problems (20) and (21) become almost equivalent. Indeed, let u_h be the function achieving the equality in (20), for some $0 \leq h < H_N$. Then, by (35), u_h is also the minimizer of $h(c)$ for $c = c_h = N - 1 - d_1(h)$ and $h(c_h) = h$ (or, equivalently, u_h achieves the equality in (5)). Furthermore, the map $[0, H_N) \ni h \mapsto c_h$ is increasing, $c_0 = -1$ and $c_{H_N} = C_N$, where C_N is as in (37).

We briefly sketch the computations to determine (explicitly) the minimizer of $h(c)$. As in [7, Section 5], we introduce an auxiliary parameter $0 \leq \alpha \leq N - 4$ and we set

$$H(\alpha) := \frac{\alpha(\alpha+4)(\alpha+4-2N)(\alpha+8-2N)}{16}. \tag{38}$$

The map $\alpha \mapsto H(\alpha)$ is increasing, $H(0) = 0$ and $H(N-4) = H_N$ so that $0 \leq H(\alpha) \leq H_N$ for all $\alpha \in [0, N-4]$. For $\alpha < N-4$, let $\gamma_N(\alpha) := \sqrt{N^2 - \alpha^2 + 2\alpha(N-4)}$ and

$$\overline{u}_\alpha(x) := |x|^{-\frac{\alpha}{2}} - |x|^{\frac{4-N+\gamma_N(\alpha)}{2}} \in H^2 \cap H_0^1(B).$$

The function $\overline{u}_\alpha$ is a *positive* solution to problem (22) with $h = H(\alpha) < H_N$ and $c = c(\alpha)$, where

$$c(\alpha) := \frac{\alpha^2 - \alpha(N-5) - N^2 + 3N - 4 + (N-3)\gamma_N(\alpha)}{\alpha + 4 - N + \gamma_N(\alpha)}. \tag{39}$$

The map $[0, N-4] \ni \alpha \mapsto c(\alpha)$ is increasing, $c(0) = -1$ and

$$C_N = c(N-4) = N - 3 - \frac{\sqrt{2(N^2 - 4N + 8)}}{2}.$$

Since the first eigenfunction $u_{h(c)}$ of problem (22) is unique (by Lemma 3), when $\Omega = B$, it must be a radial function. Furthermore, $u_{h(c)}$ turns to be the only positive eigenfunction. To see this, let $v_{\bar{h}(c)}$ be another positive eigenfunction, corresponding to some $\bar{h}(c) > h(c)$. Write (23), first with $u_{h(c)}$ and test with $v_{\bar{h}(c)}$, then with $v_{\bar{h}(c)}$ and test with $u_{h(c)}$. Subtracting, we get

$$h(c) \int_B \frac{u_{h(c)} v_{\bar{h}(c)}}{|x|^4} \, dx = \bar{h}(c) \int_B \frac{u_{h(c)} v_{\bar{h}(c)}}{|x|^4} \, dx,$$

a contradiction. By this, we conclude that $u_{h(c)} = \overline{u}_\alpha$, where $c = c(\alpha)$. Namely, $\overline{u}_\alpha$ is the minimizer of $h(c(\alpha)) = H(\alpha)$ for every $\alpha \in [0, N-4)$. In turn, this shows

Theorem 2 *For every $0 \le \alpha \le N-4$, there holds*

$$\int_B |D^2 u|^2 \, dx + c(\alpha) \int_{\partial B} u_\nu^2 \, d\sigma \ge H(\alpha) \int_B \frac{u^2}{|x|^4} \, dx \quad \textit{for all } u \in H^2 \cap H_0^1(B),$$

where $H(\alpha)$ and $c(\alpha)$ are defined in (38) *and* (39). *Furthermore, the best constant $H(\alpha)$ is attained if and only if $0 \le \alpha < N-4$, by multiples of the function*

$$\overline{u}_\alpha(x) = |x|^{-\frac{\alpha}{2}} - |x|^{\frac{4-N+\sqrt{N^2-\alpha^2+2\alpha(N-4)}}{2}}.$$

As a corollary of Theorem 2, we readily get (6) and (7). We just remark that, to get (7), one has to determine the unique solution α_N to the equation

$$c(\alpha) = 0 \quad \text{for } \alpha \in (0, N-4) \text{ and } N \ge 7.$$

By (39), we have that

$$\begin{aligned} &c(\alpha) = 0 \quad \Leftrightarrow \\ &\alpha^4 - 2(N-5)\alpha^3 - 2(5N-13)\alpha^2 + 4\big(N^2 - 7N + 8\big) + 8\big(N^2 - 3N + 2\big) = 0 \end{aligned}$$

and the above polynomial can be factorized as follows

$$\begin{aligned} &(\alpha + 1 - \sqrt{2N-1})(\alpha + 1 + \sqrt{2N-1}) \\ &\quad \times \big(\alpha - N + 4 - \sqrt{N^2 - 4N + 8}\big)\big(\alpha - N + 4 + \sqrt{N^2 - 4N + 8}\big) = 0. \end{aligned}$$

Then, since $\alpha \in (0, N-4)$ and $N \ge 7$, we obtain the unique solution $\alpha_N = \sqrt{2N-1} - 1$. Finally, $H(\alpha_N)$, with $H(\alpha)$ as in (38), is the optimal constant in (7). See also Fig. 1 for the trace of the curve $(0, N-4) \ni \alpha \mapsto (c(\alpha), H(\alpha))$ (or, equivalently, the plot of the map $(-c_0, +\infty) \ni c \mapsto h(c)$), when $N = 5$ and $N = 8$.

Appendix

Let $\Omega \subset \mathbf{R}^N (N \ge 5)$ be a bounded domain such that $0 \in \Omega$ and $\partial\Omega \in C^2$. Denote by $|\Omega|$ its N-dimensional Lebesgue measure and by $\omega_N = |B|$, where B is the unit ball. Finally, set $\gamma = j_0^2 \approx 2.4^2$, where j_0 is the first positive zero of the Bessel function J_0, and

$$A_N = A_N(\Omega) := \frac{N(N-4)}{2} \gamma \left(\frac{\omega_N}{|\Omega|}\right)^{2/N}. \tag{40}$$

Let $H_N := \frac{N^2(N-4)^2}{16}$. From [11, Theorem 2], we know that

$$\int_\Omega |\Delta u|^2\,dx \ge H_N \int_\Omega \frac{u^2}{|x|^4}\,dx + A_N \int_\Omega \frac{u^2}{|x|^2}\,dx \quad \text{for all } u \in H^2 \cap H_0^1(\Omega). \tag{41}$$

Next we prove

Proposition 3 *Let $0 < h \le H_N$ and $d_1(h)$ be the optimal constant in (20). If Ω is strictly starshaped with respect to the origin, then*

$$d_0 > d_1(h) \ge \frac{2A_N m}{MA_N + h + 4}, \tag{42}$$

where d_0 is as in (8), A_N is as in (40), $M := \max_{\partial\Omega} |x|^2$ and $m := \min_{\partial\Omega}(x \cdot \nu)$.

Proof For $0 < h < H_N$, let $u_h \in H^2 \cap H_0^1(\Omega)$ be the (positive and superharmonic) function which achieves the equality in (20). Notice that u_h solves the equation in (22) subject the conditions $u_h = 0 = \Delta u_h = d_1(h)(u_h)_\nu$ on $\partial\Omega$. By (41), we get

$$\begin{aligned} d_1(h)\int_{\partial\Omega} (u_h)_\nu^2\,d\sigma &= \int_\Omega |\Delta u_h|^2\,dx - h\int_\Omega \frac{u_h^2}{|x|^4}\,dx \\ &\ge (H_N - h)\int_\Omega \frac{u_h^2}{|x|^4}\,dx + A_N\int_\Omega \frac{u_h^2}{|x|^2}\,dx. \end{aligned} \tag{43}$$

Next, in the spirit of the computations performed in [7, Theorem 1], we deduce

$$\begin{aligned} \int_\Omega \frac{u_h^2}{|x|^2}\,dx &= \int_\Omega \big(|x|^2 u_h\big)\frac{u_h}{|x|^4}\,dx = \frac{1}{h}\int_\Omega \big(|x|^2 u_h\big)\,\Delta^2 u_h\,dx \\ &= \frac{1}{h}\int_\Omega \Delta\big(|x|^2 u_h\big)\,\Delta u_h\,dx - \frac{1}{h}\int_{\partial\Omega} |x|^2 \Delta u_h (u_h)_\nu\,d\sigma \\ &= \frac{1}{h}\int_\Omega \Delta u_h\,\big(2N u_h + 4x\cdot\nabla u_h + |x|^2\Delta u_h\big)\,dx \\ &\quad - \frac{d_1(h)}{h}\int_{\partial\Omega} |x|^2 (u_h)_\nu^2\,d\sigma. \end{aligned}$$

From [17, formula (1.3)], we have

$$\begin{aligned} \int_\Omega \Delta u_h (x\cdot\nabla u_h)\,dx &= \frac{N-2}{2}\int_\Omega |\nabla u_h|^2\,dx + \frac{1}{2}\int_{\partial\Omega} (x\cdot\nu)(u_h)_\nu^2\,d\sigma \\ &= -\frac{N-2}{2}\int_\Omega u_h \Delta u_h\,dx + \frac{1}{2}\int_{\partial\Omega} (x\cdot\nu)(u_h)_\nu^2\,d\sigma \end{aligned}$$

and we conclude

$$\begin{aligned} \int_\Omega \frac{u_h^2}{|x|^2}\,dx &= \frac{1}{h}\int_\Omega \big(4u_h\Delta u_h + |x|^2|\Delta u_h|^2\big)\,dx \\ &\quad + \frac{1}{h}\int_{\partial\Omega} \big(2(x\cdot\nu) - d_1(h)|x|^2\big)(u_h)_\nu^2\,d\sigma. \end{aligned}$$

Finally, by exploiting the Young's inequality

$$\left|\int_\Omega u_h \Delta u_h \, dx\right| \leq \frac{1}{4}\int_\Omega |x|^2 |\Delta u_h|^2 \, dx + \int_\Omega \frac{u_h^2}{|x|^2} \, dx,$$

we deduce

$$\left(1+\frac{4}{h}\right)\int_\Omega \frac{u_h^2}{|x|^2} \, dx \geq \frac{2m - Md_1(h)}{h}\int_{\partial\Omega} (u_h)_\nu^2 \, d\sigma,$$

where m and M are defined in the statement. Plugging this into (43), (42) follows for $h < H_N$.

The estimate for $d_1(H_N)$ comes by letting $h \to H_N$ in (42). Indeed, by definition of $d_1(H_N)$, we know that for all $\varepsilon > 0$ there exists $u_\varepsilon \in H^2 \cap H_0^1(\Omega) \setminus H_0^2(\Omega)$ such that

$$\frac{\int_\Omega |\Delta u_\varepsilon|^2 \, dx - H_N \int_\Omega \frac{u_\varepsilon^2}{|x|^4} \, dx}{\int_{\partial\Omega} (u_\varepsilon)_\nu^2 \, dS} < d_1(H_N) + \varepsilon.$$

Then, for all $h < H_N$ we have

$$\begin{aligned} d_1(H_N) \leq d_1(h) &\leq \frac{\int_\Omega |\Delta u_\varepsilon|^2 \, dx - H_N \int_\Omega \frac{u_\varepsilon^2}{|x|^4} \, dx}{\int_{\partial\Omega} (u_\varepsilon)_\nu^2 \, d\sigma} + (H_N - h)\frac{\int_\Omega \frac{u_\varepsilon^2}{|x|^4} \, dx}{\int_{\partial\Omega} (u_\varepsilon)_\nu^2 \, d\sigma} \\ &< d_1(H_N) + \varepsilon + C_\varepsilon (H_N - h). \end{aligned}$$

Hence,

$$\lim_{h \to H_N} d_1(h) = d_1(H_N)$$

and we conclude. □

References

1. Adams, R.A.: Sobolev Spaces. Pure and Applied Mathematics, vol. 65. Academic Press, New York (1975)
2. Adimurthi, Grossi, M., Santra, S.: Optimal Hardy-Rellich inequalities, maximum principle and related eigenvalue problem. J. Funct. Anal. **240**, 36–83 (2006)
3. Adolfsson, V.: L^2-integrability of second-order derivatives for Poisson's equation in nonsmooth domains. Math. Scand. **70**(1), 146–160 (1992)
4. Agmon, S., Douglis, A., Nirenberg, L.: Estimates near the boundary for solutions of elliptic partial differential equations satisfying general boundary conditions. II. Commun. Pure Appl. Math. **17**, 35–92 (1964)
5. Berchio, E., Gazzola, F.: Best constants and minimizers for embeddings of second order Sobolev spaces. J. Math. Anal. Appl. **320**(2), 718–735 (2006)
6. Berchio, E., Gazzola, F., Mitidieri, E.: Positivity preserving property for a class of biharmonic elliptic problems. J. Differ. Equ. **229**, 1–23 (2006)
7. Berchio, E., Cassani, D., Gazzola, F.: Hardy-Rellich inequalities with boundary remainder terms and applications. Manuscr. Math. **131**(3–4), 427–458 (2010)
8. Brezis, H., Lieb, E.: A relation between pointwise convergence of functions and convergence of functionals. Proc. Am. Math. Soc. **88**, 486–490 (1983)

9. Davies, E.B., Hinz, A.M.: Explicit constants for Rellich inequalities in $L_p(\Omega)$. Math. Z. **227**, 522–523 (1998)
10. Ferrero, A., Gazzola, G., Weth, T.: On a fourth order Steklov eigenvalue problem. Analysis **25**(4), 315–332 (2005)
11. Gazzola, F., Grunau, H.-Ch., Mitidieri, E.: Hardy inequalities with optimal constants and remainder terms. Trans. Am. Math. Soc. **356**, 2149–2168 (2004)
12. Gazzola, F., Grunau, H.-Ch., Sweers, G.: Optimal Sobolev and Hardy-Rellich constants under Navier boundary condition. Ann. Mat. Pura Appl. **189**(3), 475–486 (2010)
13. Gazzola, F., Grunau, H.-Ch., Sweers, G.: Polyharmonic Boundary Value Problems. Positivity Preserving and Nonlinear Higher Order Elliptic Equations in Bounded Domains. Lecture Notes in Mathematics, vol. 1991. Springer, Berlin (2010)
14. Grisvard, P.: Elliptic Problems in Nonsmooth Domains. Pitman, London (1985)
15. Hardy, G.H.: Notes on some points in the integral calculus. Messenger Math. **48**, 107–112 (1919)
16. Hardy, G.H., Littlewood, J.E., Pólya, G.: Inequalities. Cambridge University Press, Cambridge (1934)
17. Mitidieri, E.: A simple approach to Hardy's inequalities. Math. Notes **67**, 479–486 (2000)
18. Necas, J.: Les méthodes directes en théorie des équations elliptiques. Masson, Paris (1967) (French)
19. Rellich, F.: Halbbeschränkte Differentialoperatoren höherer Ordnung. In: Gerretsen, J.C.H. et al. (eds.) Proceedings of the International Congress of Mathematicians III, Amsterdam, 1954, pp. 243–250. Nordhoff, Groningen (1956)

Power Concavity for Solutions of Nonlinear Elliptic Problems in Convex Domains

Massimiliano Bianchini and Paolo Salani

Abstract We investigate convexity properties of solutions to elliptic Dirichlet problems in convex domains. In particular we give conditions on the operator F such that a suitable power of a positive solution u of a fully nonlinear equation $F(x,u,Du,D^2u)=0$ in a convex domain Ω, vanishing on $\partial\Omega$, is concave.

Keywords Convexity of solutions · Power concavity · Elliptic equations

1 Introduction

Let $\Omega \subset \mathbb{R}^n$ be a convex set. We will deal with problems of the following type

$$\begin{cases} F(x,u,Du,D^2u)=0 & \text{in } \Omega, \\ u=0 & \text{on } \partial\Omega, \\ u>0 & \text{in } \Omega, \end{cases} \tag{1}$$

where $F(x,t,\xi,A)$ is a real elliptic operator acting on $\mathbb{R}^n \times \mathbb{R} \times \mathbb{R}^n \times S_n$. Here Du and D^2u are the gradient and the Hessian matrix of the function u respectively, and S_n is the set of the $n \times n$ real symmetric matrices, Ω is a bounded open convex subset of $\mathbb{R}^n$.

The goal is to give assumptions that yield the concavity of some power of u. More precisely, we look for *p-concave* solutions of (1) for some $p \leq 1$; we recall hereafter the definition of p-concavity.

Definition 1 A positive function u (defined in a convex set) is said *p-concave* for some $p \neq 0$ if $\frac{p}{|p|}u^p$ is concave, while it is said *log-concave* (or *0-concave*) if $\log(u)$ is concave.

M. Bianchini · P. Salani (✉)
Dipartimento di Matematica "U. Dini", Università di Firenze, viale Morgagni 67/A, 50134 Firenze, Italy
e-mail: paolo.salani@math.unifi.it

M. Bianchini
e-mail: massimiliano.bianchini@math.unifi.it

R. Magnanini et al. (eds.), *Geometric Properties for Parabolic and Elliptic PDE's*, Springer INdAM Series 2, DOI 10.1007/978-88-470-2841-8_3,

See also the equivalent Definition 2. The case $p = 1$ corresponds to usual concavity.

Many authors investigated this question. It is for instance well known that the solution of the torsional rigidity problem (i.e. $F = \Delta u + 1$ in our notation) in a convex domain is $1/2$-concave [7] and that any positive eigenfunction of the Laplacian associated to the first Dirichlet eigenvalue in a convex domain is log-concave [3]; see also [7, 15–17] for instance. Refer to [13] for a good presentation of related problems and for a comprehensive bibliography of classical results, more recent related results (and references) are for instance in [1, 2, 5, 6, 18–21, 23, 25, 26].

In the fundamental paper [1] the authors give conditions which ensure the concavity of solutions u of (1). Recently, the method of [1] has been refined in [26], obtaining also some new rearrangement inequalities for solutions of (1) and Brunn-Minkowski inequalities for possibly related functionals. Thanks to [1], one can then investigate the p-concavity of u by writing the equation that governs u^p and proving that this equation satisfies the condition given therein. On the other hand, to choose a suitable p, to write the equation for u^p and then to verify the assumptions of [1, 26] may be sometimes hard and by no means trivial. In the present paper we show a way to track down a suitable p and retrieve the p-concavity of u directly from (1), without writing the equation governing u^p.

The method adopted here is a suitable adaptation of the one introduced in [10] to study the quasi-concavity of solutions to elliptic problems in convex rings and it makes use of the *p-concave envelope* u_p of the function u: roughly speaking, u_p is the smallest *p-concave* function greater than or equal to u. We look for conditions that imply $u = u_p$. Since we have $u_p \geq u$ by definition, we have just to take care of the reverse inequality. For this we consider operators F that satisfy the *comparison principle* for viscosity solutions and we look for conditions that force u_p to be a viscosity subsolution of (1). The main result is Theorem 3.1, where we prove that (for $p \neq 0$) the latter is true if the set

$$\left\{(x,t,A) \in \Omega \times (0,+\infty) \times S_n : F\left(x, t^{\frac{1}{p}}, t^{\frac{1}{p}-1}\theta, t^{\frac{1}{p}-3}A\right) \geq 0\right\}$$

is convex for every $\theta \in \mathbb{R}^n$.

For more details and for the corresponding assumption in case $p = 0$, see Sect. 3.

The paper is organized as follows. Section 2 contains notation and some preliminaries. In Sect. 3 we state Theorem 3.1 and give some comments. In Sect. 4 we present in detail the notion of p-concave envelope of a non-negative function and we prove some related technical lemmata. Section 5 is devoted to the proof of Theorem 3.1. Finally we give some examples of applications and some final remarks in Sect. 6.

2 Preliminaries

For $A \subset \mathbb{R}^n$, we denote by $\overline{A}$ its closure and by ∂A its boundary.

Let $n \geq 2$, for $x \in \mathbb{R}^n$ and $r > 0$, $B(x,r)$ is the Euclidean ball of radius r centered at x, i.e.

$$B(x,r) = \{z \in \mathbb{R}^n : |z - x| < r\}.$$

We denote by S_n the space of $n \times n$ real symmetric matrices and by S_n^+ and S_n^{++} the cones of nonnegative and positive definite matrices, respectively. If $A, B \in S_n$, by $A \geq 0$ (> 0) we mean that $A \in S_n^+$ (S_n^{++}) and $A \geq B$ means $A - B \geq 0$.

For a natural number m and $a = (a_1, \dots, a_m) \in \mathbb{R}^m$, by $a \geq 0$ (> 0) we mean $a_i \geq 0$ (> 0) for $i = 1, \dots, m$. Moreover we set

$$\Lambda_m = \left\{ \lambda = (\lambda_1, \dots, \lambda_m) \geq 0 : \sum_{i=1}^{m} \lambda_i = 1 \right\}.$$

By the symbol $\otimes$ we denote the direct product between vectors in $\mathbb{R}^n$, that is, for $x, y \in \mathbb{R}^n$, $x \otimes y$ is the $n \times n$ matrix with entries $(x_i y_j)$ for $i, j = 1, \dots, n$.

We will make use of basic viscosity techniques; here we recall only few notions and we refer the reader to the User's Guide [9] for more details.

The operator $F : \mathbb{R}^n \times \mathbb{R} \times \mathbb{R}^n \times S_n \to \mathbb{R}$ is said *proper* if

$$F(x, r, \xi, A) \leq F(x, s, \xi, A) \quad \text{whenever } r \geq s. \tag{2}$$

Let Γ be a convex cone in S_n, with the vertex at the origin and containing the cone of nonnegative definite symmetric matrices S_n^+. We say that F is *degenerate elliptic* in Γ if

$$F(x, u, \xi, A) \leq F(x, u, \xi, B) \quad \text{whenever } A \leq B, \ A, B \in \Gamma. \tag{3}$$

We set $\Gamma_F = \bigcup \Gamma$, where the union is extended to every cone Γ such that F is degenerate elliptic in Γ; when we say that F is degenerate elliptic, we mean F is degenerate elliptic in $\Gamma_F \neq \emptyset$.

Let u be an upper semicontinuous function and ϕ a continuous function in an open set Ω; we say that ϕ *touches u by above at $x_0 \in \Omega$* if

$$\phi(x_0) = u(x_0) \quad \text{and} \quad \phi(x) \geq u(x) \quad \text{in a neighborhood of } x_0.$$

Analogously, let u be a lower semicontinuous function and ϕ a continuous function in an open set Ω; we say that ϕ *touches u by below at $x_0 \in \Omega$* if

$$\phi(x_0) = u(x_0) \quad \text{and} \quad \phi(x) \leq u(x) \quad \text{in a neighborhood of } x_0.$$

An upper semicontinuous function u is a *viscosity subsolution* of the equation $F = 0$ if, for every admissible C^2 function ϕ touching u by above at any point $x \in \Omega$, it holds

$$F\big(x, u(x), D\phi(x), D^2\phi(x)\big) \geq 0.$$

A lower semicontinuous function u is a *viscosity supersolution* of $F = 0$ if, for every admissible C^2 function ϕ touching u by below at any point $x \in \Omega$, it holds

$$F\big(x, u(x), D\phi(x), D^2\phi(x)\big) \leq 0.$$

A *viscosity solution* is a continuous function which is a viscosity subsolution and supersolution of $F = 0$ at the same time. In our assumptions a *classical solution* is always a viscosity solution and a viscosity solution is a classical solution if it is regular enough.

The technique proposed in this paper requires the use of the comparison principle for viscosity solutions. Since we will have only to compare a viscosity subsolution with a classical solution, we will need only the following weak version of the comparison principle. Precisely, we say that the operator F satisfies *the Comparison Principle* if the following statement holds:

$$\begin{array}{l} \textit{Let } u \in C(\overline{\Omega}) \cap C^2(\Omega) \textit{ and } v \in C(\overline{\Omega}) \textit{ be respectively} \\ \textit{a classical supersolution and a viscosity subsolution} \\ \textit{of } F = 0 \textit{ such that } u \geq v \textit{ on } \partial\Omega. \textit{ Then } u \geq v \textit{ in } \Omega. \end{array} \tag{WCP}$$

We recall that $u \in C^2(\Omega)$ is a classical supersolution of $F = 0$ if $F(x, u(x), Du(x), D^2u(x)) \leq 0$ for every $x \in \Omega$. In particular, a classical solution is a classical supersolution.

Comparison Principles for viscosity solutions are an actual and deep field of investigation and it is out of our aims to give here an update picture of the state of the art, then we just refer to the User's Guide [9]. However, when one of the involved function is regular, the situation is much easier and the (WCP) is for instance satisfied if F is strictly monotone with respect to u, in other words if it is strictly proper.

3 The Main Result

For $\theta \in \mathbb{R}^n$ and $p \leq 1$ we define $G_{\theta,p} : \Omega \times (0, +\infty) \times \Gamma_F \to \mathbb{R}$ as

$$G_{\theta,p}(x, t, A) = F\big(x, t^{\frac{1}{p}}, t^{\frac{1}{p}-1}\theta, t^{\frac{1}{p}-3}A\big) \quad \text{for } p \neq 0, \tag{4}$$

and

$$G_{\theta,0}(x, t, A) = F\big(x, e^t, e^t\theta, e^t A\big). \tag{5}$$

Theorem 1 *Let Ω be an open bounded convex set, F a proper degenerate elliptic operator which satisfies the comparison principle* (WCP) *and $u \in C(\overline{\Omega}) \cap C^2(\Omega)$ a classical solution of* (1) *in Ω.*

Assume that there exists $p < 1$ such that

$$\big\{(x, t, A) \in \Omega \times (0, +\infty) \times \Gamma_F : G_{\theta,p}(x, t, A) \geq 0\big\} \quad \textit{is convex for every } \theta \in \mathbb{R}^n. \tag{6}$$

If $p > 0$, assume furthermore that

$$\liminf_{x \to x_0} \frac{\partial u(x)}{\partial \nu} > 0 \tag{7}$$

for every $x_0 \in \partial\Omega$, where ν is any inward direction of Ω at x_0.

Then u is p-concave in Ω.

The proof of Theorem 1 is given in Sect. 5.

Notice that assumption (7) is needed only for $p > 0$ and it is in general provided by a suitable version of the Hopf's Lemma.

The idea of the proof is to show that the function $v = u^p$ (or $v = \log u$ is $p = 0$) in fact coincides with its concave envelope v^* and the role of (7) is to guarantee that the *the contact set* of v (that is the set of points where v and v^* coincide) does not touch the boundary of the domain. In the case $p \leq 0$ this requirement is automatically satisfied since $|v| \to +\infty$ as $x \to \partial\Omega$; when $p \in (0, 1)$, assumption (7) implies

$$\liminf_{x \to x_0} \frac{\partial v(x)}{\partial \nu} = +\infty \tag{8}$$

for every $x_0 \in \partial\Omega$ and every inward direction ν of Ω at x_0, and the latter forces the contact set of v to stay away from $\partial\Omega$.

In fact, we can also apply our argument to the case $p = 1$, that is to prove usual concavity of solutions; in such a case however we need to assume directly (8), i.e. we have to substitute assumption (7) with the following

$$\limsup_{x \to x_0} \frac{\partial u(x)}{\partial \nu} = +\infty \quad \text{for every } x_0 \in \partial\Omega.$$

The latter coincides with the so called *zero contact angle boundary condition* which occurs in capillarity and it is a typical assumption in this kind of investigation (see for instance [16, 17]).

We finally notice that it could be even possible to remove this assumption and unify the treatment of $p > 0$ and $p \leq 0$ by considering state constraint boundary conditions, as in [1] and [26].

4 The p-Concave Envelope of a Function

Before proving Theorem 1, we need some preliminary definitions and results. First of all we recall the notion of p-means; for more details we refer to [11].

Given $2 \leq m \in \mathbb{N}$, $a = (a_1, \dots, a_m) > 0$, $\lambda \in \Lambda_m$ and $p \in [-\infty, +\infty]$, the quantity

$$M_p(a, \lambda) = \begin{cases} [\lambda_1 a_1^p + \lambda_2 a_2^p + \cdots + \lambda_m a_m^p]^{1/p} & \text{for } p \neq -\infty, 0, +\infty \\ \max\{a_1, \dots, a_m\} & p = +\infty \\ a_1^{\lambda_1} \cdots a_m^{\lambda_m} & p = 0 \\ \min\{a_1, a_2, \dots, a_m\} & p = -\infty \end{cases} \tag{9}$$

is the (λ-weighted) p-mean of a. For $a \geq 0$, we define $M_p(a, \lambda)$ as above if $p \geq 0$ and we set $M_p(a, \lambda) = 0$ if $p < 0$ and $a_i = 0$ for some $i = 1, \dots, m$.

A simple consequence of Jensen's inequality is that, for a fixed $0 \leq a \in \mathbb{R}^m$ and $\lambda \in \Lambda_m$,

$$M_p(a, \lambda) \leq M_q(a, \lambda) \quad \text{if } p \leq q. \tag{10}$$

Moreover, it is easily seen that

$$\lim_{p\to+\infty} M_p(a,\lambda)=\max\{a_1,\dots,a_m\} \tag{11}$$

and

$$\lim_{p\to-\infty} M_p(a,\lambda)=\min\{a_1,\dots,a_m\}. \tag{12}$$

Notice that the definition of p-concavity can be now equivalently restated in the following way.

Definition 2 A non-negative function u defined in a convex set is said *p-concave* for some $p\in[-\infty,+\infty)$ if

$$u\big((1-\mu)x_1+\mu x_2\big)\ge M_p\big(\big(u(x_1),u(x_2)\big),(1-\mu,\mu)\big)$$
$$\text{for every } x_1,x_2\in\Omega,\ \mu\in[0,1]. \tag{13}$$

For $p=1$ we have usual concavity and for $p=0$ we have log-concavity, as already said in the introduction; $-\infty$-concave functions are usually said *quasi-concave*. Due to (10) we have that the p-concavity property is monotone, in the sense that if u is q-concave for some q, then it is p-concave for every $p\le q$.

Let us fix $\lambda\in\Lambda_{n+1}$ and $p\in[-\infty,+\infty]$.

Definition 3 Let $\Omega\subset\mathbb{R}^n$ be an open bounded convex set and $u\in C(\overline{\Omega})$, $u>0$ in Ω; the (p,λ)-*envelope* of u is the function $u_{p,\lambda}:\overline{\Omega}\to\mathbb{R}$ defined as follows

$$u_{p,\lambda}(x)=\sup\left\{M_p\big(u(x_1),\dots,u(x_{n+1}),\lambda\big):x_i\in\overline{\Omega},\ i=1,\dots,n+1,\right.$$
$$\left.x=\sum_{i=1}^{n+1}\lambda_i x_i\right\}. \tag{14}$$

Although the above definition is given also for $p=\pm\infty$, throughout we consider only $p\in(-\infty,+\infty)$, if not otherwise specified.

Notice that, as $\overline{\Omega}$ is compact and M_p is continuous, the supremum in (14) is in fact a maximum. Hence, for every $\bar{x}\in\overline{\Omega}$, there exist $x_{1,p},\dots,x_{n+1,p}\in\overline{\Omega}$ such that

$$\bar{x}=\sum_{i=1}^{n+1}\lambda_i x_{i,p},\quad u_{p,\lambda}(\bar{x})=\begin{cases}\left(\sum_{i=1}^{n+1}\lambda_i u(x_{i,p})^p\right)^{1/p} & \text{if } p\neq 0,\\ \prod_{i=1}^{n+1}u(x_i)^{\lambda_i} & \text{if } p=0.\end{cases} \tag{15}$$

An immediate consequence of the definition is that

$$u_{p,\lambda}(x)\ge u(x),\quad \forall x\in\overline{\Omega},\ p\in[-\infty,+\infty]. \tag{16}$$

Moreover, from (10), we have

$$u_{p,\lambda}(x)\le u_{q,\lambda}(x),\quad \text{for } p\le q,\ x\in\Omega. \tag{17}$$

Notice that if the points $x_{i,p} \in \Omega$ in (15) lie in the interior of Ω for $i = 1, \dots, n+1$, then by the Lagrange Multipliers Theorem we have

$$u(x_{1,p})^{p-1} Du(x_{1,p}) = \cdots = u(x_{n+1,p})^{p-1} Du(x_{n+1,p}). \tag{18}$$

We also set

$$u_p(x) = \sup\{u_{p,\lambda}(x) : \lambda \in \Lambda_{n+1}\}.$$

We notice that the above supremum is in fact a maximum too in our assumptions and that u_p is the smallest *p-concave* function greater than or equal to u by Carathéodory's Theorem, see [24, Theorem 17.1 and Corollary 17.1.5].

Lemma 1 *Let $p < 1$, $\lambda \in \Lambda_{n+1}$, $\Omega \subset \mathbb{R}^n$ an open bounded convex set, $u \in C(\overline{\Omega})$ such that $u = 0$ on $\partial\Omega$ and $u > 0$ in Ω. Then $u_{p,\lambda} \in C(\overline{\Omega})$ and*

$$u_{p,\lambda} > 0 \quad in\ \Omega, \qquad u_{p,\lambda} = 0 \quad on\ \partial\Omega. \tag{19}$$

Proof The proof of (19) is almost straightforward and for $p \le 0$ has been already explicitly given in [10, Lemma 4.1]. The case $p \in (0, 1)$ is completely analogous, but for the sake of completeness we give hereafter a sketch of the proof. That $u_{p,\lambda} > 0$ in Ω comes from (16), while $u_{p,\lambda} = 0$ on $\partial\Omega$ is a simple consequence of the convexity of Ω. Regarding continuity, by definition $u^p_{p,\lambda}$ is the supremal convolution of $n+1$ copies of u^p, then we can refer to [27, Corollary 2.1] to get $u^p_{p,\lambda} \in C(\Omega)$, or we can simply observe that it is the supremum of a family of uniformly continuous functions, each of these sharing the same modulus of continuity, then $u^p_{p,\lambda} \in C(\overline{\Omega})$. □

Lemma 2 *Let $p \le 0$, $\lambda \in \Lambda_{n+1}$, $\Omega \subset \mathbb{R}^n$ an open bounded convex set, $u \in C(\overline{\Omega})$ such that $u = 0$ on $\partial\Omega$, $u > 0$ on Ω. If $x \in \Omega$, then the point $x_{i,p}$ defined by (15) belongs to Ω for $i = 1, \dots, n+1$.*

Proof Since $x \in \Omega$, then $u(x) > 0$ and the thesis follows immediately from (16) and the definition of p-means with $p \le 0$ (with some null a_i). □

Lemma 3 *Let $0 < p < 1$, $\lambda \in \Lambda_{n+1}$, $\Omega \subset \mathbb{R}^n$ an open bounded convex set, $u \in C^1(\Omega) \cap C(\overline{\Omega})$ such that $u = 0$ on $\partial\Omega$, $u > 0$ on Ω. Assume that (7) holds true.*

If $x \in \Omega$, then the points $x_{1,p}, \dots, x_{n+1,p}$ defined by (15) belong to Ω.

Proof By contradiction, assume that (up to a relabeling) $x_1 \in \partial\Omega$ (and $\lambda_1 > 0$). Notice that at least one of the x_i's must be in the interior of Ω and $\lambda_i > 0$, otherwise $u_{p,\lambda}(x) = 0$, while $u_{p,\lambda}(x) \ge u(x) > 0$. Then let $x_{n+1} \in \Omega$ (with $\lambda_{n+1} > 0$), set $v = u^p$ and

$$a = |Dv(x_{n+1})| = pu(x_{n+1})^{p-1} |Du(x_{n+1})|.$$

By the regularity of u, we have

$$|Dv| = pu^{p-1}|Du| < a + 1 \quad \text{in } B(x_{n+1}, r_{n+1}) \subset \Omega \tag{20}$$

for $r_{n+1} > 0$ small enough.

On the other hand, the direction

$$\nu = (x_{n+1} - x_1)/|x_{n+1} - x_1|$$

is pointing inward at x_1 and by assumption (7) we get (8), whence

$$\frac{\partial v}{\partial \nu} > a + 1 \quad \text{in } \Omega \cap B(x_1, r_1) \tag{21}$$

for $r_1 > 0$ small enough.

Next we take $\rho < \min\{\lambda_1 r_1, \lambda_{n+1} r_{n+1}\}$ and we consider the points

$$\begin{aligned} \tilde{x}_1 &= x_1 + \frac{\rho}{\lambda_1}\nu, \\ \tilde{x}_i &= x_i \quad \text{for } i = 2, \dots, n, \\ \tilde{x}_{n+1} &= x_{n+1} - \frac{\rho}{\lambda_{n+1}}\nu. \end{aligned}$$

We have

$$\tilde{x}_1 \in B(x_1, r_1), \qquad \tilde{x}_{n+1} \in B(x_{n+1}, r_{n+1})$$

and

$$x = \sum_{i=1}^{n+1} \lambda_i \tilde{x}_i. \tag{22}$$

Then by (20) and (21) we get

$$\begin{aligned} u(\tilde{x}_1)^p &= v(\tilde{x}_1) > v(x_1) + (a+1)\frac{\rho}{\lambda_1} = (a+1)\frac{\rho}{\lambda_1}, \\ u(\tilde{x}_{n+1})^p &= v(\tilde{x}_{n+1}) \geq v(x_{n+1})^p - (a+1)\frac{\rho}{\lambda_{n+1}} = u(x_{n+1})^p - (a+1)\frac{\rho}{\lambda_{n+1}}, \end{aligned}$$

whence

$$\begin{aligned} &\left(\sum_{i=1}^{n+1} \lambda_i u(\tilde{x}_i)^p\right)^{1/p} \\ &> \left(\lambda_1 (a+1)\frac{\rho}{\lambda_1} + \sum_{i=2}^{n} \lambda_i u(x_i)^p + \lambda_{n+1} u(x_{n+1})^p - \lambda_{n+1}(a+1)\frac{\rho}{\lambda_{n+1}}\right)^{1/p} \\ &= \left(\sum_{i=1}^{n+1} \lambda_i u(x_i)^p\right)^{1/p} = u_{p,\lambda}(x) \end{aligned}$$

which contradicts the definition of $u_{p,\lambda}$, due to (22). □

5 Proof of Theorem 1

Lemma 4 *Let $\Omega \subset \mathbb{R}^n$ be an open bounded convex set and F a degenerate elliptic operator. Assume that there exists $p \leq 0$, such that* (6) *holds. Let $u \in C(\overline{\Omega}) \cap C^2(\Omega)$ be a classical solution of* (1) *in Ω. Then $u_{p,\lambda}$ is a viscosity subsolution of* (1) *for every $\lambda \in \Lambda_{n+1}$.*

Proof Let u and Ω be as in the statement and let us fix $\lambda \in \Lambda_{n+1}$. The proof follows the steps of [10] and the strategy is the following: for every $\bar{x} \in \Omega$, we construct a $C^2(\Omega)$ function $\varphi_{p,\lambda}$ which touches the (p,λ)-envelope $u_{p,\lambda}$ of u by below at $\bar{x}$ and such that

$$F\big(\bar{x}, \varphi_{p,\lambda}(\bar{x}), D\varphi_{p,\lambda}(\bar{x}), D^2\varphi_{p,\lambda}(\bar{x})\big) \geq 0. \tag{23}$$

Clearly this implies that $u_{p,\lambda}$ is a viscosity subsolution of (1); indeed every test function ϕ touching $u_{p,\lambda}$ at $\bar{x}$ from above must also touch $\varphi_{p,\lambda}$ at $\bar{x}$ from above, then

$$\phi(\bar{x}) = \varphi_{p,\lambda}(\bar{x}), \qquad D\phi(\bar{x}) = D\varphi_{p,\lambda}(\bar{x}) \quad \text{and} \quad D^2\phi(\bar{x}) \geq D^2\varphi_{p,\lambda}(\bar{x})$$

and the conclusion follows from the ellipticity of F.

Let us start with the case $p < 0$ and consider $\bar{x} \in \Omega$. By (15) and Lemma 2, there exist $x_{1,p}, \dots, x_{n+1,p} \in \Omega$ satisfying (15) and such that (18) holds.

Now we introduce the function $\varphi_{p,\lambda} : B(\bar{x}, r) \to \mathbb{R}$, for a small enough $r > 0$, defined as follows

$$\begin{aligned} \varphi_{p,\lambda}(x) = \big[&\lambda_1 u\big(x_{1,p} + a_{1,p}(x - \bar{x})\big)^p + \cdots \\ &+ \lambda_{n+1} u\big(x_{n+1,p} + a_{n+1,p}(x - \bar{x})\big)^p\big]^{1/p} \end{aligned} \tag{24}$$

where

$$a_{i,p} = \frac{u(x_{i,p})^p}{u_{p,\lambda}(\bar{x})^p}, \qquad \text{for } i = 1, \dots, n+1. \tag{25}$$

The following facts trivially hold:

(A) $\sum_{i=1}^{n+1} \lambda_i a_{i,p} = 1$ by (15);
(B) $x = \sum_{i=1}^{n+1} \lambda_i (x_{i,p} + a_{i,p}(x - \bar{x}))$ for every $x \in B(\bar{x}, r)$, thanks to (A) and the first equation in (15);
(C) $\varphi_{p,\lambda}(\bar{x}) = u_{p,\lambda}(\bar{x})$;
(D) $\varphi_{p,\lambda}(x) \leq u_{p,\lambda}(x)$ in $B(\bar{x}, r)$ (this follows from (B) and from the definition of $u_{p,\lambda}$).

In particular, (C) and (D) say that $\varphi_{p,\lambda}$ touches $u_{p,\lambda}$ from below at $\bar{x}$.

A straightforward calculation yields

$$D\varphi_{p,\lambda}(\bar{x}) = \varphi_{p,\lambda}(\bar{x})^{1-p} \sum_{i=1}^{n+1} \lambda_i u(x_{i,p})^{p-1} a_{i,p} Du(x_{i,p}),$$

and

$$\begin{aligned} D^2\varphi_{p,\lambda}(\bar{x}) = {} & (1-p)\varphi_{p,\lambda}(\bar{x})^{-1} D\varphi_{p,\lambda}(\bar{x}) \otimes D\varphi_{p,\lambda}(\bar{x}) \\ & - (1-p)\varphi_{p,\lambda}(\bar{x})^{1-p} \sum_{i=1}^{n+1} \lambda_i u(x_{i,p})^{p-2} a_{i,p}^2 Du(x_{i,p}) \otimes Du(x_{i,p}) \\ & + \varphi_{p,\lambda}(\bar{x})^{1-p} \sum_{i=1}^{n+1} \lambda_i u(x_{i,p})^{p-1} a_{i,p}^2 D^2 u(x_{i,p}). \end{aligned}$$

Then, by (18), (25), (26) and the definition of $\varphi_{p,\lambda}$, we get

$$\begin{aligned} D\varphi_{p,\lambda}(\bar{x}) &= \varphi_{p,\lambda}(\bar{x})^{1-p}u(x_{i,p})^{p-1}Du(x_{i,p})\sum_{i=1}^{n+1}\lambda_i\frac{u(x_{i,p})^p}{\varphi_{p,\lambda}(\bar{x})^p} \\ &= \varphi_{p,\lambda}(\bar{x})^{1-p}u(x_{i,p})^{p-1}Du(x_{i,p}) \quad \text{for } i=1,\ldots,n+1 \end{aligned} \tag{26}$$

and

$$\begin{aligned} D^2\varphi_{p,\lambda}(\bar{x}) = &\sum_{i=1}^{n+1}\lambda_i\frac{u(x_{i,p})^{3p-1}}{\varphi_{p,\lambda}(\bar{x})^{3p-1}}D^2u(x_{i,p}) \\ &+(1-p)\varphi_{p,\lambda}(\bar{x})^{-1}\left[1-\varphi_{p,\lambda}(\bar{x})^{-p}\sum_{i=1}^{n+1}\lambda_i u(x_{i,p})^p\right] \\ &\times D\varphi_{p,\lambda}(\bar{x})\otimes D\varphi_{p,\lambda}(\bar{x}). \end{aligned}$$

The quantity in square brackets is equal to 0 by (C) and (15), then

$$D^2\varphi_{p,\lambda}(\bar{x}) = \sum_{i=1}^{n+1}\lambda_i\frac{u(x_{i,p})^{3p-1}}{\varphi_{p,\lambda}(\bar{x})^{3p-1}}D^2u(x_{i,p}). \tag{27}$$

Since u is a classical solution of (1), it follows that

$$\begin{aligned} &G_{\theta,p}\big(x_{i,p}, u(x_{i,p})^p, u(x_{i,p})^{3p-1}D^2u(x_{i,p})\big) \\ &\quad = F\big(x_{i,p}, u(x_{i,p}), Du(x_{i,p}), D^2u(x_{i,p})\big) = 0, \end{aligned}$$

for $i=1,\ldots,n+1$, where

$$\theta = \varphi_{p,\lambda}(\bar{x})^{p-1}D\varphi_{p,\lambda}(\bar{x}),$$

that is

$$\begin{aligned} &\big(x_{i,p}, u(x_{i,p})^p, u(x_{i,p})^{3p-1}D^2u(x_{i,p})\big) \\ &\quad \in \big\{(x,t,A)\in\Omega\times(0,+\infty)\times\Gamma_F : G_{\theta,p}(x,t,A)=0\big\} \end{aligned} \tag{28}$$

for $i=1,\ldots,n+1$. Then from (6) we have

$$\begin{aligned} &\left(\sum_{i=1}^{n+1}\lambda_i x_{i,p}, \sum_{i=1}^{n+1}\lambda_i u(x_{i,p})^p, \sum_{i=1}^{n+1}\lambda_i u(x_{i,p})^{3p-1}D^2u(x_{i,p})\right) \\ &\quad \in \big\{(x,t,A) : G_{\theta,p}(x,t,A)\geq 0\big\}, \end{aligned}$$

that is

$$G_{\theta,p}\left(\sum_{i=1}^{n+1}\lambda_i x_{i,p}, \sum_{i=1}^{n+1}\lambda_i u(x_{i,p})^p, \sum_{i=1}^{n+1}\lambda_i u(x_{i,p})^{3p-1}D^2u(x_{i,p})\right)\geq 0,$$

i.e.

$$G_{\theta,p}\big(\bar{x}, \varphi_{p,\lambda}(\bar{x})^p, \varphi_{p,\lambda}(\bar{x})^{3p-1}D^2\varphi_{p,\lambda}(\bar{x})\big)\geq 0,$$

whence (23) by the definition of $G_{\theta,p}$.

The case $p = 0$ is similar to the case $p < 0$, the only differences consisting in that we set

$$\varphi_{0,\lambda} := \exp\bigl(\lambda_1 \log u(x_{1,0} + x - \bar{x}) + \cdots + \lambda_{n+1} \log u(x_{n+1,0} + x - \bar{x})\bigr),$$

which means $a_{i,0} = 1$ for $i = 1, \dots, n+1$ and $\theta = \frac{D\varphi_{0,\lambda}(\bar{x})}{\varphi_{0,\lambda}(\bar{x})}$, and that in place of (28) we notice that the points $(x_i, \log u(x_i), u(x_i)^{-1} D^2 u(x_i))$, $i = 1, \dots, n+1$, belong to the 0-level set of $G_{\theta,0}$. □

Lemma 5 *Let $\Omega \subset \mathbb{R}^n$ be an open convex set of class C^1, $\lambda \in \Lambda_{n+1}$, F a degenerate elliptic operator, $p \in (0, 1)$, such that* (6) *holds. Let $u \in C(\overline{\Omega}) \cap C^2(\Omega)$ be a classical solution of* (1) *such that* (7) *holds. Then $u_{p,\lambda}$ is a viscosity subsolution of* (1).

Proof The proof of this lemma is the same as the proof of the previous lemma using Lemma 3 instead of Lemma 2. □

With these lemmata at hands, the proof of Theorem 1 is now very easy.

Proof of Theorem 1 Under the assumptions of the theorem we can apply the previous lemmata to obtain that $u_{p,\lambda}$ is a viscosity subsolution of (1) for every $\lambda \in \Lambda_{n+1}$. Then by the Comparison Principle (WCP) we have $u \geq u_{p,\lambda}$ in Ω, while, by construction, we have $u \leq u_{p,\lambda}$ in Ω. Hence $u = u_{p,\lambda}$ in Ω for every $\lambda \in \Lambda_{n+1}$, which finally implies $u = u_p$ and the proof is complete. □

6 Examples

For the sake of completeness, we give some examples of operators satisfying (6).

In all the following examples the function f is nonnegative.

The first one is the q-Laplacian for $q > 1$ (including in particular the Laplacian for $q = 2$):

$$F\bigl(x, u, Du, D^2u\bigr) = \Delta_q u + f(x, u, Du),$$

satisfies (6) for some $p \in (-\infty, 1]$ if the function

$$g(x, t, A) = \begin{cases} t^{q+1-\frac{q-1}{p}} f(x, t^{\frac{1}{p}}, t^{\frac{1}{p}-1}\theta) & \text{if } p \neq 0 \\ e^{t(1-q)} f(x, e^t, e^t\theta) & \text{if } p = 0 \end{cases} \tag{29}$$

is concave in $\Omega \times [0, +\infty) \times S_n$ for every fixed $\theta \in \mathbb{R}^n$.

For instance, when

$$f(x, u, Du) = cu^\alpha$$

for some $c > 0$ and $\alpha \leq q - 1$, (29) means

$$p \leq \frac{q - 1 - \alpha}{q},$$

then we can say that the solution of (1) is $\frac{(q-1-\alpha)}{q}$-concave.

In particular, when $q=2$ and $f\equiv 1$, i.e. in the case of the torsional rigidity, we get that the solution u of (1) is $\frac{1}{2}$-concave, as it is well known, while for a generic $q>1$ we get that the solution of

$$\begin{cases} \Delta_q u = -1 & \text{in } \Omega, \\ u = 0 & \text{on } \partial\Omega \end{cases}$$

is $\frac{q-1}{q}$-concave. Another particular instance is the first (positive) Dirichlet eigenfunction of the q-Laplacian, that is the solution of

$$\begin{cases} \Delta_q u = -\lambda_1 u & \text{in } \Omega, \\ u = 0 & \text{on } \partial\Omega, \\ u > 0 & \text{in } \Omega. \end{cases}$$

In this case condition (29) is satisfied up to $p=(q-2)/q$. Notice that in this case we cannot trust on the comparison principle; on the other hand Lemma 4 tells us that $u_{p,\lambda}$ satisfies

$$\begin{cases} \Delta_q u_{p,\lambda} \geq -\lambda_1 u_{p,\lambda} & \text{in } \Omega, \\ u_{p,\lambda} = 0 & \text{on } \partial\Omega, \\ u_{p,\lambda} > 0 & \text{in } \Omega \end{cases}$$

and this yields $u = u_{p,\lambda}$ thanks to the variational characterization of u.

On the other hand, the Laplacian and the q-Laplacian have been deeply investigated and the results above stated are mostly already known, see the bibliography.

In order to show some explicit examples of new applications, hereafter we consider Dirichlet problems for the Finsler Laplacian $\Delta_H u$ and the Pucci's Extremal Operator $\mathcal{M}^-_{\lambda,\Lambda}u$.

We recall that the *Finsler Laplacian* $\Delta_H u$ of a regular function u is defined as follows

$$\Delta_H u = \operatorname{div}\big(H(Du)\nabla_\xi H(Du)\big),$$

where $H(\xi)$ is a given norm in $\mathbb{R}^n$, that is a nonnegative centrally symmetric 1-homogeneous convex function (or, if you prefer, the support function of a centrally symmetric convex body), and ∇_ξ denotes the gradient with respect to the variable $\xi \in \mathbb{R}^n$. For more detail, please refer for instance to [8, 12]. Then

$$F\big(x,u,Du,D^2u\big) = \Delta_H u + f(x,u,Du)$$

satisfies (6) for some $p \in (-\infty, 1]$ if the function

$$g(x,t,A) = \begin{cases} t^{3-\frac{1}{p}} f(x, t^{\frac{1}{p}}, t^{\frac{1}{p}-1}\theta) & \text{if } p \neq 0 \\ e^{-t} f(x, e^t, e^t\theta) & \text{if } p = 0 \end{cases} \tag{30}$$

is concave in $\Omega \times [0,+\infty) \times S_n$ for every fixed $\theta \in \mathbb{R}^n$. Notice that this is the same condition as for the Laplacian. In particular the solution u of

$$\begin{cases} \Delta_H u = -1 & \text{in } \Omega, \\ u = 0 & \text{on } \partial\Omega \end{cases}$$

is $\frac{1}{2}$-concave. While a positive eigenfunction u corresponding to the first eigenvalue $\lambda_{H,1}$ of the operator $\Delta_H u$ (i.e. $f(x,u,Du) = \lambda_{H,1} u$) in a convex domain Ω is log-concave, as already proved in [14].

We finally recall that the *Pucci's Extremal Operators* were introduced by C. Pucci in [22] and they are perturbations of the usual Laplacian. Precisely, given two numbers $0 < \lambda \leq \Lambda$ and a real symmetric $n \times n$ matrix M, whose eigenvalues are $e_i = e_i(M)$, for $i = 1, \dots, n$, the Pucci's extremal operators are

$$\mathcal{M}^+_{\lambda,\Lambda}(M) = \Lambda \sum_{e_i>0} e_i + \lambda \sum_{e_i<0} e_i \tag{31}$$

and

$$\mathcal{M}^-_{\lambda,\Lambda}(M) = \lambda \sum_{e_i>0} e_i + \Lambda \sum_{e_i<0} e_i. \tag{32}$$

We recall that $\mathcal{M}^+_{\lambda,\Lambda}$ and $\mathcal{M}^-_{\lambda,\Lambda}$ are uniformly elliptic and positively homogeneous of degree 1; moreover $\mathcal{M}^+_{\lambda,\Lambda}$ is convex, while $\mathcal{M}^-_{\lambda,\Lambda}$ is concave over S_n (see [4] for instance). Our results then can be applied to the equation

$$F\big(x, u, Du, D^2u\big) = \mathcal{M}^-_{\lambda,\Lambda}\big(D^2u\big) + f(x, u, Du),$$

if (30) holds for some $p \in (-\infty, 1]$. We notice that in fact exactly the same conclusion holds for every elliptic equation of the type

$$F\big(D^2u\big) + f(x, u, Du) = 0,$$

where $F : S_n \to \mathbb{R}$ is concave and positively 1-homogeneous.

Acknowledgements We thank two unknown referees for their careful reading and for several useful suggestions that helped us to improve the paper.

References

1. Alvarez, O., Lasry, J.M., Lions, P.L.: Convex viscosity solutions and state constraints. J. Math. Pures Appl. **76**, 265–288 (1997)
2. Bian, B., Guan, P.: A microscopic convexity principle for nonlinear partial differential equations. Invent. Math. **177**, 307–335 (2009)
3. Brascamp, H.J., Lieb, E.H.: On extensions of the Brunn-Minkowski and Prékopa-Leindler theorems, including inequalities for log-concave functions, and with an application to the diffusion equation. J. Funct. Anal. **22**, 366–389 (1976)
4. Cabré, X., Caffarelli, L.A.: Fully Nonlinear Elliptic Equations. AMS Colloquium Publications, vol. 43. AMS, Providence (1995)
5. Caffarelli, L.A., Guan, P., Ma, X.N.: A constant rank theorem for solutions of fully nonlinear elliptic equations. Commun. Pure Appl. Math. **60**, 1769–1791 (2007)
6. Caffarelli, L.A., Friedman, A.: Convexity of solutions of semilinear elliptic equations. Duke Math. J. **52**, 431–456 (1985)
7. Caffarelli, L.A., Spruck, J.: Convexity of solutions to some classical variational problems. Commun. Partial Differ. Equ. **7**, 1337–1379 (1982)
8. Cianchi, A., Salani, P.: Overdetermined anisotropic elliptic problems. Math. Ann. **345**, 859–881 (2009)

9. Crandall, M.G., Ishii, H., Lions, P.L.: User's guide to viscosity solution of second order elliptic PDE. Bull. Am. Math. Soc. **27**, 1–67 (1992)
10. Cuoghi, P., Salani, P.: Convexity of level sets for solutions to nonlinear elliptic problems in convex rings. Electron. J. Differ. Equ. **124**, 12 pp. (2006) (electronic)
11. Hardy, G.H., Littlewood, J.E., Pólya, G.: Inequalities. Cambridge University Press, Cambridge (1959)
12. Ferone, V., Kawohl, B.: Remarks on a Finsler Laplacian. Proc. Am. Math. Soc. **137**, 247–253 (2009)
13. Kawohl, B.: Rearrangements and Convexity of Level Sets in P.D.E. Lectures Notes in Mathematics, vol. 1150. Springer, Berlin (1985)
14. Kawohl, B., Novaga, M.: The p-Laplace eigenvalue problem as $p \to 1$ and Cheeger sets in a Finsler metric. J. Convex Anal. **15**(3), 623–634 (2008)
15. Kennington, A.U.: Power concavity and boundary value problems. Indiana Univ. Math. J. **34**(3), 687–704 (1985)
16. Korevaar, N.J.: Capillary surface convexity above convex domains. Indiana Univ. Math. J. **32**(1), 73–81 (1983)
17. Korevaar, N.J.: Convex solutions to nonlinear elliptic and parabolic boundary value problems. Indiana Univ. Math. J. **32**(4), 603–614 (1983)
18. Korevaar, N.J., Lewis, J.L.: Convex solutions of certain elliptic equations have constant rank Hessians. Arch. Ration. Mech. Anal. **97**(1), 19–32 (1987)
19. Juutinen, P.: Concavity maximum principle for viscosity solutions of singular equations. Nonlinear Differ. Equ. Appl. **17**, 601–618 (2010)
20. Liu, P., Ma, X.N., Xu, L.: A Brunn-Minkowski inequality for the Hessian eigenvalue in three-dimensional convex domain. Adv. Math. **225**, 1616–1633 (2010)
21. Ma, X.N., Xu, L.: The convexity of solution of a class Hessian equation in bounded convex domain in $\mathbb{R}^3$. J. Funct. Anal. **255**, 1713–1723 (2008)
22. Pucci, C.: Operatori ellittici estremanti. Ann. Mat. Pura Appl. **72**(4), 141–170 (1966)
23. Sakaguchi, S.: Concavity properties of solutions to some degenerate quasilinear elliptic Dirichlet problems. Ann. Sc. Norm. Super. Pisa, Cl. Sci. (4) **14**, 403–421 (1987)
24. Rockafellar, R.T.: Convex Analysis. Princeton Landmarks in Mathematics. Princeton University Press, Princeton (1970)
25. Salani, P.: Convexity of solutions and Brunn-Minkowski inequalities for Hessian equations in R^3. Adv. Math. **229**, 1924–1948 (2012)
26. Salani, P.: Mean-width rearrangements, Brunn-Minkowski inequalities and convexity for elliptic equations (in preparation)
27. Strömberg, T.: The operation of infimal convolution. Diss. Math. (Rozprawy Mat.) **352**, 58 pp. (1996)

The Heart of a Convex Body

Lorenzo Brasco and Rolando Magnanini

Abstract We investigate some basic properties of the *heart* $\heartsuit(\mathcal{K})$ of a convex set $\mathcal{K}$. It is a subset of $\mathcal{K}$, whose definition is based on mirror reflections of Euclidean space, and is a non-local object. The main motivation of our interest for $\heartsuit(\mathcal{K})$ is that this gives an estimate of the location of the hot spot in a convex heat conductor with boundary temperature grounded at zero. Here, we investigate on the relation between $\heartsuit(\mathcal{K})$ and the mirror symmetries of $\mathcal{K}$; we show that $\heartsuit(\mathcal{K})$ contains many (geometrically and physically) relevant points of $\mathcal{K}$; we prove a simple geometrical lower estimate for the diameter of $\heartsuit(\mathcal{K})$; we also prove an upper estimate for the area of $\heartsuit(\mathcal{K})$, when $\mathcal{K}$ is a triangle.

Keywords Convex bodies · Hot spots · Critical points · Shape optimization

1 Introduction

Let $\mathcal{K}$ be a convex body in the Euclidean space $\mathbb{R}^N$, that is $\mathcal{K}$ is a compact convex set with non-empty interior. In [1] we defined the *heart* $\heartsuit(\mathcal{K})$ of $\mathcal{K}$ as follows. Fix a unit vector $\omega \in \mathbb{S}^{N-1}$ and a real number λ; for each point $x \in \mathbb{R}^N$, let $T_{\lambda,\omega}(x)$ denote the reflection of x in the hyperplane $\pi_{\lambda,\omega}$ of equation $\langle x, \omega\rangle = \lambda$ (here, $\langle x, \omega\rangle$ denotes the usual scalar product of vectors in $\mathbb{R}^N$); then set

$$\mathcal{K}_{\lambda,\omega} = \{x \in \mathcal{K} : \langle x, \omega\rangle \geq \lambda\}$$

(see Fig. 1). The heart of $\mathcal{K}$ is thus defined as

$$\heartsuit(\mathcal{K}) = \bigcap_{\omega\in\mathbb{S}^{N-1}} \{\mathcal{K}_{-\lambda,-\omega} : T_{\lambda,\omega}(\mathcal{K}_{\lambda,\omega}) \subset \mathcal{K}\}.$$

L. Brasco (✉)
Laboratoire d'Analyse, Topologie, Probabilités, Aix-Marseille Université, CMI 39, Rue Frédéric Joliot Curie, 13453 Marseille Cedex 13, France
e-mail: brasco@cmi.univ-mrs.fr

R. Magnanini
Dipartimento di Matematica "U. Dini", Università di Firenze, viale Morgagni 67/A, 50134 Firenze, Italy
e-mail: magnanin@math.unifi.it

R. Magnanini et al. (eds.), *Geometric Properties for Parabolic and Elliptic PDE's*, Springer INdAM Series 2, DOI 10.1007/978-88-470-2841-8_4,

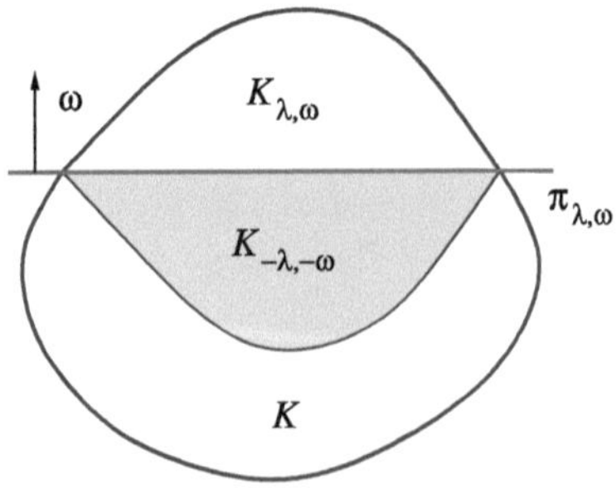

Fig. 1 The sets $\mathcal{K}_{\lambda,\omega}$ and $\mathcal{K}_{-\lambda,-\omega}$

Our interest in $\heartsuit(\mathcal{K})$ was motivated in [1] in connection to the problem of locating the (unique) point of maximal temperature—the *hot spot*—in a convex heat conductor with boundary temperature grounded at zero. There, by means of A.D. Aleksandrov's reflection principle, we showed that $\heartsuit(\mathcal{K})$ must contain the hot spot at each time and must also contain the maximum point of the first Dirichlet eigenfunction of the Laplacian, which is known to control the asymptotic behavior of temperature for large times. By the same arguments, we showed in [1] that $\heartsuit(\mathcal{K})$ must also contain the maximum point of positive solutions of nonlinear equations belonging to a quite large class. By these observations, the set $\heartsuit(\mathcal{K})$ can be viewed as a geometrical means to estimate the positions of these important points.

Another interesting feature of $\heartsuit(\mathcal{K})$ is the non-local nature of its definition. We hope that the study of $\heartsuit(\mathcal{K})$ can help, in a relatively simple setting, to develop techniques that may be useful in the study of other objects and properties of non-local nature, which have lately raised interest in the study of partial differential equations.

A further reason of interest is that the shape of $\heartsuit(\mathcal{K})$ seems to be related to the mirror symmetry of $\mathcal{K}$. By means of a numerical algorithm, developed in [1], that (approximately) constructs $\heartsuit(\mathcal{K})$ for any given convex polyhedron $\mathcal{K}$, one can observe that relationship—and other features of $\heartsuit(\mathcal{K})$—and raise some questions.

1. We know that, if $\mathcal{K}$ has a hyperplane of symmetry, then $\heartsuit(\mathcal{K})$ is contained in that hyperplane; is the converse true?
2. How small $\heartsuit(\mathcal{K})$ can be? Can we estimate from below the ratio between the diameters of $\heartsuit(\mathcal{K})$ and $\mathcal{K}$?
3. How big $\heartsuit(\mathcal{K})$ can be? Can we estimate from above the ratio between the volumes of $\heartsuit(\mathcal{K})$ and $\mathcal{K}$?

The purpose of this note is to collect results in that direction.

In Sect. 2, we give a positive answer to question (i) (see Theorem 1).

In Sect. 3, we start by showing that many relevant points related to a convex set lie in its heart. For instance, we shall prove that, besides the center of mass $M_{\mathcal{K}}$ of $\mathcal{K}$ (as seen in [1, Proposition 4.1]), $\heartsuit(\mathcal{K})$ must also contain the center $C_{\mathcal{K}}$ of the smallest ball containing $\mathcal{K}$—the so-called *circumcenter*—and the center of mass of the set of all the *incenters* of $\mathcal{K}$,

$$\mathcal{M}(\mathcal{K}) = \left\{x \in \mathcal{K} : \mathrm{dist}(x, \partial\mathcal{K}) = r_{\mathcal{K}}\right\};$$

here $r_{\mathcal{K}}$ is the *inradius* of $\mathcal{K}$, i.e. the radius of the largest balls contained in $\mathcal{K}$. This information gives a simple estimate from below of the diameter of $\heartsuit(\mathcal{K})$, thus partially answering to question (ii) (see Theorem 2).

By further exploring in this direction, we prove that $\heartsuit(\mathcal{K})$ must also contain the points minimizing each p-moment of the set $\mathcal{K}$ (see Sect. 3.2) and other more general moments associated to $\heartsuit(\mathcal{K})$. As a consequence of this general result, we relate $\heartsuit(\mathcal{K})$ to a problem in spectral optimization considered in [5] and show that $\heartsuit(\mathcal{K})$ must contain the center of a ball realizing the so-called *Fraenkel asymmetry* (see [4] and Sect. 3 for a definition).

Finally, in Sect. 4, we begin an analysis of problem (iii). Therein, we discuss the shape optimization problem (10) and prove that an optimal shape does not exist in the subclass of triangles. In fact, in Theorem 5 we show that

$$\left|\heartsuit(\mathcal{K})\right| < \frac{3}{8}|\mathcal{K}| \quad \text{for every triangle } \mathcal{K};$$

the constant $3/8$ is not attained but is only approached by choosing a sequence of obtuse triangles.

2 Dimension of the Heart and Symmetries

Some of the results in Sect. 2.1 where proved in [1] but, for the reader's convenience, we reproduce them here.

2.1 Properties of the Maximal Folding Function

The *maximal folding function* $\mathcal{R}_{\mathcal{K}} : \mathbb{S}^{N-1} \to \mathbb{R}$ of a convex body $\mathcal{K} \subset \mathbb{R}^N$ was defined in [1] by

$$\mathcal{R}_{\mathcal{K}}(\omega) = \min\{\lambda \in \mathbb{R} : T_{\lambda,\omega}(\mathcal{K}_{\lambda,\omega}) \subseteq \mathcal{K}\}, \quad \omega \in \mathbb{S}^{N-1}.$$

The heart of $\mathcal{K}$ can be defined in terms of $\mathcal{R}_{\mathcal{K}}$:

$$\heartsuit(\mathcal{K}) = \{x \in \mathcal{K} : \langle x, \omega\rangle \le \mathcal{R}_{\mathcal{K}}(\omega), \text{ for every } \omega \in \mathbb{S}^{N-1}\}. \tag{1}$$

It is important to remark that we have

$$\max_{x \in \heartsuit(\mathcal{K})} \langle x, \omega\rangle \le \mathcal{R}_{\mathcal{K}}(\omega), \quad \omega \in \mathbb{S}^{N-1}$$

and in general the two terms do not coincide: in other words, $\mathcal{R}_{\mathcal{K}}$ does not coincide with the support function of $\heartsuit(\mathcal{K})$ (see [1, Example 4.8]).

Lemma 1 *The maximal folding function $\mathcal{R}_{\mathcal{K}} : \mathbb{S}^{N-1} \to \mathbb{R}$ is lower semicontinuous. In particular, $\mathcal{R}_{\mathcal{K}}$ attains its minimum on $\mathbb{S}^{N-1}$.*

Proof Fix $\lambda \in \mathbb{R}$ and let $\omega_0 \in \{\omega \in \mathbb{S}^{N-1} : \mathcal{R}_{\mathcal{K}}(\omega) > \lambda\}$. Then

$$\left|T_{\lambda,\omega_0}(\mathcal{K}_{\lambda,\omega_0}) \cap \left(\mathbb{R}^N \setminus \mathcal{K}\right)\right| > 0.$$

The continuity of the function $\omega \mapsto |T_{\lambda,\omega}(\mathcal{K}_{\lambda,\omega}) \cap (\mathbb{R}^N \setminus \mathcal{K})|$ implies that, for every ω in some neighborhood of ω_0, we have that $\mathcal{R}_{\mathcal{K}}(\omega) > \lambda$. □

Remark 1 In general $\mathcal{R}_{\mathcal{K}}$ is not continuous on $\mathbb{S}^{N-1}$: it sufficient to take a rectangle $\mathcal{K} = [-a, a] \times [-b, b]$ and observe that we have $\mathcal{R}_{\mathcal{K}}(1, 0) = 0$, since $\mathcal{K}$ is symmetric with respect to the y-axis but, for a sufficiently small ϑ, $\mathcal{R}_{\mathcal{K}}(\cos\vartheta, \sin\vartheta) = a\cos\vartheta - b\sin\vartheta$, so that

$$\lim_{\vartheta\to 0} \mathcal{R}_{\mathcal{K}}(\vartheta) = a > \mathcal{R}_{\mathcal{K}}(1, 0).$$

Observe that here the lack of continuity of the maximal folding function is not due to the lack of smoothness of the boundary of $\mathcal{K}$, but rather to the presence of non-strictly convex subsets of $\partial\mathcal{K}$.

Proposition 1 *Let $\mathcal{K} \subset \mathbb{R}^N$ be a convex body and define its* center of mass *by*

$$M_{\mathcal{K}} = \frac{1}{|\mathcal{K}|}\int_{\mathcal{K}} y\, dy.$$

Then we have that

$$\mathcal{R}_{\mathcal{K}}(\omega) \geq \langle M_{\mathcal{K}}, \omega\rangle, \quad \text{for every } \omega \in \mathbb{S}^{N-1}, \tag{2}$$

and the equality sign can hold for some $\omega \in \mathbb{S}^{N-1}$ if and only if $\mathcal{K}$ is ω-symmetric, i.e. if $T_{\lambda,\omega}(\mathcal{K}) = \mathcal{K}$ for some $\lambda \in \mathbb{R}$. In particular, $M_{\mathcal{K}} \in \heartsuit(\mathcal{K})$.

Proof Set $\lambda = \mathcal{R}_{\mathcal{K}}(\omega)$ and consider the set $\Omega = \mathcal{K}_{\lambda,\omega} \cup T_{\lambda,\omega}(\mathcal{K}_{\lambda,\omega})$; by the definition of center of mass and since Ω is symmetric with respect to the hyperplane $\pi_{\lambda,\omega}$, we easily get that

$$\begin{aligned} |\mathcal{K}|\big[\mathcal{R}_{\mathcal{K}}(\omega) - \langle M_{\mathcal{K}}, \omega\rangle\big] &= \int_{\mathcal{K}} \big[\mathcal{R}_{\mathcal{K}}(\omega) - \langle y, \omega\rangle\big]\, dy \\ &= \int_{\mathcal{K}\setminus\Omega} \big[\mathcal{R}_{\mathcal{K}}(\omega) - \langle y, \omega\rangle\big]\, dy. \end{aligned}$$

Observe that the last integral contains a positive quantity to be integrated on the region $\mathcal{K} \setminus \Omega$: this already shows (2). Moreover, the same identity implies:

$$\mathcal{R}_{\mathcal{K}}(\omega) - \langle M_{\mathcal{K}}, \omega\rangle = 0 \quad \Longleftrightarrow \quad |\mathcal{K} \setminus \Omega| = 0,$$

and the latter condition is equivalent to say that $\mathcal{K}$ is ω-symmetric. Finally, by combining (2) with the definition (1) of $\heartsuit(\mathcal{K})$, it easily follows $M_{\mathcal{K}} \in \heartsuit(\mathcal{K})$. □

2.2 *On the Mirror Symmetries of* $\mathcal{K}$

A first application of Proposition 1 concerns the relation between $\heartsuit(\mathcal{K})$ and the mirror symmetries of $\mathcal{K}$.

Theorem 1 *Let $\mathcal{K} \subset \mathbb{R}^N$ be a convex body.*

(i) *If there exist k $(1 \le k \le N)$ independent directions $\omega_1, \dots, \omega_k \in \mathbb{S}^{N-1}$ such that $\mathcal{K}$ is ω_j-symmetric for $j = 1, \dots, k$, then $\mathcal{R}_{\mathcal{K}}(\omega_j) = \langle M_{\mathcal{K}}, \omega_j \rangle$ for $j = 1, \dots, k$ and*

$$\heartsuit(\mathcal{K}) \subseteq \bigcap_{j=1}^{k} \pi_{\mathcal{R}_{\mathcal{K}}(\omega_j), \omega_j}.$$

In particular, the co-dimension of $\heartsuit(\mathcal{K})$ is at least k.

(ii) *If $\heartsuit(\mathcal{K})$ has dimension k $(1 \le k \le N-1)$, then there exists at least a direction $\theta \in \mathbb{S}^{N-1}$ such that $\mathcal{R}_{\mathcal{K}}(\theta) = \langle M_{\mathcal{K}}, \theta \rangle$, and $\mathcal{K}$ is θ-symmetric.*

Proof (i) The assertion is a straightforward consequence of Proposition 1.

(ii) By Lemma 1, the function $\omega \mapsto \mathcal{R}_{\mathcal{K}}(\omega) - \langle M_{\mathcal{K}}, \omega \rangle$ attains its minimum for some $\theta \in \mathbb{S}^{N-1}$. Set $r = \mathcal{R}_{\mathcal{K}}(\theta) - \langle M_{\mathcal{K}}, \theta \rangle$ and suppose that $r > 0$.

Then, for every $x \in B(M_{\mathcal{K}}, r)$, we have that

$$\langle x, \omega \rangle = \langle M_{\mathcal{K}}, \omega \rangle + \langle x - M_{\mathcal{K}}, \omega \rangle < \langle M_{\mathcal{K}}, \omega \rangle + r \le \mathcal{R}_{\mathcal{K}}(\omega) - r + r = \mathcal{R}_{\mathcal{K}}(\omega),$$

for every $\omega \in \mathbb{S}^{N-1}$, and hence $x \in \heartsuit(\mathcal{K})$ by (1). Thus, $B(x, r) \subset \heartsuit(\mathcal{K})$—a contradiction to the fact that $1 \le k \le N-1$. Hence, $\mathcal{R}_{\mathcal{K}}(\theta) = \langle M_{\mathcal{K}}, \theta \rangle$ and $\mathcal{K}$ is θ-symmetric by Proposition 1. □

Remark 2 It is clear that the dimension of the heart only gives information on the *minimal* number of symmetries of a convex body: the example of a ball is quite explicative.

We were not able to prove the following result:

if $\heartsuit(\mathcal{K})$ has co-dimension m $(1 \le m \le N)$, then there exist at least m independent directions $\theta_1, \dots, \theta_N \in \mathbb{S}^{N-1}$ such that $\mathcal{R}_{\mathcal{K}}(\theta_j) = \langle M_{\mathcal{K}}, \theta_j \rangle$, $j = 1, \dots, m$, and $\mathcal{K}$ is θ_j-symmetric for $j = 1, \dots, m$.

We leave it as a conjecture.

3 Relevant Points Contained in the Heart

In this section, we will show that many relevant points of a convex set are contained in its heart (e.g. the incenter and the circumcenter, besides the center mass). This fact will give us a means to estimate from below the diameter of the heart.

3.1 An Estimate of the Heart's Diameter

Proposition 2 *Let $\mathcal{K} \subset \mathbb{R}^N$ be a convex body. Then its circumcenter $C_{\mathcal{K}}$ belongs to $\heartsuit(\mathcal{K})$.*

Proof Suppose that $C_{\mathcal{K}} \notin \heartsuit(\mathcal{K})$; projection on the set $\heartsuit(\mathcal{K})$; then there is a half-space H^+ such that the hyperplane $\Pi = \partial H^+$ separates $C_{\mathcal{K}}$ and $\heartsuit(\mathcal{K})$, and for which the reflection in Π of $H^+ \cap \mathcal{K}$ is contained in $\mathcal{K}$, by the definition of $\heartsuit(\mathcal{K})$. Hence this is contained in the smallest ball $B_R(C_{\mathcal{K}})$ containing $\mathcal{K}$. Thus, $\mathcal{K}$ must be contained in the set $B_R(C_{\mathcal{K}}) \cap B'_R(C_{\mathcal{K}})$, where $B'_R(C_{\mathcal{K}})$ is the reflection of $B_R(C_{\mathcal{K}})$ in the hyperplane Π.

This is a contradiction, since the smallest ball containing $\mathcal{K}$ would have a radius strictly smaller than R. □

We now consider the *incenters* of $\mathcal{K}$: these are the centers of the balls of largest radius $r_{\mathcal{K}}$ inscribed in $\mathcal{K}$. Needless to say, a convex body may have many incenters. We start with the simpler case of a convex body with a unique incenter.

Proposition 3 *Let $\mathcal{K} \subset \mathbb{R}^N$ be a convex body; if its incenter $I_{\mathcal{K}}$ is unique, then $I_{\mathcal{K}} \in \heartsuit(\mathcal{K})$. In particular, $I_{\mathcal{K}} \in \heartsuit(\mathcal{K})$ if $\mathcal{K}$ is strictly convex.*

Proof Consider the unique maximal ball $B(I_{\mathcal{K}}, r_{\mathcal{K}})$ inscribed in $\mathcal{K}$ and suppose that $I_{\mathcal{K}} \notin \heartsuit(\mathcal{K})$; this implies that there exists $\omega \in \mathbb{S}^{N-1}$ such that

$$\mathcal{R}_{\mathcal{K}}(\omega) - \langle I_{\mathcal{K}}, \omega \rangle < 0.$$

Set $\lambda = \mathcal{R}_{\mathcal{K}}(\omega)$ and define $I'_{\mathcal{K}} = T_{\lambda,\omega}(I_{\mathcal{K}})$; then $I'_{\mathcal{K}} \neq I_{\mathcal{K}}$ and $\langle I'_{\mathcal{K}}, \omega \rangle < \langle I_{\mathcal{K}}, \omega \rangle$. Now, the half-ball $B^+ = \{x \in B(I_{\mathcal{K}}, r_{\mathcal{K}}) : \langle x, \omega \rangle \geq \lambda\}$ and its reflection $T_{\lambda,\omega}(B^+)$ in the hyperplane $\pi_{\lambda,\omega}$ are contained in $\mathcal{K}$, since B^+ is contained in the maximal cap $\mathcal{K}_{\lambda,\omega}$.

This fact implies in particular that the reflection of the whole ball B is contained in $\mathcal{K}$: but the latter is still a maximal ball of radius $r_{\mathcal{K}}$, with center $I'_{\mathcal{K}}$ different from $I_{\mathcal{K}}$. This is a contradiction, since it violates the assumed uniqueness of the incenter. □

To treat the general case, we need the following simple result.

Lemma 2 *Let $\mathcal{K} \subset \mathbb{R}^N$ be a convex body and let us set*

$$\mathcal{M}(\mathcal{K}) = \{x \in \mathcal{K} : \operatorname{dist}(x, \partial\mathcal{K}) = r_{\mathcal{K}}\}.$$

Then $\mathcal{M}(\mathcal{K})$ is a closed convex set with $|\mathcal{M}(\mathcal{K})| = 0$; in particular, the dimension of $\mathcal{M}(\mathcal{K})$ is at most $N - 1$.

Proof The quasi-convexity[1] of $\mathrm{dist}(x, \partial\mathcal{K})$ (being $\mathcal{K}$ convex) immediately implies that $\mathcal{M}(\mathcal{K})$ is convex. For the reader's convenience, here we give a proof anyway. Let us take $x, z \in \mathcal{M}(\mathcal{K})$ two distinct points, then by definition of inradius we have

$$B(x, r_{\mathcal{K}}) \cup B(z, r_{\mathcal{K}}) \subset \mathcal{K}.$$

Since $\mathcal{K}$ is convex, it must contain the convex hull of $B(x, r_{\mathcal{K}}) \cup B(z, r_{\mathcal{K}})$, as well: hence, for every $t \in [0, 1]$ we have

$$B\big((1-t)x + tz, r_{\mathcal{K}}\big) \subset \mathcal{K},$$

that is $(1-t)x + tz \in \mathcal{M}(\mathcal{K})$, which proves the convexity of $\mathcal{M}(\mathcal{K})$.

Now, suppose that $|\mathcal{M}(\mathcal{K})| > 0$; since $\mathcal{M}(\mathcal{K})$ is convex, it must contain a ball $B(x, \varrho)$. The balls of radius $r_{\mathcal{K}}$ having centers on $\partial B(x, \varrho/2)$ are all contained in $\mathcal{K}$, so that their whole union is contained in $\mathcal{K}$ as well. Observing that this union is given by $B(x, \varrho/2 + r_{\mathcal{K}})$, we obtain the desired contradiction, since we violated the maximality of $r_{\mathcal{K}}$. □

Proposition 4 *Let $\mathcal{K} \subset \mathbb{R}^N$ be a convex body and let us suppose that $\mathcal{M}(\mathcal{K})$ has dimension $k \in \{1, \dots, N-1\}$.*

Then the center of mass of $\mathcal{M}(\mathcal{K})$, defined by

$$I_{\mathcal{M}} = \frac{\int_{\mathcal{M}(\mathcal{K})} y\, d\mathcal{H}^k(y)}{\mathcal{H}^k(\mathcal{M}(\mathcal{K}))},$$

belongs to $\heartsuit(\mathcal{K})$. Here, $\mathcal{H}^k$ denotes the standard k-dimensional Hausdorff measure.

Proof The proof is based on the observation that

$$\mathcal{R}_{\mathcal{M}(\mathcal{K})}(\omega) \le \mathcal{R}_{\mathcal{K}}(\omega), \quad \omega \in \mathbb{S}^{N-1}, \tag{3}$$

where $\mathcal{R}_{\mathcal{M}(\mathcal{K})}$ is the maximal folding function of $\mathcal{M}(\mathcal{K})$, thought as a subset of $\mathbb{R}^N$. Assuming (3) to be true, we can use the definition of heart and Proposition 1 to obtain that the center of mass $I_{\mathcal{M}}$ belongs to $\heartsuit(\mathcal{M}(\mathcal{K}))$ and hence to $\heartsuit(\mathcal{K})$, which would conclude the proof.

Now, suppose by contradiction that there is an $\omega \in \mathbb{S}^{N-1}$ such that $\mathcal{R}_{\mathcal{M}(\mathcal{K})}(\omega) > \mathcal{R}_{\mathcal{K}}(\omega)$, and set $\lambda = \mathcal{R}_{\mathcal{K}}(\omega)$, as usual. Then, there exists $x \in \mathcal{M}(\mathcal{K})$ with $\langle x, \omega\rangle \ge \lambda$ such that its reflection $x^\lambda = T_{\lambda,\omega}(x)$ in the hyperplane $\pi_{\lambda,\omega}$ falls outside $\mathcal{M}(\mathcal{K})$: this would imply in particular that $B(x^\lambda, r_{\mathcal{K}}) \not\subset \mathcal{K}$. Observe that by definition of $\mathcal{M}(\mathcal{K})$, the ball $B(x, r_{\mathcal{K}})$ lies inside $\mathcal{K}$, so that the cap $B(x, r_{\mathcal{K}}) \cap \{\langle y, \omega\rangle \ge \lambda\}$ is reflected in $\mathcal{K}$; thus, as before, we obtain that the union of this cap and its reflection is contained in $\mathcal{K}$. This shows that the ball $B(x^\lambda, r_{\mathcal{K}})$ is contained in $\mathcal{K}$, thus giving a contradiction. □

The result here below follows at once from Propositions 1, 2 and 4.

---

[1] This means that the superlevel sets of the function are convex.

Theorem 2 *Let $\mathcal{K} \subset \mathbb{R}^N$ be a convex body, then*
$$\operatorname{diam}\big[\heartsuit(\mathcal{K})\big] \geq \max\big(|M_{\mathcal{K}} - C_{\mathcal{K}}|, |C_{\mathcal{K}} - I_{\mathcal{M}}|, |I_{\mathcal{M}} - M_{\mathcal{K}}|\big).$$

Remark 3 Notice that, when $\heartsuit(\mathcal{K})$ degenerates to a single point, then clearly
$$\heartsuit(\mathcal{K}) = \{M_{\mathcal{K}}\} = \{I_{\mathcal{K}}\} = \{C_{\mathcal{K}}\}.$$

Needless to say, it may happen that the three points $M_{\mathcal{K}}$, $I_{\mathcal{K}}$ and $C_{\mathcal{K}}$ coincide, but $\heartsuit(\mathcal{K})$ is not a point. For example, take an ellipse parametrized in polar coordinates as
$$E = \big\{(\varrho, \vartheta) : 0 \leq \varrho \leq \sqrt{a^2 \cos^2 \vartheta + b^2 \sin^2 \vartheta},\ \vartheta \in [-\pi, \pi]\big\}$$
and a π-periodic, smooth non-negative function η on $[-\pi, \pi]$, having its support in two small neighborhoods of $-3\pi/4$ and $\pi/4$. Then, if ε is sufficiently small, the deformed set
$$E_\varepsilon = \big\{(\varrho, \vartheta) : 0 \leq \varrho \leq \sqrt{a^2 \cos^2 \vartheta + b^2 \sin^2 \vartheta} - \varepsilon\eta(\vartheta),\ \vartheta \in [-\pi, \pi]\big\}$$
is still convex and centrally symmetric. Moreover, it is easy to convince oneself that $\{M_{E_\varepsilon}\} = \{I_{E_\varepsilon}\} = \{C_{E_\varepsilon}\} = \{(0,0)\}$, whereas, by Theorem 1, $\heartsuit(E_\varepsilon)$ is not a point, since E_ε has no mirror symmetries.

3.2 The p-Moments of $\mathcal{K}$ and More

We recall that the point $M_{\mathcal{K}}$ can also be characterized as the unique point in $\mathcal{K}$ which minimizes the function
$$x \mapsto \int_{\mathcal{K}} |x - y|^2\, dy, \quad x \in \mathcal{K},$$
that can be viewed as the moment of inertia (or 2-moment) of $\mathcal{K}$ about the point x. In this subsection, we will extend the results of Sect. 3.1 to more general moments of $\mathcal{K}$, that include as special cases the p-moments $\int_{\mathcal{K}} |x - y|^p\, dy$.

We first establish a preliminary lemma.

Lemma 3 *Let $\varphi, \psi : [0, \infty) \to \mathbb{R}$ be, respectively, an increasing and a decreasing function and suppose that ψ is also integrable in $[0, \infty)$. Define the two functions*
$$F(t) = \int_a^b \varphi\big(|t - s|\big)\, ds, \quad t \in \mathbb{R},$$
and
$$G(t) = \int_{-\infty}^a \psi\big(|t - s|\big)\, ds + \int_b^\infty \psi\big(|t - s|\big)\, ds, \quad t \in \mathbb{R},$$
where a and b are two numbers with $0 \leq a \leq b$.

Then, both F and G attain their minimum at the midpoint $(a+b)/2$ of $[a, b]$. Moreover, F and G are convex in $[a, b]$.

If φ is strictly increasing (resp. ψ is strictly decreasing), then the minimum point of F (resp. G) is unique.

Proof (i) We easily get that

$$F(a+b-t) = F(t) \quad \text{for every } t \in \mathbb{R};$$

this means that the graph of F in $[a,b]$ is symmetric with respect to the line $t = (a+b)/2$. Moreover, since F can be rewritten as

$$F(t) = F\big((a+b)/2\big) + \int_{(b-a)/2}^{t-a} \varphi(s)\,ds - \int_{b-t}^{(b-a)/2} \varphi(s)\,ds,$$

we infer that F is increasing for $t \geq (a+b)/2$. This shows that F attains its minimum in $[a,b]$ at the midpoint. Clearly, if φ is strictly increasing, then F is also strictly increasing for $t \geq (a+b)/2$ and the minimum point is unique.

The convexity of F in $[a,b]$ follows from the identity

$$F(t) = \Phi(t-a) + \Phi(b-t), \quad t \in [a,b],$$

where

$$\Phi(t) = \int_0^t \varphi(s)\,ds$$

is a convex function, being a primitive of an increasing function.

(ii) It is enough to rewrite the function G as follows

$$G(t) = \int_{\mathbb{R}} \psi\big(|t-s|\big)\,ds - \int_a^b \psi\big(|t-s|\big)\,ds, \quad t \in [a,b];$$

then we notice that the first term is a constant, while the second one behaves as F, thanks to the first part of this lemma, since the function $-\psi$ is increasing. □

The following result generalizes one in [7] in the case of $\mathcal{K}$ convex.

Theorem 3 *Let $\mathcal{K} \subset \mathbb{R}^N$ be a convex body and let $\varphi : [0,\infty) \to \mathbb{R}$ be an increasing function. Define the function*

$$\mu_\varphi(x) = \int_{\mathcal{K}} \varphi\big(|x-y|\big)\,dy, \quad x \in \mathbb{R}^N,$$

and the set

$$\mathbf{m}(\mu_\varphi) = \big\{x \in \mathbb{R}^N : \mu_\varphi(x) = \min \mu_\varphi\big\}.$$

Then

$$\mathbf{m}(\mu_\varphi) \cap \heartsuit(\mathcal{K}) \neq \varnothing. \tag{4}$$

Proof We shall refer to $\mu_\varphi(x)$ as the *φ-moment* of $\mathcal{K}$ about the point x. First of all, we observe that μ_φ is lower semicontinuous thanks to Fatou's Lemma and that

$$\inf_{\mathcal{K}} \mu_\varphi = \inf_{\mathbb{R}^N} \mu_\varphi,$$

so that the minimum of μ_φ is attained at some point belonging to $\mathcal{K}$, i.e. $\varnothing \neq \mathbf{m}(\mu_\varphi) \subset \mathcal{K}$.

We first prove (4) when φ is strictly increasing. Let $x \in \mathcal{K}$ be a minimum point of μ_φ. If $x \notin \heartsuit(\mathcal{K})$, then there exists a direction $\omega \in \mathbb{S}^{N-1}$ such that $\mathcal{R}_{\mathcal{K}}(\omega) < \langle x, \omega\rangle$. We set for simplicity $\lambda = \mathcal{R}_{\mathcal{K}}(\omega)$ and consider the hyperplane $\pi = \pi_{\lambda,\omega}$, so that

$$x \in \mathcal{K}_{\lambda,\omega} \quad \text{and} \quad \mathcal{T}_{\lambda,\omega}(\mathcal{K}_{\lambda,\omega}) \subset \mathcal{K}.$$

Modulo a rotation, we can always assume that $\omega = (1, 0, \dots, 0)$ and the hyperplane π has the form $\{x \in \mathbb{R}^N : x_1 = \lambda\}$.

Now, we define the symmetric set $\Omega = \mathcal{K}_{\lambda,\omega} \cup \mathcal{T}_{\lambda,\omega}(\mathcal{K}_{\lambda,\omega})$, which can be written as

$$\Omega = \{(y_1, y') \in \mathcal{K} : y' \in \mathcal{K} \cap \pi,\ \lambda - a(y') \le y_1 \le \lambda + a(y')\}.$$

Consider the projection $z = (\lambda, x')$ of x in π and observe that $z \in \Omega$, thanks to the convexity of $\mathcal{K}$. For $y \in \mathcal{K} \setminus \Omega$, we have that $|z - y| < |x - y|$. Thus

$$\int_{\mathcal{K}\setminus\Omega} \varphi(|x - y|)\, dy \ge \int_{\mathcal{K}\setminus\Omega} \varphi(|z - y|)\, dy, \tag{5}$$

since φ is increasing. Moreover, by Fubini's theorem, we compute:

$$\int_\Omega \varphi(|x - y|)\, dy = \int_{\mathcal{K}\cap\pi} \left\{ \int_{\lambda-a(y')}^{\lambda+a(y')} \varphi\left(\sqrt{|x' - y'|^2 + |x_1 - y_1|^2}\right) dy_1 \right\} dy'.$$

We now apply Lemma 3 with the choice $t \mapsto \varphi(\sqrt{|x' - y'|^2 + t^2})$, a strictly increasing function. Thus, we can infer that the last integral is strictly larger than

$$\int_{\mathcal{K}\cap\pi} \left\{ \int_{\lambda-a(y')}^{\lambda+a(y')} \varphi\left(\sqrt{|x' - y'|^2 + |\lambda - y_1|^2}\right) dy_1 \right\} dy' = \int_\Omega \varphi(|z - y|)\, dy.$$

With the aid of (5), we then conclude that $\mu_\varphi(x) > \mu_\varphi(z)$, which gives a contradiction. We observe in passing that we have proved something more; namely, we showed that $\mathbf{m}(\mu_\varphi) \subseteq \heartsuit(\mathcal{K})$.

If φ is only increasing, we approximate it by the following sequence of strictly increasing functions

$$\varphi_n(t) = \varphi(t) + \frac{1}{n} t^2, \quad t \in [0, \infty),\ n \in \mathbb{N}.$$

Let $x_n \in \mathbf{m}(\mu_{\varphi_n}) \subseteq \heartsuit(\mathcal{K})$ be a sequence of minimizers of μ_{φ_n}, then by compactness of $\heartsuit(\mathcal{K})$ they convergence (up to a subsequence) to a point $x_0 \in \heartsuit(\mathcal{K})$. For every $x \in \mathbb{R}^N$, by Fatou's Lemma we have

$$\mu_\varphi(x) = \lim_{n\to\infty} \mu_{\varphi_n}(x) \ge \liminf_{n\to\infty} \mu_{\varphi_n}(x_n) \ge \mu_\varphi(x_0),$$

which implies that $x_0 \in \mathbf{m}(\mu_\varphi)$. This concludes the proof. □

The analogous of Theorem 2 is readily proved.

Theorem 4 *Let $\mathcal{K}$ be a convex body. Then the convex hull of the set*

$$\bigcup \{\mathbf{m}(\mu_\varphi) \cap \heartsuit(\mathcal{K}) : \varphi \text{ is increasing on } [0, \infty)\},$$

is contained in $\heartsuit(\mathcal{K})$.

Remark 4 By similar arguments, we can prove that, if ψ is decreasing and the function

$$\nu_\psi(x) = \int_{\mathbb{R}^N\setminus\mathcal{K}} \psi\big(|x-y|\big)\,dy,$$

is finite for every $x \in \mathcal{K}$, then $\mathbf{m}(\nu_\psi) \cap \heartsuit(\mathcal{K}) \neq \varnothing$.

Particularly interesting are the cases where $\varphi(t) = t^p$ with $p > 0$ and $\psi(t) = t^{-p}$ with $p > N$.

Corollary 1 *Consider the functions*

$$\mu_p(x) = \int_{\mathcal{K}} |x-y|^p\,dy, \quad x \in \mathcal{K},$$

for $p > 0$ *or*

$$\nu_p(x) = \int_{\mathbb{R}^N\setminus\mathcal{K}} |x-y|^{-p}\,dy, \quad x \in \mathcal{K},$$

for $p > N$.

Then, their minimum points belong to $\heartsuit(\mathcal{K})$.

Remark 5 Propositions 1, 2 and 3 can be re-proved by means of Corollary 1 by choosing $p = 2$ or, respectively, by taking limits as $p \to \infty$.

Notice, in fact, that

$$\lim_{p\to\infty} \mu_p(x)^{1/p} = \max_{y\in\mathcal{K}} |x-y|$$

and

$$\lim_{p\to\infty} \nu_p(x)^{-1/p} = \min_{y\in\mathcal{K}} |x-y|.$$

Hence the *circumradius* $\rho_\mathcal{K}$ and inradius $r_\mathcal{K}$ are readily obtained as

$$\rho_\mathcal{K} = \min_{x\in\mathcal{K}} \max_{y\in\partial\mathcal{K}} |x-y| = \lim_{p\to+\infty} \min_{x\in\mathcal{K}} \mu_p(x)^{1/p},$$

and

$$r_\mathcal{K} = \max_{x\in\mathcal{K}} \min_{y\in\partial\mathcal{K}} |x-y| = \lim_{p\to+\infty} \max_{x\in\mathcal{K}} \nu_p(x)^{-1/p}.$$

These observations quite straightforwardly imply that $C_\mathcal{K}$ and $I_\mathcal{K}$ belong to $\heartsuit(\mathcal{K})$.

A final remark concerns the case $p = 0$. It is well-known that

$$\lim_{p\to 0^+} \left(\frac{\mu_p(x)}{|\mathcal{K}|}\right)^{1/p} = \exp\left\{\int_\mathcal{K} \log|x-y|\,\frac{dy}{|\mathcal{K}|}\right\} = \exp\big\{\mu_{\log}(x)/|\mathcal{K}|\big\},$$

that can be interpreted as the *geometric mean* of the function $y \mapsto |x-y|$ on $\mathcal{K}$; needless to say, the set of its minimum points intersects $\heartsuit(\mathcal{K})$.

3.3 On Fraenkel's Asymmetry

Lemma 4 *Let $\mathcal{K} \subset \mathbb{R}^N$ be a convex body. For $r > 0$, define the function*

$$\gamma(x) = \left|\mathcal{K} \cap B(x,r)\right|, \quad x \in \mathbb{R}^N.$$

Then γ is log-concave and, if $\mathbf{M}(\gamma) = \{x \in \mathbb{R}^N : \gamma(x) = \max \gamma\}$, then

$$\mathbf{M}(\gamma) \cap \heartsuit(\mathcal{K}) \neq \varnothing. \tag{6}$$

Proof The log-concavity of G is a consequence of Prékopa–Leindler's inequality, that we recall here for the reader's convenience: let $0 < t < 1$ and let f, g and h be nonnegative integrable functions on $\mathbb{R}^N$ satisfying

$$h\big((1-t)x + ty\big) \geq f(x)^{1-t} g(y)^t, \quad \text{for every } x, y \in \mathbb{R}^N; \tag{7}$$

then

$$\int_{\mathbb{R}^N} h(x)\,dx \geq \left(\int_{\mathbb{R}^N} f(x)\,dx\right)^{1-t} \left(\int_{\mathbb{R}^N} g(x)\,dx\right)^t. \tag{8}$$

(For a proof and a discussion on the links between (8) and the Brunn-Minkowski inequality, the reader is referred to [3].)

Indeed, we pick two points $z, w \in \mathbb{R}^N$ and a number $t \in (0,1)$, and apply Prékopa–Leindler's inequality to the triple of functions

$$f = 1_{\mathcal{K} \cap B(z,r)}, \quad g = 1_{\mathcal{K} \cap B(w,r)}, \quad h = 1_{\mathcal{K} \cap B((1-t)z+tw,r)};$$

then, (7) is readily satisfied. Thus, (8) easily implies

$$\begin{aligned} \gamma\big((1-t)z + tw\big) &= \left|\mathcal{K} \cap B\big((1-t)z + tw, r\big)\right| \\ &\geq \left|\mathcal{K} \cap B(z,r)\right|^{1-t} \left|\mathcal{K} \cap B(w,r)\right|^t = \gamma(z)^{1-t} \gamma(w)^t, \end{aligned}$$

and, by taking the logarithm on both sides, we get the desired convexity. A straightforward consequence is that the set $\mathbf{M}(\gamma)$ is convex.

Once again, the validity of (6) will be a consequence of the inequality

$$\mathcal{R}_{\mathbf{M}(\gamma)}(\omega) \leq \mathcal{R}_{\mathcal{K}}(\omega), \quad \text{for every } \omega \in \mathbb{S}^{N-1}. \tag{9}$$

By contradiction: let us suppose that there exist $\omega \in \mathbb{S}^{N-1}$ and $x \in \mathbf{M}(\gamma)$ such that

$$\mathcal{R}_{\mathcal{K}}(\omega) < \mathcal{R}_{\mathbf{M}(\gamma)}(\omega) \leq \langle x, \omega \rangle.$$

In particular, this implies that the point $x^\lambda = T_{\lambda,\omega}(x)$, with $\lambda = \mathcal{R}_{\mathcal{K}}(\omega)$, does not belong to $\mathbf{M}(\gamma)$—i.e. the reflection of x with respect to the hyperplane $\pi_{\lambda,\omega}$ falls outside $\mathbf{M}(\gamma)$.

We set for brevity $B = B(x,r)$ and $B^\lambda = B(x^\lambda, r)$, and we again consider the ω-symmetric set $\Omega = \mathcal{K}_{\lambda,\omega} \cup T_{\lambda,\omega}(\mathcal{K}_{\lambda,\omega}) \subseteq \mathcal{K}$. Then, observe that

$$\begin{aligned} B \cap (\mathcal{K} \setminus \Omega) &= \big\{x \in B : \langle x, \omega \rangle < \lambda\big\} \cap (\mathcal{K} \setminus \Omega) \\ &\subseteq \big(B \cap B^\lambda\big) \cap (\mathcal{K} \setminus \Omega) \subseteq B^\lambda \cap (\mathcal{K} \setminus \Omega), \end{aligned}$$

which implies that $|B \cap (\mathcal{K} \setminus \Omega)| \le |B^\lambda \cap (\mathcal{K} \setminus \Omega)|$. Also, notice that since Ω is symmetric in the hyperplane $\pi_{\lambda,\omega}$ and $B^\lambda = T_{\lambda,\omega}(B)$, we have that $|B^\lambda \cap \Omega| = |B \cap \Omega|$.

By using these informations and the maximality of x, we can infer that

$$\gamma(x^\lambda) = \left|\mathcal{K} \cap B^\lambda\right| = \left|\Omega \cap B^\lambda\right| + \left|(\mathcal{K} \setminus \Omega) \cap B^\lambda\right|$$
$$\ge |\Omega \cap B| + \left|(\mathcal{K} \setminus \Omega) \cap B\right| = |\Omega \cap B| = \gamma(x),$$

that is x^λ is also a maximum point, i.e. $x^\lambda \in \mathbf{M}(\gamma)$—a contradiction. □

As a consequence of this lemma, we obtain a result concerning the so-called *Fraenkel asymmetry* of $\mathcal{K}$:

$$\mathcal{A}(\mathcal{K}) = \min_{x \in \mathbb{R}^N} \frac{|\mathcal{K} \Delta B(x, r^*_{\mathcal{K}})|}{|\mathcal{K}|},$$

where Δ denotes the symmetric difference of the two sets and the radius $r^*_{\mathcal{K}}$ is determined by $|B(x, r^*_{\mathcal{K}})| = |\mathcal{K}|$. This is a measure of how a set is far from being spherically symmetric and was introduced in [4]; we refer the reader to [2] for a good account on $\mathcal{A}(\mathcal{K})$.

Corollary 2 *Let $\mathcal{K} \subset \mathbb{R}^N$ be a convex body. Then $\mathcal{A}(\mathcal{K})$ is attained for at least one ball centered at a point belonging to $\heartsuit(\mathcal{K})$.*

Proof It is sufficient to observe that $|\mathcal{K} \Delta B(x, r^*_{\mathcal{K}})| = 2(|\mathcal{K}| - |\mathcal{K} \cap B(x, r^*_{\mathcal{K}})|)$, since $|B(x, r^*_{\mathcal{K}})| = |\mathcal{K}|$, and hence

$$\frac{|\mathcal{K} \Delta B(x, r^*_{\mathcal{K}})|}{|\mathcal{K}|} = 2\left(1 - \frac{\gamma(x)}{|\mathcal{K}|}\right).$$

Thus, $\mathcal{A}(\mathcal{K})$ is attained by points that maximize γ; hence, Lemma 4 provides the desired conclusion. □

Remark 6 Observe in particular that if $\mathcal{K}$ has N hyperplanes of symmetry, then an optimal ball can be placed at their intersection. However, in general, even under this stronger assumption, such optimal ball is not unique. For example, take the rectangle $Q_\varepsilon = [-\pi/4\varepsilon, \pi/4\varepsilon] \times [-\varepsilon, \varepsilon]$ with $0 < \varepsilon < \pi/4$; any unit ball centered at a point in the segment $(-\pi/4\varepsilon + 1, \pi/4\varepsilon - 1) \times \{0\}$ realizes the Fraenkel asymmetry $\mathcal{A}(Q_\varepsilon)$. Thus, in general *it is not true that all optimal balls are centered in the heart.*

Remark 7 The following problem in spectral optimization was considered in [5]: given a (convex) set $\mathcal{K} \subset \mathbb{R}^N$ and a radius $0 < r < r_{\mathcal{K}}$, find the ball $B(x_0, r) \subset \mathcal{K}$ which maximizes the quantity

$$\lambda_1\big(\mathcal{K} \setminus B(x, r)\big)$$

as a function of x: here, $\lambda_1(\Omega)$ stands for the first Dirichlet-Laplacian eigenvalue of a set Ω. By considerations similar to the ones used in this section and remarks contained in [5, Theorem 2.1], it can be proved that $x_0 \in \heartsuit(\mathcal{K})$.

4 Estimating the Volume of the Heart

In this section, we begin an analysis of the following problem in shape optimization:

$$\text{maximize the ratio } \frac{|\heartsuit(\mathcal{K})|}{|\mathcal{K}|} \text{ among all convex bodies } \mathcal{K} \subset \mathbb{R}^N; \tag{10}$$

solving (10) would give an answer to question (iii) in the introduction. Since this ratio is scaling invariant, (10) is equivalent to the following problem:

$$\text{maximize the ratio } \frac{|\heartsuit(\mathcal{K})|}{|\mathcal{K}|} \text{ among all convex bodies } \mathcal{K} \subset [0,1]^N; \tag{11}$$

here, $[0, 1]^N$ is the unit cube in $\mathbb{R}^N$.

We notice that the class of the competing sets in problem (11) is relatively compact in the topology induced by the *Hausdorff distance* (see [6, Chap. 2])—the most natural topology when one deals with the constraint of convexity. This fact implies, in particular, that any maximizing sequence $\{\mathcal{K}_n\}_{n\in\mathbb{N}} \subset [0,1]^N$ of convex bodies converges—up to a subsequence—to a compact convex set $\mathcal{K} \subset [0,1]^N$.

However, there are two main obstructions to the existence of a maximizing set for (11): (a) in general, the limit set $\mathcal{K}$ may not be a convex body, i.e. $\mathcal{K}$ could have empty interior; in other words, maximizing sequences could "collapse" to a lower dimensional object; (b) it is not clear whether the shape functional $\mathcal{K} \mapsto |\heartsuit(\mathcal{K})|$ is upper semicontinuous or not in the aforementioned topology.

The next example assures that the foreseen semicontinuity property *fails to be true* in general.

Example 1 Let $Q = [-2,2] \times [-1,1]$ and take the points

$$p^1_\varepsilon = (1, 1+\varepsilon) \quad \text{and} \quad p^2_\varepsilon = (-2-\varepsilon, 1/2),$$

and define Q_ε as the convex hull of $Q \cup \{p^1_\varepsilon,\ p^2_\varepsilon\}$. As ε vanishes, $\heartsuit(Q_\varepsilon)$ shrinks to the quadrangle having vertices

$$(0,0) \quad (1,0) \quad (1/2,1/2) \quad \text{and} \quad (0,1/2),$$

while clearly $\heartsuit(Q) = \{0\}$: indeed, observe that due to the presence of the new corners p^1_ε and p^2_ε, it is no more possible to use $\{x = 0\}$ or $\{y = 0\}$ as maximal axis of reflection in the directions $\mathbf{e}_1 = (1,0)$ or $\mathbf{e}_2 = (0,1)$, respectively. In particular, we get that

$$0 = \big|\heartsuit(Q)\big| < \lim_{\varepsilon\to 0^+} \big|\heartsuit(Q_\varepsilon)\big|.$$

The situation is illustrated in Fig. 2.

These considerations show that the existence of a solution of (10) is not a trivial issue. Indeed, we are able to show that *an optimal shape does not exist* in the class of triangles. This is the content of the main result of this section.

Fig. 2 The heart of Q_ε

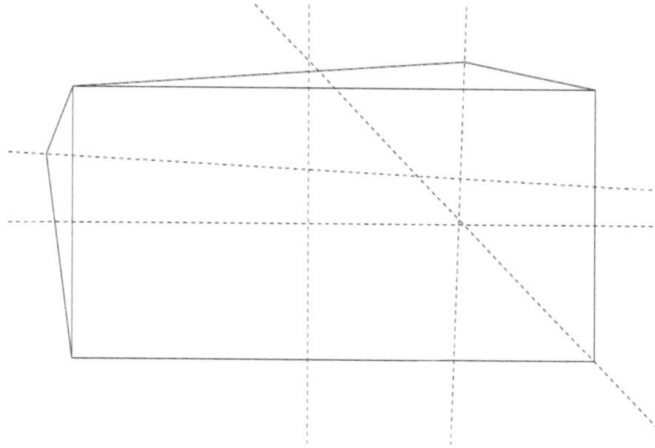

Theorem 5 *It holds that*

$$\sup\left\{\frac{|\heartsuit(\mathcal{K})|}{|\mathcal{K}|} : \mathcal{K} \textit{ is a triangle}\right\} = \frac{3}{8},$$

and the supremum is attained by a sequence of obtuse triangles.

The proof of Theorem 5 is based on the following lemma, in which we exactly determine $\heartsuit(\mathcal{K})$, when $\mathcal{K}$ is a triangle.

Lemma 5 *Let $\mathcal{K}$ be a triangle. Then the following assertions hold:*

(i) *if $\mathcal{K}$ is acute, $\heartsuit(\mathcal{K})$ is contained in the triangle formed by the segments joining the midpoints of the sides of $\mathcal{K}$; also, $\heartsuit(\mathcal{K})$ equals the quadrangle $\mathcal{Q}$ formed by the bisectors of the smallest and largest angles and the axes of the shortest and longest sides of $\mathcal{K}$;*

(ii) *if $\mathcal{K}$ is obtuse, $\heartsuit(\mathcal{K})$ is contained in the parallelogram whose vertices are the midpoints mentioned in* (i) *and the vertex of the smallest angle in $\mathcal{K}$; also, $\heartsuit(\mathcal{K})$ equals the polygon $\mathcal{P}$ formed by the largest side of $\mathcal{K}$ and the bisectors and axes mentioned in* (i); *$\mathcal{P}$ may be either a pentagon or a quadrangle.*

Proof Observe that bisectors of angles and axes of sides are admissible axes of reflection. If $\mathcal{K}$ is acute, $C_\mathcal{K}$ and $I_\mathcal{K}$ fall in its interior and are the intersection of the axes and bisectors, respectively. If $\mathcal{K}$ is obtuse, $I_\mathcal{K}$ still falls in the interior of $\mathcal{K}$, while $C_\mathcal{K}$ is the midpoint of the largest side of $\mathcal{K}$ and is no longer the intersection of the axes. These remarks imply that $\heartsuit(\mathcal{K}) \subseteq \mathcal{Q}$ in case (i) and $\heartsuit(\mathcal{K}) \subseteq \mathcal{P}$ in case (ii); also $C_\mathcal{K}, I_\mathcal{K} \in \mathcal{Q} \cap \heartsuit(\mathcal{K})$ and $C_\mathcal{K}, I_\mathcal{K} \in \mathcal{P} \cap \heartsuit(\mathcal{K})$.

The segments specified in (i) are also admissible axes of reflection if $\mathcal{K}$ is acute; thus, the inclusion in the triangle mentioned in (i) easily follows. If $\mathcal{K}$ is obtuse, only the segment joining the midpoints of the smallest and intermediate side is an axis of reflection. However, we can still claim that $\heartsuit(\mathcal{K})$ is contained in the parallelogram mentioned in (ii), since $C_\mathcal{K}$ is now the midpoint of the largest side from which one of the axes is issued: thus, $\heartsuit(\mathcal{K})$ must stay below that axis.

Fig. 3 The heart of a triangle

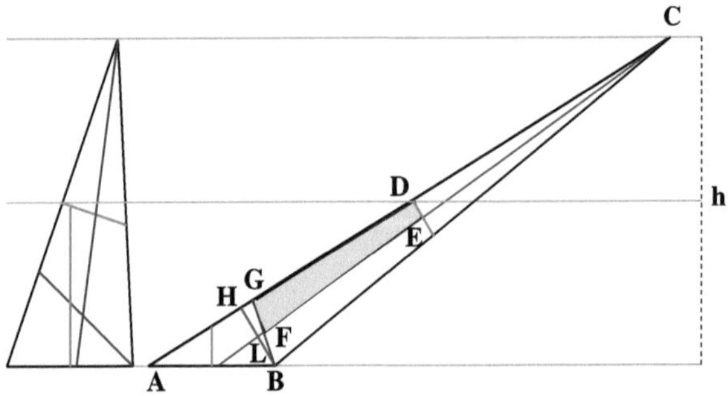

Now, if $\heartsuit(\mathcal{K})$ were smaller than $\mathcal{Q}$ (or $\mathcal{P}$), then there would be an axis of reflection that cuts off one of the vertices of $\mathcal{Q}$ (or $\mathcal{P}$) different from $C_{\mathcal{K}}$ and $I_{\mathcal{K}}$ (that always belong to $\heartsuit(\mathcal{K})$). In any case, such an axis would violate the maximality that axes of sides and bisectors of angles enjoy with respect of reflections. □

We are now ready to prove Theorem 5.

Proof First of all, thanks to the inclusion mentioned in Lemma 5, we get $|\heartsuit(\mathcal{K})| < 1/4|\mathcal{K}|$ when $\mathcal{K}$ is acute. Thus, we can restrict ourselves to the case of $\mathcal{K}$ obtuse.

Here, we refer to Fig. 3. We observe that, by what we proved in Lemma 5, $\heartsuit(\mathcal{K})$ is always contained in the quadrangle $DEFG$ (when the angle in B is much larger than $\pi/2$, $DEFG$ and $\heartsuit(\mathcal{K})$ coincide), which is contained in the trapezoid $DELH$. Thus, $|\heartsuit(\mathcal{K})| \le |DELH|$; hence it is enough to prove that, if the angle in B increases, $|DELH|$ increases and both ratios $|\heartsuit(\mathcal{K})|/|\mathcal{K}|$ and $|DELH|/|\mathcal{K}|$ tend to 3/8.

We proceed to compute $|\heartsuit(\mathcal{K})|$, when the angle in B is large. We fix a base and a height of $\mathcal{K}$: as a base we choose the smallest side and we suppose it has length b; h will denote the length of its corresponding height. In this way, $|\mathcal{K}| = bh/2$.

In Fig. 3, the lines through the points B and G, and C and L bisect the angles in B and C, respectively. The line through D and E is the only axis that contributes to form $\heartsuit(\mathcal{K})$, that equals the quadrangle $DEFG$; thus, $\heartsuit(\mathcal{K})$ is obtained as

$$\heartsuit(\mathcal{K}) = T_1 \setminus (T_2 \cup T_3),$$

where the T_i's are triangles:

$$T_1 = CBG, \qquad T_2 = CBF, \qquad T_3 = CED.$$

We place the origin of Cartesian axes in A and set $B = (b, 0)$; we also set $C = (t, h)$. Finally, we denote by α, β and γ the respective measures of the angles in A, B and C.

Trigonometric formulas imply that

$$\begin{aligned}
|T_1| &= \frac{1}{2}\big[h^2 + (t-b)^2\big]\frac{\sin(\beta/2)\sin(\gamma)}{\sin(\beta/2+\gamma)}, \\
|T_2| &= \frac{1}{2}\big[h^2 + (t-b)^2\big]\frac{\sin(\beta/2)\sin(\gamma/2)}{\sin(\beta/2+\gamma/2)}, \qquad (12) \\
|T_3| &= \frac{1}{8}\big[h^2 + t^2\big]\tan(\gamma/2),
\end{aligned}$$

where the angles β and γ are related by the *theorem of sines*:

$$\frac{h}{\sqrt{h^2+t^2}\sqrt{h^2+(t-b)^2}} = \frac{\sin(\beta)}{\sqrt{h^2+t^2}} = \frac{\sin(\gamma)}{b}.$$

The area of *DELH* is readily computed as

$$|DELH| = \frac{1}{2}\frac{b^2h^2}{(h^2+t^2)\tan(\gamma/2)} - \frac{1}{8}\left(h^2+t^2\right)\tan(\gamma/2). \tag{13}$$

Now, observe that this quantity increases with t, since it is the composition of two decreasing functions: $s \mapsto b^2h^2/(2s) - s/8$ and $t \mapsto (h^2+t^2)\tan(\gamma/2)$.

As $t \to \infty$, $|\mathcal{K}|$ does not change, the angle γ vanishes and the angle β tends to π; moreover, we have that

$$t\sin(\beta) \to h, \qquad t^2\sin(\gamma) \to bh \quad \text{as } t \to \infty.$$

Formulas (12) then yield:

$$|T_1| \to \frac{1}{2}bh = |\mathcal{K}|, \qquad |T_2| \to \frac{1}{4}bh = \frac{1}{2}|\mathcal{K}|, \qquad |T_3| \to \frac{1}{16}bh = \frac{1}{8}|\mathcal{K}|.$$

Thus, since $|\heartsuit(\mathcal{K})| = |T_1| - |T_2| - |T_3|$, we have that $|\heartsuit(\mathcal{K})| \to \frac{3}{8}|\mathcal{K}|$; by (13), $|DELH| \to \frac{3}{8}|\mathcal{K}|$ as well.

The proof is complete. □

Remark 8 Thus, Theorem 5 sheds some light on problem (10). In fact, observe that the maximizing sequence, once properly re-scaled, gives a maximizing sequence for the equivalent problem (11), that precisely collapses to a one-dimensional object.

Numerical evidence based on the algorithm developed in [1] suggests that, the more $\mathcal{K}$ is *round*, the more $\heartsuit(\mathcal{K})$ is small compared to $\mathcal{K}$. We conjecture that

$$\sup\left\{\frac{|\heartsuit(\mathcal{K})|}{|\mathcal{K}|} : \mathcal{K} \subset \mathbb{R}^2 \text{ is a convex body}\right\} = \frac{3}{8},$$

and the supremum is realized by a sequence of obtuse triangles.

Acknowledgements The authors warmly thank Prof. J. O'Hara, for the reference [7], and his PhD student Shigehiro Sakata, who pointed out a gap in a preliminary version of the proof of Theorem 3.

The first author has been partially supported by the ERC Advanced Grant No. 226234. The second author has been supported by GNAMPA-INdAM and the PRIN-MIUR grant "Proprietà e metodi geometrici nelle equazioni alle derivate parziali, disuguaglianze di Sobolev e convessità". Both authors gratefully acknowledge the Banff International Research Station and its facilities, where this work started.

References

1. Brasco, L., Magnanini, R., Salani, P.: The location of the hot spot in a grounded convex conductor. Indiana Univ. Math. J. **60**, 633–660 (2011)

2. Fusco, N., Maggi, F., Pratelli, A.: The sharp quantitative isoperimetric inequality. Ann. Math. (2) **168**(3), 941–980 (2008)
3. Gardner, R.J.: The Brunn-Minkowski inequality. Bull. Am. Math. Soc. **39**, 355–405 (2002)
4. Hall, R.R., Hayman, W.K., Weitsman, A.W.: On asymmetry and capacity. J. Anal. Math. **56**, 87–123 (1991)
5. Harrell, E.M., Kröger, P., Kurata, K.: On the placement of an obstacle or a well so as to optimize the fundamental eigenvalue. SIAM J. Math. Anal. **33**, 240–259 (2001)
6. Henrot, A., Pierre, M.: Variation et optimisation de formes. Une analyse géométrique. [Shape variation and optimization. A geometric analysis]. Mathématiques & Applications [Mathematics & Applications], vol. 48. Springer, Berlin (2005) (French)
7. O'Hara, J.: Renormalization of potentials and generalized centers. Adv. Appl. Math. **48**, 365–392 (2012). See also arXiv:1008.2731

A Viscosity Equation for Minimizers of a Class of Very Degenerate Elliptic Functionals

Giulio Ciraolo

Abstract We consider the functional

$$J(v) = \int_{\Omega} [f(|\nabla v|) - v] dx,$$

where Ω is a bounded domain and $f : [0, +\infty) \to \mathbb{R}$ is a convex function vanishing for $s \in [0, \sigma]$, with $\sigma > 0$. We prove that a minimizer u of J satisfies an equation of the form

$$\min\big(F\big(\nabla u, D^2 u\big), |\nabla u| - \sigma\big) = 0$$

in the viscosity sense.

Keywords Nonlinear degenerate elliptic operators · Viscosity solutions · Torsion problem

1 Introduction

Let Ω be a bounded domain in $\mathbb{R}^N$, $N \geq 2$, with boundary $\partial\Omega$ of class $C^{2,\alpha}$, with $0 < \alpha < 1$. We consider the variational problem

$$\inf\big\{J(v) : v \in W_0^{1,\infty}(\Omega)\big\}, \quad \text{where } J(v) = \int_{\Omega} \big[f\big(|\nabla v|\big) - v\big] dx; \tag{1}$$

here, the function $f : [0, +\infty) \to \mathbb{R}$ is convex, monotone, nondecreasing and we assume that there exists $\sigma > 0$ such that

$$f \in C^1\big([0, +\infty)\big) \cap C^3\big((\sigma, +\infty)\big); \tag{2a}$$

$$f(0) = 0 \quad \text{and} \quad \lim_{s \to +\infty} \frac{f(s)}{s} = +\infty; \tag{2b}$$

$$f'(s) = 0 \quad \text{for every } 0 \leq s \leq \sigma; \tag{2c}$$

$$f''(s) > 0 \quad \text{for } s > \sigma. \tag{2d}$$

G. Ciraolo (✉)
Dipartimento di Matematica e Informatica, Università di Palermo, Via Archirafi 34, 90123 Palermo, Italy
e-mail: g.ciraolo@math.unipa.it

R. Magnanini et al. (eds.), *Geometric Properties for Parabolic and Elliptic PDE's*, Springer INdAM Series 2, DOI 10.1007/978-88-470-2841-8_5,

Functionals of this kind occur in the study of complex-valued solutions of the *eikonal* equation (see [6] and [16–19]), as well as in the study of problems linked to traffic congestion (see [2]) and in variational problems which are relaxations of non-convex ones (see [5] and [10]). We have in mind the following two main examples of a function f:

$$f(s) = \begin{cases} 0, & 0 \leq s \leq 1, \\ \frac{1}{2}[s\sqrt{s^2-1} - \log(s + \sqrt{s^2-1})], & s > 1, \end{cases} \tag{3}$$

which arises from the study of complex-valued solutions of the eikonal equation, and

$$f(s) = \begin{cases} \frac{1}{q}(s-1)^q, & s > 1, \\ 0, & 0 \leq s \leq 1, \end{cases} \tag{4}$$

$q > 1$, which is linked to traffic congestion problems.

Since f vanishes in the interval $[0, \sigma]$, problem (1) is strongly degenerate and, as far as we know, few studies have been done. Besides the papers cited before, we mention [1] and [20] where regularity issues were tackled.

In this paper, we shall prove that the minimizer u of (1) satisfies an equation of the form

$$\min\big(F(\nabla u, D^2 u), |\nabla u| - \sigma\big) = 0 \tag{5}$$

in the viscosity sense (see Theorems 1 and 2 for the meaning of F).

Our strategy is to approximate J by a sequence of less degenerating functionals so that the minimizers of the corresponding variational problems converge uniformly to u; this is done in Sect. 2. Then, the machinery of viscosity equations applies and, in Sect. 3, we prove that u satisfies (5). To prove Theorems 1 and 2, which are our main results, we make use of techniques which have been used in the context of the ∞-Laplace operator (see for instance [3, 13, 14]).

2 Preliminary Results

We start by recalling some well-known facts. Since Ω is bounded and $\partial\Omega$ is of class $C^{2,\alpha}$ then the following *uniform exterior sphere condition* holds: there exists $\rho > 0$ such that for every $x_0 \in \partial\Omega$ there exists a ball $B_\rho(y)$ of radius ρ centered at $y = y(x_0) \in \mathbb{R}^N \setminus \overline{\Omega}$ such that $\overline{B_\rho(y)} \cap \overline{\Omega} = \overline{B_\rho(y)} \cap \partial\Omega$ and $x_0 \in \partial B_\rho(y)$.

Notice that, since f satisfies (2a)–(2d), the functional J is differentiable and a critical point u of J satisfies the problem

$$\begin{cases} -\operatorname{div}\left(\dfrac{f'(|\nabla u|)}{|\nabla u|}\nabla u\right) = 1, & \text{in } \Omega, \\ u = 0, & \text{on } \partial\Omega, \end{cases} \tag{6}$$

in the weak sense, i.e.

$$\int_\Omega \frac{f'(|\nabla u|)}{|\nabla u|}\nabla u \cdot \nabla\phi\, dx = \int_\Omega \phi\, dx, \quad \text{for every } \phi \in C_0^1(\Omega). \tag{7}$$

It will be useful in the sequel to have at hand the solution of (6) when Ω is the ball of given radius R (centered at the origin): it is given by

$$u_R(x) = \int_{|x|}^{R} g'\left(\frac{s}{N}\right) ds, \tag{8}$$

where

$$g(t) = \sup\{st - f(s) : s \geq 0\}$$

is the Fenchel conjugate of f (see for instance [9] and [11]).

It is clear that, when $\sigma = 0$ (1) has a unique solution, since f is strictly convex. When $\sigma > 0$, the uniqueness of a minimizer for (1) is proved in [7].

In this section we shall approximate the functional J by a sequence of strictly convex functionals

$$J_n(v) = \int_\Omega \left[f_n(|\nabla v|) - v \right] dx, \tag{9}$$

$n \in \mathbb{N}$, which are less degenerating than J (see Proposition 1 for the assumptions on the functions f_n) and prove some uniform bounds for the minimizers u_n of

$$\inf\{J_n(v) : v \in W_0^{1,\infty}(\Omega)\}. \tag{10}$$

Notice that, if $f_n \in C^1([0,+\infty)) \cap C^3((0,+\infty))$ satisfies (2b) and it is such that $f_n'(0) = 0$ and $f_n''(s) > 0$ for $s > 0$, then the minimizer u_n of (9) is unique and satisfies

$$\int_\Omega \frac{f_n'(|\nabla u_n|)}{|\nabla u_n|} \nabla u_n \cdot \nabla \phi \, dx = \int_\Omega \phi \, dx, \quad \text{for every } \phi \in C_0^1(\Omega). \tag{11}$$

We shall say that $w \in W^{1,\infty}(\Omega)$ is a subsolution of (11) if

$$\int_\Omega \frac{f_n'(|\nabla w|)}{|\nabla w|} \nabla w \cdot \nabla \phi \, dx \leq \int_\Omega \phi \, dx, \quad \text{for every } \phi \in C_0^1(\Omega) \text{ with } \phi \geq 0,$$

and that $w \in W^{1,\infty}(\Omega)$ is a supersolution of (11) if

$$\int_\Omega \frac{f_n'(|\nabla w|)}{|\nabla w|} \nabla w \cdot \nabla \phi \, dx \geq \int_\Omega \phi \, dx, \quad \text{for every } \phi \in C_0^1(\Omega) \text{ with } \phi \geq 0.$$

Let u_n and v_n be a subsolution and a supersolutions of (11), respectively. Then, the following *weak comparison principle* holds: if $u_n \leq v_n$ on $\partial\Omega$ then $u_n \leq v_n$ in $\overline{\Omega}$ (see Lemma 3.7 in [11]).

It will be useful to define the following P-function (see [11]):

$$P_n(x) = \Phi_n\left(|\nabla u_n(x)|\right) + \frac{2}{N} u_n(x), \quad x \in \overline{\Omega}, \tag{12}$$

where

$$\Phi_n(t) = 2 \int_0^t s f_n''(s) \, ds. \tag{13}$$

To avoid heavy notations, in Lemmas 1 and 2 we drop the dependence on n.

Lemma 1 *Let* $f \in C^1([0,+\infty)) \cap C^3((0,+\infty))$ *be such that* $f'(0)=0$ *and* $f''(s)>0$ *for* $s>0$ *and let* u *be the solution of* (1). *Then,* $|\nabla u|$ *attains its maximum on the boundary of* Ω *and the following estimate holds*:

$$\big|\nabla u(x)\big| \le M, \quad x \in \overline{\Omega}, \tag{14}$$

with

$$M = g'\left(\frac{\rho}{N-1}\left(e^{\frac{(N-1)R^*}{\rho}} - 1\right)\right),$$

where g *is the Fenchel conjugate of* f, $R^* = \sup\{|x-y| : x, y \in \partial\Omega\}$ *and* ρ *is the radius of the uniform exterior sphere.*

Furthermore,

$$0 \le u(x) \le \min\left(\int_0^{R^*} g'\left(\frac{s}{N}\right) ds, \frac{N}{2}\Phi(M)\right) \quad x \in \overline{\Omega}. \tag{15}$$

Proof Since u is a minimizer of J, it is easy to show that $u \ge 0$. Being R^* the diameter of Ω, there exist a ball of radius R^* that contains Ω (we can assume that such ball is centered at the origin). Since $u_{R^*}(x) \ge 0$ for $x \in \partial\Omega$, the weak comparison principle implies that

$$u(x) \le u_{R^*}(x) \quad \text{for every } x \in \overline{\Omega}.$$

From $u_{R^*}(x) \le u_{R^*}(0)$, $x \in B_{R^*}$ and from (8), we have

$$u(x) \le \int_0^{R^*} g'\left(\frac{s}{N}\right) ds, \tag{16}$$

for every $x \in \overline{\Omega}$.

Now, we consider the P-function given by (12). As proved in Lemma 3.2 in [11], P attains its maximum on the boundary of Ω and thus

$$P(x) \le \max_{\partial\Omega} P = \max_{\partial\Omega} \Phi\big(|\nabla u|\big), \quad x \in \overline{\Omega}.$$

Since Φ is strictly increasing, then we get

$$\max_{\overline{\Omega}} \big|\nabla u(x)\big| = \max_{\partial\Omega} \big|\nabla u(x)\big|, \tag{17}$$

i.e. $|\nabla u|$ attains its maximum on the boundary of Ω.

Following [12], we construct a barrier function for u which will give us an upper bound for $|\nabla u|$ on the boundary of Ω. Let $x_0 \in \partial\Omega$ be fixed and let $B_\rho(y(x_0))$ be the ball in the exterior sphere condition. Set

$$\delta(x) = \operatorname{dist}\big(x, \partial B_\rho\big(y(x_0)\big)\big),$$

and let $w(x) = \psi(\delta(x))$ be a function depending only on the distance from $\partial B_\rho(y(x_0))$; we have

$$\operatorname{div}\left\{f'\big(|\nabla w|\big)\frac{\nabla w}{|\nabla w|}\right\} = \psi''\big(\delta(x)\big)f''\big(\psi'\big(\delta(x)\big)\big) + f'\big(\psi'\big(\delta(x)\big)\big)\Delta\delta(x). \tag{18}$$

Since

$$\left|\Delta\delta(x)\right| = \frac{N-1}{|x-y|} \leq \frac{N-1}{\rho},$$

from (18) we obtain

$$\operatorname{div}\left\{f'\left(|\nabla w|\right)\frac{\nabla w}{|\nabla w|}\right\} + 1 \leq \psi''\left(\delta(x)\right)f''\left(\psi'\left(\delta(x)\right)\right) + \frac{N-1}{\rho}f'\left(\psi'\left(\delta(x)\right)\right) + 1. \tag{19}$$

By choosing

$$\psi(t) = \int_0^t g'\left(\frac{\rho}{N-1}\left(e^{\frac{N-1}{\rho}(R^*-s)} - 1\right)\right) ds,$$

the right hand side of (19) vanishes and thus w is a supersolution of (7). Notice that $\psi'(t) > 0$ for $t > 0$ and then $\psi(t) > 0$ for $t > 0$. Since $x \in \Omega$ implies that $\operatorname{dist}(x, \partial B_\rho(x_0)) > 0$, we have that $w(x) \geq 0$ for $x \in \overline{\Omega}$. The weak comparison principle yields $u(x) \leq w(x)$ in $\overline{\Omega}$. Since $x_0 \in \partial\Omega$ is arbitrary, we obtain

$$\left|\nabla u(x)\right| \leq g'\left(\frac{\rho}{N-1}\left(e^{\frac{(N-1)R^*}{\rho}} - 1\right)\right),$$

for any $x \in \partial\Omega$. According to (17) the same estimate holds in the whole of Ω and (14) holds.

Notice that from (12)

$$u(x) \leq \frac{N}{2}P(x), \quad x \in \overline{\Omega};$$

since P attains its maximum on the boundary of Ω and from (14) we have that

$$u(x) \leq \frac{N}{2}\Phi(M)$$

which, together with (16), implies (15). □

We denote by $H_{\partial\Omega}(x)$ the mean curvature of $\partial\Omega$ at the point $x \in \partial\Omega$ and set

$$H^*_{\partial\Omega} = \min_{x\in\partial\Omega} H_{\partial\Omega}(x).$$

In the following lemma, we give a further bound for u and $|\nabla u|$ in the case that the mean curvature of $\partial\Omega$ attains a positive minimum.

Lemma 2 *Let f be as in Lemma* 1 *and assume that $H^*_{\partial\Omega} > 0$. Then,*

$$u(x) \leq \frac{N}{2}\Phi\left(g'\left(\frac{1}{NH^*_{\partial\Omega}}\right)\right) \tag{20}$$

and

$$\left|\nabla u(x)\right| \leq g'\left(\frac{1}{NH^*_{\partial\Omega}}\right) \tag{21}$$

for every $x \in \overline{\Omega}$, where Φ is given by (13) *and g is the Fenchel conjugate of f.*

Proof Since $|\nabla u| > 0$ on $\partial\Omega$ (see Lemma 2.7 in [7]), Eq. (7) is nondegenerate in a neighborhood of $\partial\Omega$; from standard elliptic regularity theory (see [21] and [12]), we know that $u \in C^{2,\alpha}(\overline{\Omega} \setminus \{x : \nabla u \neq 0\})$ for some $\alpha \in (0,1)$, and then (7) can be written pointwise on $\partial\Omega$ as

$$f''\big(|u_\nu(x)|\big)u_{\nu\nu}(x) - (N-1)f'\big(u_\nu(x)\big)H_{\partial\Omega}(x) = -1;$$

here, ν denotes the exterior unit normal to $\partial\Omega$, $u_\nu = \nabla u \cdot \nu$ and $u_{\nu\nu} = (D^2 u)\nu \cdot \nu$. From Lemma 3.3 in [11], we know that

$$Nf'\big(|\nabla u(x)|\big)H_{\partial\Omega}(x) \le 1,$$

for every $x \in \partial\Omega$ and, since g' is nondecreasing, then

$$|\nabla u(x)| \le g'\left(\frac{1}{NH^*_{\partial\Omega}}\right),$$

for every $x \in \partial\Omega$. Since Φ is nondecreasing and P (given by (12)) attains its maximum on $\partial\Omega$, from $u = 0$ on $\partial\Omega$ we obtain

$$P(x) \le \Phi\left(g'\left(\frac{1}{NH^*_{\partial\Omega}}\right)\right) \tag{22}$$

for every $x \in \Omega$. From (12) and (22) we conclude. □

Notice that, when Ω is a ball, (21) is optimal.

Proposition 1 *Let* $(f_n)_{n\in\mathbb{N}}$ *be such that:*

(i) $f_n \in C^1([0,+\infty)) \cap C^3((0,+\infty))$;
(ii) f_n *converges uniformly to* f *on the compact sets contained in* $[0,+\infty)$;
(iii) $f_n'(0) = 0$, *the functions* f_n' *decrease to* f' *in* $[0,+\infty)$ *and* f_n' *converges uniformly to* f' *on the compact sets contained in* $[0,+\infty)$;
(iv) $f_n''(t) > 0$ *for* $t > 0$.

Let u *(resp.* u_n*) be the solution of* (1) *for* J *(resp. of* (10) *for* J_n*). Then*

(a) u_n *is a minimizing sequence for* J *and* $J_n(u_n) \to J(u)$;
(b) u_n *and* ∇u_n *are uniformly bounded and (up to a subsequence)* $(u_n)_{n\in\mathbb{N}}$ *tends to* u *in the sup norm topology and* u *satisfies estimates* (14) *and* (15) *almost everywhere in* Ω.

Proof Since $J_n \to J$ uniformly (a) is standard. Since the sequence $(f_n')_{n\in\mathbb{N}}$ is decreasing in n, then g_n' is increasing in n and converges pointwise to g' (here, we denote by g and g_n the Fenchel conjugates of f and f_n, respectively). Thus, $g_n(t) \le g(t)$ and $g_n'(t) \le g'(t)$ for every $t \in [0,+\infty)$ and (b) follows by Lemma 1 and an application of Ascoli-Arzelà's theorem. □

3 Viscosity Euler-Lagrange Equation

In this section we prove that the solution u of (1) satisfies an equation of the form (5) in the viscosity sense. Firstly, we do it for $f \in C^2((0,+\infty)) \cup C^3((\sigma,+\infty))$ and then we deal with the case that f is not twice differentiable at $s=\sigma$.

Consider a sequence of approximating functions $\{f_n\}_{n\in\mathbb{N}}$ satisfying (i)–(iv) in Proposition 1. The minimizer u_n for (10) satisfies

$$-\operatorname{div}\frac{f_n'(|\nabla u_n|)}{|\nabla u_n|}=1,$$

in weak sense. Assume for a moment that u_n is regular enough so that we can differentiate, then u_n satisfies

$$-\frac{|\nabla u_n| f_n''(|\nabla u_n|)-f_n'(|\nabla u_n|)}{|\nabla u_n|^3}\Delta_\infty u_n-\frac{f_n'(|\nabla u_n|)}{|\nabla u_n|}\Delta u_n=1,$$

where

$$\Delta_\infty u=\sum_{i,j=1}^{N}\frac{\partial^2 u}{\partial x_i\partial x_j}\frac{\partial u}{\partial x_i}\frac{\partial u}{\partial x_j}.$$

Since this equation is fully nonlinear and degenerate elliptic, it makes sense to define and study its viscosity solutions (see [4]).

Let $P\in\mathbb{R}^N$ and $X\in\mathcal{S}^N$, where $\mathcal{S}^N$ is the space of real-valued $N\times N$ symmetric matrices. Consider the function

$$F_n(P,X):=\begin{cases}-\frac{|P|f_n''(|P|)-f_n'(|P|)}{|P|^3}P\cdot XP-\frac{f_n'(|P|)}{|P|}\operatorname{tr}(X)-1, & P\neq 0,\\ -1, & P=0.\end{cases}\tag{23}$$

Notice that, if

$$\lim_{s\to 0^+}\frac{sf_n''(s)-f_n'(s)}{s^3}=0,\quad\text{and}\quad \lim_{s\to 0^+}\frac{f_n'(s)}{s}=0,\tag{24}$$

then F_n is continuous. For future use, we shall assume that the sequence $\{f_n\}_{n\in\mathbb{N}}$ is such that

$$\lim_{n\to+\infty}\frac{sf_n''(s)-f_n'(s)}{s^3}=0\quad\text{and}\quad \lim_{n\to+\infty}\frac{f_n'(s)}{s}=0\tag{25}$$

uniformly on the compact sets of $[0,\sigma)$; here, thanks to (24), the functions in the limits are understood as continuously extended to 0 at $s=0$.

We shall introduce the definition of viscosity solution of an equation of the form $F(\nabla v,D^2v)=0$ (see [14]).

Definition An upper semicontinuous function u defined in Ω is a *viscosity subsolution* of

$$F\big(\nabla v,D^2v\big)=0,\tag{26}$$

$x \in \Omega$, if, whenever $x_0 \in \Omega$ and $\phi \in C^2(\Omega)$ are such that $u(x_0) = \phi(x_0)$ and $u(x) < \phi(x)$ if $x \neq x_0$, then

$$F\big(\nabla\phi(x_0), D^2\phi(x_0)\big) \le 0.$$

A lower semicontinuous function u defined in Ω is a *viscosity supersolution* of (26) if, whenever $x_0 \in \Omega$ and $\phi \in C^2(\Omega)$ are such that $u(x_0) = \phi(x_0)$ and $u(x) > \phi(x)$ if $x \neq x_0$, then

$$F\big(\nabla\phi(x_0), D^2\phi(x_0)\big) \ge 0.$$

Finally, $u \in C^0(\Omega)$ is a *viscosity solution* of (26) if it is both a viscosity subsolution and a viscosity supersolution of (26).

Lemma 3 *Let u_n be the minimizer of J_n, where $f_n \in C^1([0,+\infty)) \cup C^3((0,+\infty))$ satisfies* (24) *and is such that $f_n''(s) > 0$ for $s > 0$. Then u_n is a viscosity solution of* (26), *with $F = F_n$ and F_n given by* (23).

Proof Notice that, since f_n satisfies (24), then F_n is continuous. We present the details for the case of supersolutions. Let $x_0 \in \Omega$ and $\phi \in C^2(\Omega)$ be such that $u_n(x_0) = \phi(x_0)$ and $u_n(x) > \phi(x)$ for $x \neq x_0$. Assume that $\nabla\phi(x_0) \neq 0$; we have to show that

$$\begin{aligned} &-\frac{|\nabla\phi(x_0)| f_n''(|\nabla\phi(x_0)|) - f_n'(|\nabla\phi(x_0)|)}{|\nabla\phi(x_0)|^3}\Delta_\infty\phi(x_0) \\ &-\frac{f_n'(|\nabla\phi(x_0)|)}{|\nabla\phi(x_0)|}\Delta\phi(x_0) - 1 \ge 0. \end{aligned}$$

By contradiction, suppose that this is not the case. By continuity, there exists $r > 0$ small enough such that

$$-\frac{|\nabla\phi(x)| f_n''(|\nabla\phi(x)|) - f_n'(|\nabla\phi(x)|)}{|\nabla\phi(x)|^3}\Delta_\infty\phi(x) - \frac{f_n'(|\nabla\phi(x)|)}{|\nabla\phi(x)|}\Delta\phi(x) < 1,$$

for any $|x - x_0| < r$. Let $m = \inf\{u_n(x) - \phi(x) : \ |x - x_0| = r\}$ and set $\eta = \phi + \frac{1}{2}m$. Since $m > 0$ then $\eta < u_n$ on $\partial B_r(x_0)$, $\eta(x_0) > u_n(x_0)$ and

$$-\frac{|\nabla\eta(x)| f_n''(|\nabla\eta(x)|) - f_n'(|\nabla\eta(x)|)}{|\nabla\eta(x)|^3}\Delta_\infty\eta(x) - \frac{f_n'(|\nabla\eta(x)|)}{|\nabla\eta(x)|}\Delta\eta(x) < 1,$$

for any $|x - x_0| < r$. By multiplying by $(\eta - u_n)^+$, integrating in $B_r(x_0)$ and using an integration by parts, we have

$$\int_{\{\eta > u_n\}} f_n'\big(|\nabla\eta|\big)\frac{\nabla\eta}{|\nabla\eta|}\cdot\nabla(\eta - u_n)\,dx < \int_{\{\eta > u_n\}} (\eta - u_n)\,dx. \tag{27}$$

Notice that, since $\eta(x_0) > u_n(x_0)$ and $\eta - u_n$ is continuous, the Lebesgue measure of $\{\eta > u_n\}$ is strictly positive. The function $(\eta - u_n)^+$ extended to zero outside $B_r(x_0)$ can be used as a test function in (11):

$$\int_{\{\eta > u_n\}} f_n'\big(|\nabla u_n|\big)\frac{\nabla u_n}{|\nabla u_n|}\cdot\nabla(\eta - u_n)\,dx = \int_{\{\eta > u_n\}} (\eta - u_n)\,dx. \tag{28}$$

Subtracting (28) from (27) we have

$$\int_{\{\eta>u_n\}} \left[f_n'(|\nabla\eta|)\frac{\nabla\eta}{|\nabla\eta|} - f_n'(|\nabla u_n|)\frac{\nabla u_n}{|\nabla u_n|} \right] \cdot \nabla(\eta - u_n)\,dx < 0. \tag{29}$$

Since

$$\begin{aligned}
&\left[f_n'(|\nabla\eta|)\frac{\nabla\eta}{|\nabla\eta|} - f_n'(|\nabla u_n|)\frac{\nabla u_n}{|\nabla u_n|} \right] \cdot \nabla(\eta - u_n) \\
&\quad = f_n'(|\nabla\eta|)|\nabla\eta| + f_n'(|\nabla u_n|)|\nabla u_n| \\
&\qquad - f_n'(|\nabla\eta|)\frac{\nabla\eta}{|\nabla\eta|}\cdot\nabla u_n - f_n'(|\nabla u_n|)\frac{\nabla u_n}{|\nabla u_n|}\cdot\nabla\eta,
\end{aligned}$$

Cauchy-Schwarz inequality yields

$$\begin{aligned}
&\left[f_n'(|\nabla\eta|)\frac{\nabla\eta}{|\nabla\eta|} - f_n'(|\nabla u_n|)\frac{\nabla u_n}{|\nabla u_n|} \right] \cdot \nabla(\eta - u_n) \\
&\quad \geq \big(f_n'(|\nabla\eta|) - f_n'(|\nabla u_n|)\big)\big(|\nabla\eta| - |\nabla u_n|\big);
\end{aligned}$$

from the convexity of f_n we obtain

$$\left[f_n'(|\nabla\eta|)\frac{\nabla\eta}{|\nabla\eta|} - f_n'(|\nabla u_n|)\frac{\nabla u_n}{|\nabla u_n|} \right] \cdot \nabla(\eta - u_n) \geq 0, \tag{30}$$

which gives the desired contradiction on account of (29).

To complete the proof that u_n is a viscosity supersolution of (26), we shall prove that if ϕ is a test function touching u_n at x_0 from below, then $\nabla\phi(x_0) \neq 0$ (i.e. the set of test functions touching u_n from below with vanishing gradient is the empty set).

By contradiction, let us assume that $\phi \in C^2(\Omega)$ is such that $u_n(x_0) = \phi(x_0)$, $u_n(x) > \phi(x)$ for $x \neq x_0$ and $\nabla\phi(x_0) = 0$. Thus, there exists $c > 0$ and $r_1 > 0$ such that $u_n(x) > \phi(x) > \psi(x)$ for $0 < |x - x_0| < r_1$, where

$$\psi(x) = -c|x - x_0|^2 + u_n(x_0).$$

We notice that ψ is of class C^2 and satisfies $F_n(\nabla\psi(x), D^2\psi(x)) < 0$ for every x in some ball of radius r_2 centered at x_0, i.e. there exists $r_2 > 0$ such that ψ is a strict classical subsolution of $F_n(Dv, D^2v) = 0$ in $B_{r_2}(x_0)$.

Let $r = \min(r_1, r_2)/2$, $m = \inf\{u_n(x) - \phi(x) : |x - x_0| = r\}$ and set $\eta = \phi + \frac{1}{2}m$. Notice that $\eta < u_n$ on $\partial B_r(x_0)$, $\eta(x_0) > u_n(x_0)$ and $F_n(\nabla\eta, D^2\eta) < 0$ in $B_r(x_0)$. As done in the first part of the proof, we use the function $(\eta - u_n)^+$ extended to zero outside $B_r(x_0)$ as a test function in (11) and we obtain (29); then, from (30) we get a contradiction. Thus, the set of test functions touching u_n from below with vanishing gradient is the empty set and hence u_n is a viscosity supersolution of (26).

To prove that u_n is a subsolution of (26), we first consider a test function ϕ touching u_n at x_0 from above with $\nabla\phi(x_0) \neq 0$. This case in analogous to the supersolution case. The case $\nabla\phi(x_0) = 0$ is simpler than before, since in this case $F_n(\nabla\phi(x_0), D^2\phi(x_0)) \leq 0$ is straightforwardly satisfied. □

Theorem 1 *Let u be the minimizer of* (1), *with f satisfying* (2a)–(2d) *and $f \in C^2((0,+\infty))$. Assume that there exists a sequence $\{f_n\}_{n\in\mathbb{N}}$ satisfying* (i)–(iv) *in Proposition* 1, (24), (25) *and such that f_n'' converges to f'' uniformly on the compact sets contained in* $(0,+\infty)$.

Then, u is a viscosity solution of

$$\min\left(-\frac{|\nabla u| f''(|\nabla u|) - f'(|\nabla u|)}{|\nabla u|^3}\Delta_\infty u - \frac{f'(|\nabla u|)}{|\nabla u|}\Delta u - 1, |\nabla u| - \sigma\right) = 0. \tag{31}$$

Proof Let $\{f_n\}_{n\in\mathbb{N}}$ be an approximating sequence of the function f satisfying (i)–(iv) in Proposition 1, (24), (25) and such that f_n'' converges to f'' uniformly on the compact sets contained in $(0,+\infty)$. We refer to Theorem 3 for the existence of such a sequence in some relevant cases. From Proposition 1, we can assume that u_n converges to u uniformly as n tends to infinity. By using a standard argument from the theory of viscosity solutions (see [8] and [15]), we shall prove that u is a viscosity supersolution and subsolution of (31). The two proofs are not symmetric and we prove firstly that u is a viscosity supersolution and then that it is also a viscosity subsolution.

Assume ϕ is a smooth function touching u from below at $\hat{x} \in \Omega$, i.e., $u(\hat{x}) = \phi(\hat{x})$ and $u(x) > \phi(x)$ for any $x \neq \hat{x}$. Since u_n is a viscosity solution of (26) and u_n converges uniformly to u, there exist $\{x_n\}_{n\in\mathbb{N}} \subset \Omega$ such that

(i) for any $x \in \Omega$, $u_n(x_n) - \phi(x_n) \leq u_n(x) - \phi(x)$;
(ii) x_n tends to $\hat{x}$ as n tends to infinity;

see for instance [14] p. 95. Being u_n a viscosity supersolution of (26), we can conclude that

$$F_n\big(\nabla\phi(x_n), D^2\phi(x_n)\big) \geq 0.$$

Let assume that $|\nabla\phi(\hat{x})| < \sigma$; since ϕ is of class C^2 and from (ii), there exists $\delta > 0$ such that $|\nabla\phi(x_n)| \leq \sigma - \delta$ for n large enough. By taking the limit as $n \to \infty$ and from (25) we get a contradiction. Thus, we may exclude that $|\nabla\phi(\hat{x})| < \sigma$.

Now assume that $|\nabla\phi(\hat{x})| \geq \sigma$. Since f_n' and f_n'' converge uniformly on compact sets to f' and f'', respectively, by taking the limit as $n \to \infty$ we get that both

$$-\frac{|\nabla\phi(\hat{x})| f''(|\nabla\phi(\hat{x})|) - f'(|\nabla\phi(\hat{x})|)}{|\nabla\phi(\hat{x})|^3}\Delta_\infty\phi(\hat{x}) - \frac{f'(|\nabla\phi(\hat{x})|)}{|\nabla\phi(\hat{x})|}\Delta\phi(\hat{x}) - 1 \geq 0,$$

and

$$\big|\nabla\phi(\hat{x})\big| - \sigma \geq 0$$

are satisfied. Hence the claim is proven.

Now, we prove that u is a viscosity subsolution of (31). Assume ϕ is a smooth function such that $u(\hat{x}) = \phi(\hat{x})$ and $u(x) < \phi(x)$ for any $x \neq \hat{x}$. As claimed at the previous case, there exists a sequence $\{x_n\}_{n\in\mathbb{N}}$ such that

(i) $u_n(x_n) - \phi(x_n) \geq u_n(x) - \phi(x)$;
(ii) x_n tends to $\hat{x}$ as n tends to infinity.

If $|\nabla\phi(\hat{x})| \leq \sigma$, then obviously

$$\min\Bigg(-\frac{|\nabla\phi(\hat{x})|f''(|\nabla\phi(\hat{x})|) - f'(|\nabla\phi(\hat{x})|)}{|\nabla\phi(\hat{x})|^3}\Delta_\infty\phi(\hat{x}) - \frac{f'(|\nabla\phi(\hat{x})|)}{|\nabla\phi(\hat{x})|}\Delta\phi(\hat{x}) - 1, \\ \big|\nabla\phi(\hat{x})\big| - \sigma\Bigg) \leq 0$$

holds. In case $|\nabla\phi(\hat{x})| > \sigma$, then $|\nabla\phi(x_n)| \geq \sigma + \delta$ for some $\delta > 0$ and for any n large enough. Since u_n is a viscosity subsolution of (26), then we have $F_n(\nabla\phi(x_n), D^2\phi(x_n)) \leq 0$. Since f_n and its first and second derivatives converges uniformly as $n \to +\infty$, by taking the limit leads to

$$-\frac{|\nabla\phi(\hat{x})|f''(|\nabla\phi(\hat{x})|) - f'(|\nabla\phi(\hat{x})|)}{|\nabla\phi(\hat{x})|^3}\Delta_\infty\phi(\hat{x}) - \frac{f'(|\nabla\phi(\hat{x})|)}{|\nabla\phi(\hat{x})|}\Delta\phi(\hat{x}) - 1 \leq 0,$$

which completes the proof. □

Now, we assume that f satisfies (2a)–(2d) and

$$\lim_{s\to\sigma^+} f''(s) = +\infty. \tag{32}$$

Thus, $f \notin C^2((0, +\infty))$ (i.e. f is not twice differentiable at $s = \sigma$). Since it is not possible to choose f_n such that f_n'' converges uniformly to f'', we cannot proceed as in Theorem 1.

Let

$$a(s) = \begin{cases} \frac{f'(s)}{sf''(s)}, & s > \sigma, \\ 0, & 0 \leq s \leq \sigma, \end{cases} \tag{33}$$

and

$$b(s) = \begin{cases} \frac{s^2}{f''(s)}, & s > \sigma, \\ 0, & 0 \leq s \leq \sigma. \end{cases} \tag{34}$$

Notice that $a, b \in C^0([0, +\infty))$. Analogously, we define

$$a_n(s) = \frac{f_n'(s)}{sf_n''(s)} \quad \text{and} \quad b_n(s) = \frac{s^2}{f_n''(s)}, \tag{35}$$

for $s > 0$.

Theorem 2 *Let u be the minimizer of* (1), *with f satisfying* (2a)–(2d) *and* (32). *Assume that there exists a sequence $\{f_n\}_{n\in\mathbb{N}}$ satisfying* (i)–(iv) *in Proposition* 1, (24) *and* (25). *Let a, b, a_n, b_n be defined by* (33)–(35) *and assume that f_n is such that a_n and b_n converge uniformly to a and b in the compact sets contained in* $(0, +\infty)$ *and* $(\sigma, +\infty)$, *respectively.*

Then u is a viscosity solution of

$$\min\big(-\big[1 - a\big(|\nabla u|\big)\big]\Delta_\infty u - |\nabla u|^2 a\big(|\nabla u|\big)\Delta u - b\big(|\nabla u|\big), \big|\nabla u(x)\big| - \sigma\big) = 0. \tag{36}$$

Proof The proof splits in two parts. First we prove that u is a viscosity supersolution, then that it is also a subsolution. The earlier is slightly more involved and we deal with it first. Notice that the existence of the sequence $\{f_n\}_{n\in\mathbb{N}}$ is proved in Theorem 3 for some relevant cases.

The function u is a viscosity supersolution of (36).

Assume ϕ is a smooth function touching u from below at $\hat{x} \in \Omega$, i.e., $u(\hat{x}) = \phi(\hat{x})$ and $u(x) > \phi(x)$ for any $x \neq \hat{x}$. Recall that u_n is a viscosity solution of (26) and that, from Proposition 1, we can assume that u_n converges uniformly to u as n tends to infinity. Thus, there exist $\{x_n\}_{n\in\mathbb{N}} \subset \Omega$ such that for any $x \in \Omega$, $u_n(x_n) - \phi(x_n) \le u_n(x) - \phi(x)$ and x_n tends to $\hat{x}$ as n tends to infinity.

Since u_n is a viscosity supersolution of (26), we can conclude that

$$F_n\big(\nabla\phi(x_n), D^2\phi(x_n)\big) \ge 0. \tag{37}$$

Let assume that $|\nabla\phi(\hat{x})| < \sigma$. As done in the proof of Theorem 1, we get a contradiction and we may exclude that $|\nabla\phi(\hat{x})| < \sigma$.

Now assume that $|\nabla\phi(\hat{x})| > \sigma$. Hence, we may assume that $|\nabla\phi(x_n)| > \sigma$ (at least for n large). By multiplying both sides of (37) by

$$\frac{|\nabla\phi(x_n)|^2}{f_n''(|\nabla\phi(x_n)|)},$$

we have

$$-\big[1 - a_n\big(|\nabla\phi(x_n)|\big)\big]\Delta_\infty\phi(x_n) \\ - \big|\nabla\phi(x_n)\big|^2 a_n\big(|\nabla\phi(x_n)|\big)\Delta\phi(x_n) - b_n\big(|\nabla\phi(x_n)|\big) \ge 0.$$

From the uniform convergence of a_n and b_n and by taking the limit as $n \to \infty$, we get that both

$$-\big[1 - a\big(|\nabla\phi(\hat{x})|\big)\big]\Delta_\infty\phi(\hat{x}) - \big|\nabla\phi(\hat{x})\big|^2 a\big(|\nabla\phi(\hat{x})|\big)\Delta\phi(\hat{x}) - b\big(|\nabla\phi(\hat{x})|\big) \ge 0,$$

and

$$\big|\nabla\phi(\hat{x})\big| - \sigma \ge 0$$

are satisfied.

It remains to consider the case $|\nabla\phi(\hat{x})| = \sigma$. Since we do not have the uniform convergence of b_n to b in a neighborhood of σ, we must proceed in a different way. By contradiction, let us assume that u is not a viscosity supersolution of (36). For what we have proven in the first part of the proof, there exist $\hat{x} \in \Omega$ and a smooth function ϕ touching u from below at $\hat{x} \in \Omega$ with $|\nabla\phi(\hat{x})| = \sigma$ such that

$$\min\Big(-\big[1 - a\big(|\nabla\phi(\hat{x})|\big)\big]\Delta_\infty\phi(\hat{x}) - \big|\nabla\phi(\hat{x})\big|^2 a\big(|\nabla\phi(\hat{x})|\big)\Delta\phi(\hat{x}) - b\big(|\nabla\phi(\hat{x})|\big), \\ \big|\nabla\phi(\hat{x})\big| - \sigma\Big) < 0.$$

Since $|\nabla\phi(\hat{x})| = \sigma$, then $a(\sigma) = b(\sigma) = 0$ and the above inequality yields

$$-\Delta_\infty\phi(\hat{x}) < 0. \tag{38}$$

As done before, there exists $\{x_n\}_{n\in\mathbb{N}} \subset \Omega$ such that for any $x \in \Omega$, $u_n(x_n) - \phi(x_n) \le u_n(x) - \phi(x)$ and x_n tends to $\hat{x}$ as n tends to infinity. Notice that u_n is a viscosity supersolution of (26) and then

$$
\begin{aligned}
&-\frac{f_n''(|\nabla\phi(x_n)|)}{|\nabla\phi(x_n)|^2}\Delta_\infty\phi(x_n) \\
&\ge 1 + \frac{f_n'(|\nabla\phi(x_n)|)}{|\nabla\phi(x_n)|}\Delta\phi(x_n) - \frac{f_n'(|\nabla\phi(x_n)|)}{|\nabla\phi(x_n)|^3}\Delta_\infty\phi(x_n).
\end{aligned}
$$

Since x_n converges to $\hat{x}$ and ϕ is of class C^2, then $|\nabla\phi(x_n)|$ converges to σ and, from (38), $\Delta_\infty\phi(x_n) > 0$ for n large enough. The uniform convergence of f_n' to f' yields the following contradiction:

$$
\frac{1}{2} \le -\frac{f''(|\nabla\phi(x_n)|)}{|\nabla\phi(x_n)|^2}\Delta_\infty\phi(x_n) < 0,
$$

for n large enough. Hence the claim is proven.

The function u is a viscosity subsolution of (36).

Assume ϕ is a smooth function such that $u(\hat{x}) = \phi(\hat{x})$ and $u(x) < \phi(x)$ for any $x \ne \hat{x}$. Thus, there exists a sequence $\{x_n\}_{n\in\mathbb{N}}$ such that $u_n(x_n) - \phi(x_n) \ge u_n(x) - \phi(x)$ and x_n tends to $\hat{x}$ as n tends to infinity.

If $|\nabla\phi(\hat{x})| \le \sigma$, then obviously

$$
\begin{aligned}
\min\Big(&-\big[1 - a\big(|\nabla\phi(\hat{x})|\big)\big]\Delta_\infty\phi(\hat{x}) - \big|\nabla\phi(\hat{x})\big|^2 a\big(|\nabla\phi(\hat{x})|\big)\Delta\phi(\hat{x}) - b\big(|\nabla\phi(\hat{x})|\big), \\
&\big|\nabla\phi(\hat{x})\big| - \sigma\Big) \le 0
\end{aligned}
$$

holds. In case $|\nabla\phi(\hat{x})| > \sigma$, from the fact that u_n is a viscosity subsolution of (26), we can conclude (carrying out the same algebraic manipulation showed at the previous step)

$$
\begin{aligned}
&-\big[1 - a_n\big(|\nabla\phi(x_n)|\big)\big]\Delta_\infty\phi(x_n) \\
&\quad - \big|\nabla\phi(x_n)\big|^2 a_n\big(|\nabla\phi(x_n)|\big)\Delta\phi(x_n) - b_n\big(|\nabla\phi(x_n)|\big) \le 0.
\end{aligned}
$$

Taking the limit leads to the desired conclusion. □

Remark 1 It is of interest to have an analogue of Theorem 2 when f satisfies

$$
0 < \lim_{s\to\sigma^+} f''(s) < +\infty.
$$

This case can be studied by using an argument analogue to the one used in the proof of Theorem 2 and under the additional assumption

$$
\Big(\limsup_{n\to\infty} f_n''\Big)(\sigma) \le \lim_{s\to\sigma^+} f''(s).
$$

We will not write in this paper the details of the proof. We just mention that if f is given by (4) with $q = 2$, then the approximating sequence given in the proof of the following Theorem satisfies this additional assumption.

Theorems 1 and 2 require the existence of a sequence $\{f_n\}_{n\in\mathbb{N}}$ which satisfies several assumptions. In the following theorem, we construct an explicit example.

Theorem 3 *Let f satisfy* (2a)–(2d) *and let a be given by* (33). *Assume that there exists $\tilde{\sigma} > \sigma$ such that $a(s)$ is nondecreasing for $s \in [\sigma, \tilde{\sigma}]$. Then there exists $\{f_n\}_{n\in\mathbb{N}}$ which satisfies the assumptions required in Theorems* 1 *and* 2.

Proof We construct an explicit example. A convenient way to construct the sequence $\{f_n\}_{n\in\mathbb{N}}$ is to modify f' only in the interval $(0, \sigma + \varepsilon)$, with $\varepsilon > 0$ small enough. We define

$$f'_\varepsilon(s) = \begin{cases} f'(\sigma+\varepsilon)[2(\frac{s}{\sigma+\varepsilon})^{p_\varepsilon} - (\frac{s}{\sigma+\varepsilon})^{q_\varepsilon}], & 0 \le s \le \sigma+\varepsilon, \\ f'(s), & s > \sigma+\varepsilon, \end{cases}$$

with

$$p_\varepsilon = \frac{(\sigma+\varepsilon)f''(\sigma+\varepsilon)}{f'(\sigma+\varepsilon)}\left(1 + \frac{1}{2}\omega_\varepsilon\right),$$

and

$$q_\varepsilon = \frac{(\sigma+\varepsilon)f''(\sigma+\varepsilon)}{f'(\sigma+\varepsilon)}(1 + \omega_\varepsilon),$$

where,

$$\omega_\varepsilon = \sqrt{2\left[1 - \frac{f'''(\sigma+\varepsilon)f'(\sigma+\varepsilon)}{f''(\sigma+\varepsilon)^2} - \frac{f'(\sigma+\varepsilon)}{(\sigma+\varepsilon)f''(\sigma+\varepsilon)}\right]}.$$

Since $a(s)$ is nondecreasing in $[\sigma, \tilde{\sigma}]$, then the same holds for $\log a(s)$. Thus,

$$0 \le \frac{d}{ds}\log a(s) = \frac{f''(s)}{f'(s)} - \frac{1}{s} - \frac{f'''(s)}{f''(s)}.$$

By multiplying by $f'(s)/f''(s)$ we get

$$\frac{f'(s)}{sf''(s)} + \frac{f'(s)f'''(s)}{f''(s)^2} \le 1,$$

for $s \in (\sigma, \tilde{\sigma})$, which implies that ω_ε is well-defined.

Tedious but straightforward computations show that $f_\varepsilon \in C^3(0,+\infty) \cap C^1([0,+\infty))$. Since we modified f' only on a compact set, it is easy to show the uniform convergence of f_n and f'_n to f and f', respectively.

Notice that

$$f''_\varepsilon(s) = \begin{cases} \frac{f'(\sigma+\varepsilon)}{\sigma+\varepsilon}[2p_\varepsilon(\frac{s}{\sigma+\varepsilon})^{p_\varepsilon-1} - q_\varepsilon(\frac{s}{\sigma+\varepsilon})^{q_\varepsilon-1}], & 0 \le s \le \sigma+\varepsilon, \\ f''(s), & s > \sigma+\varepsilon; \end{cases}$$

since $2p_\varepsilon > q_\varepsilon$ and $p_\varepsilon < q_\varepsilon$, then $f''_\varepsilon > 0$.

Notice that we have

$$\lim_{s\to\sigma^+} a(s) = 0. \tag{39}$$

Indeed, assume by contradiction that there exists $\alpha > 0$ such that $a(s) \to 1/\alpha$ as $s \to \sigma^+$. Since $a(s)$ is nondecreasing, then $a(s) \geq 1/\alpha$ for any $s \in [\sigma, \tilde{\sigma}]$, which implies that

$$\frac{d}{ds} \log f'(s) \leq \frac{d}{ds} \log s^\alpha, \quad s \in (\sigma, \tilde{\sigma}).$$

By integrating both sides of the above inequality from s to $\tilde{\sigma}$ and after simple manipulations, we obtain that

$$f'(s) \geq f'(\tilde{\sigma}) \left(\frac{s}{\tilde{\sigma}} \right)^\alpha,$$

for any $s \in (\sigma, \tilde{\sigma})$. By taking the limit as $s \to \sigma^+$, we obtain $f'(\sigma) > 0$, a contradiction. Thus, (39) holds.

Since $p_\varepsilon, q_\varepsilon \geq (\sigma + \varepsilon) f''(\sigma + \varepsilon)/f'(\sigma + \varepsilon)$, from (39) we have

$$\lim_{\varepsilon \to 0^+} p_\varepsilon = +\infty \quad \text{and} \quad \lim_{\varepsilon \to 0^+} q_\varepsilon = +\infty. \tag{40}$$

Assume that ε is small enough so that p_ε and q_ε are greater than 3; then

$$\frac{s f_\varepsilon''(s) - f_\varepsilon'(s)}{s^3} = f'(\sigma + \varepsilon) \left[\frac{2(p_\varepsilon - 1)}{(\sigma + \varepsilon)^{p_\varepsilon}} s^{p_\varepsilon - 3} - \frac{(q_\varepsilon - 1)}{(\sigma + \varepsilon)^{q_\varepsilon}} s^{q_\varepsilon - 3} \right],$$
$$0 < s \leq \sigma + \varepsilon,$$

and

$$\frac{f_\varepsilon'(s)}{s} = f'(\sigma + \varepsilon) \left[\frac{2p_\varepsilon}{(\sigma + \varepsilon)^{p_\varepsilon}} s^{p_\varepsilon - 1} - \frac{q_\varepsilon}{(\sigma + \varepsilon)^{q_\varepsilon}} s^{q_\varepsilon - 1} \right], \quad 0 < s \leq \sigma + \varepsilon.$$

From (40), we get (24) and (25).

We notice that

$$a_\varepsilon(s) = \frac{f'(\sigma + \varepsilon)}{(\sigma + \varepsilon) f''(\sigma + \varepsilon)} \cdot \frac{2(\frac{s}{\sigma+\varepsilon})^{p_\varepsilon} - (\frac{s}{\sigma+\varepsilon})^{q_\varepsilon}}{2(\frac{s}{\sigma+\varepsilon})^{p_\varepsilon} - (\frac{s}{\sigma+\varepsilon})^{q_\varepsilon} + \omega_\varepsilon[(\frac{s}{\sigma+\varepsilon})^{p_\varepsilon} - (\frac{s}{\sigma+\varepsilon})^{q_\varepsilon}]},$$

for $0 \leq s \leq \sigma + \varepsilon$ and $a_\varepsilon(s) = a(s)$ for $s > \sigma + \varepsilon$. Thus,

$$\sup_{s \in \mathbb{R}} \left| a_\varepsilon(s) - a(s) \right| \leq \frac{f'(\sigma + \varepsilon)}{(\sigma + \varepsilon) f''(\sigma + \varepsilon)};$$

by (39), we obtain that a_ε converges uniformly to a.

Since $f_\varepsilon''(s) = f''(s)$ for $s \geq \sigma + \varepsilon$, it is clear that b_n converges uniformly to b in the compact sets contained in $(\sigma, +\infty)$. □

Remark 2 We notice when f is given by (3) or (4), then a satisfies the assumptions of Theorem 3. Indeed, if f is given by (3), then we have

$$a(s) = \begin{cases} 0, & 0 \leq s \leq 1, \\ 1 - \frac{1}{s^2}, & s > 1, \end{cases}$$

and then $a(s)$ is nondecreasing.

When f is given by (4), then

$$a(s) = \begin{cases} 0, & 0 \le s \le 1, \\ \frac{1}{q-1}(1 - \frac{1}{s}), & s > 1, \end{cases}$$

which is a nondecreasing function.

Example 1 Let f be given by (3). Then

$$f'(s) = \sqrt{(s^2-1)^+}$$

and a and b in (33) and (34) read as

$$a(s) = \begin{cases} 0, & 0 \le s \le 1, \\ 1 - \frac{1}{s^2}, & s > 1, \end{cases}$$

and

$$b(s) = s\sqrt{(s^2-1)^+}.$$

We notice that, working as in the proof of Theorem 2, we can prove that u satisfies other equations in the viscosity sense which are of the same form as (5). For instance, let $a^* > 0$ be such that $a(s) < a^*$ for any $s \ge 0$; then it can be shown that u satisfies

$$\min\left\{-\left[1 + \frac{1-a^*}{a^* - a(|\nabla u|)}\right]\Delta_\infty u - \frac{|\nabla u|^2 a(|\nabla u|)}{a^* - a(|\nabla u|)}\Delta u - \frac{b(|\nabla u|)}{a^* - a(|\nabla u|)},\right.$$
$$\left.\big|\nabla u(x)\big| - \sigma\right\} = 0, \qquad (41)$$

in the viscosity sense. If f is given by (3), we can choose $a^* = 1$ and (41) reads as

$$\min\Big(-\Delta_\infty u - |\nabla u|^2\big(|\nabla u|^2 - 1\big)^+\Delta u - |\nabla u|^3\sqrt{\big(|\nabla u|^2-1\big)^+}, \big|\nabla u(x)\big| - 1\Big) = 0.$$

Acknowledgements The author wishes to thank Rolando Magnanini and Simone Cecchini for the many helpful discussions. The author is grateful for the careful and thoughtful comments of the referees which led to substantial improvements over a first version of this paper.

References

1. Brasco, L.: Global L^∞ gradient estimates for solutions to a certain degenerate elliptic equation. Nonlinear Anal. **74**, 516–531 (2011)
2. Brasco, L., Carlier, G., Santambrogio, F.: Congested traffic dynamics, weak flows and very degenerate elliptic equations. J. Math. Pures Appl. **93**, 652–671 (2010)
3. Buttazzo, G., Kawohl, B.: Overdetermined boundary value problems for the ∞-Laplacian. Int. Math. Res. Not. **2**, 237–247 (2011)
4. Cabré, X., Caffarelli, L.: Fully Nonlinear Elliptic Equations. Colloquium Publications, vol. 43. American Mathematical Society, Providence (1995)

5. Carstensen, C., Müller, S.: Local stress regularity in scalar nonconvex variational problems. SIAM J. Math. Anal. **34**, 495–509 (2002)
6. Cecchini, S., Magnanini, R.: Critical points of solutions of degenerate elliptic equations in the plane. Calc. Var. Partial Differ. Equ. **39**(1–2), 121–138 (2010)
7. Ciraolo, G., Magnanini, R., Sakaguchi, S.: Symmetry of minimizers with a level surface parallel to the boundary. Preprint. http://arxiv.org/abs/1203.5295 (2012)
8. Crandall, M.G., Ishii, H., Lions, P.L.: User's guide to viscosity solutions of second order partial differential equations. Bull. Am. Math. Soc. **27**(1), 1–67 (1992)
9. Crasta, G.: Existence, uniqueness and qualitative properties of minima to radially symmetric non-coercive non-convex variational problems. Math. Z. **235**, 569–589 (2000)
10. Fonseca, I., Fusco, N., Marcellini, P.: An existence result for a nonconvex variational problem via regularity. ESAIM Control Optim. Calc. Var. **7**, 69–95 (2002)
11. Fragalà, I., Gazzola, F., Kawohl, B.: Overdetermined problems with possibly degenerate ellipticity, a geometric approach. Math. Z. **254**, 117–132 (2006)
12. Gilbarg, D., Trudinger, N.S.: Elliptic Partial Differential Equations of Second Order. Springer, Berlin (1977)
13. Jensen, R.: Uniqueness of Lipschitz's extensions: minimizing the sup norm of the gradient. Arch. Ration. Mech. Anal. **123**, 51–74 (1993)
14. Juutinen, P., Lindqvist, P., Manfredi, J.J.: The ∞-eigenvalue problem. Arch. Ration. Mech. Anal. **148**(2), 89–105 (1999)
15. Koike, S.: A Beginner's Guide to the Theory of Viscosity Solutions. MSJ Mem., vol. 13. Mathematical Society of Japan, Tokyo (2004)
16. Magnanini, R., Talenti, G.: On complex-valued solutions to a 2D eikonal equation. Part one: qualitative properties. In: Nonlinear Partial Differential Equations. Contemporary Mathematics, vol. 283, pp. 203–229. American Mathematical Society, Providence (1999)
17. Magnanini, R., Talenti, G.: On complex-valued solutions to a 2D eikonal equation. Part two: existence theorems. SIAM J. Math. Anal. **34**, 805–835 (2003)
18. Magnanini, R., Talenti, G.: On complex-valued solutions to a 2D eikonal equation. Part three: analysis of a Backlund transformation. Appl. Anal. **85**(1–3), 249–276 (2006)
19. Magnanini, R., Talenti, G.: On complex-valued 2D eikonals. Part four: continuation past a caustic. Milan J. Math. **77**(1), 1–66 (2009)
20. Santambrogio, F., Vespri, V.: Continuity in two dimensions for a very degenerate elliptic equation. Nonlinear Anal. **73**, 3832–3841 (2010)
21. Tolksdorf, P.: On the Dirichlet problem for quasilinear equations in domains with conical boundary points. Commun. Partial Differ. Equ. **8**, 773–817 (1983)

Kato's Inequality in the Half Space: An Alternative Proof and Relative Improvements

Adele Ferone

Abstract In this paper we give an alternative proof of the optimal Kato's inequality in the half space. The approach is based on a very classical method of Calculus of Variation due to Weirstrass (and developed by Hilbert) that usually is considered to prove that the solutions of the Euler Lagrange equation associated to a functional are, in fact, extremals. In this paper we will show how this method is well suited also to functionals that have no extremals. Moreover, we will present a class of inequalities that interpolate Kato's inequality and Hardy's inequality in the half space.

Keywords Kato's inequality · Hardy's inequality · Remainder terms · Optimal constants

1 Introduction

Sobolev spaces play a fundamental rule in the study of differential and integral operators especially for their embedding characteristics. Most of the results assert that, if Ω is an open set of $\mathbb{R}^n$ with smooth boundary, then $W^{1,p}(\Omega)$ is embedded into some Lebesgue spaces $L^q(\Omega)$ or $L^q(\partial\Omega)$, with $q > p$.

The standard Sobolev inequality states that $W^{1,2}(\mathbb{R}^n)$ is continuously embedded into $L^{2^*}(\mathbb{R}^n)$, $2^* = \frac{2n}{n-2}$, and

$$\sqrt{(n-2)\pi n}\left[\frac{\Gamma(n/2)\Gamma(1+n/2)}{\Gamma(n)\Gamma(1+n/2)}\right]^{1/n} \|u\|_{L^{2^*}(\mathbb{R}^n)} \le \|\nabla u\|_{L^2(\mathbb{R}^n)}, \tag{1}$$

where, here and in the following, Γ is the usual Gamma function defined as $\Gamma(s) = \int_0^\infty t^{s-1}\exp(-t)\,dt$. The constant in (1) is optimal and was given independently by Aubin and Talenti in [6, 26].

The Hardy inequality

$$\frac{(n-2)^2}{4}\int_{\mathbb{R}^n}\frac{u^2(y)}{|y|^2}\,dy \le \int_{\mathbb{R}^n}|\nabla u|^2(y)\,dy, \quad u \in W^{1,2}(\mathbb{R}^n) \tag{2}$$

A. Ferone (✉)
Dipartimento di Matematica, Seconda Università di Napoli, Viale Lincoln 5, 81100 Caserta, Italy
e-mail: adele.ferone@unina2.it

R. Magnanini et al. (eds.), *Geometric Properties for Parabolic and Elliptic PDE's*, Springer INdAM Series 2, DOI 10.1007/978-88-470-2841-8_6,

strengthens (1) in as much as it states that if u has a distributional gradient in $L^2(\mathbb{R}^n)$ then it has an higher summability, indeed it belongs to the Lorentz space $L^{2^*,2}(\mathbb{R}^n)$ and $L^{2^*,2}(\mathbb{R}^n) \subsetneq L^{2^*}(\mathbb{R}^n)$. The constant that appear in (2) is optimal but never attained as demonstrated by sequences obtained on truncating functions having the form

$$\psi_a(x) = a|x|^{\frac{2-n}{2}} \quad \text{for } x \in \mathbb{R}^n,$$

with $a \in \mathbb{R} \setminus \{0\}$, at levels $1/k$ and k, and then letting $k \to \infty$. The optimality of the constant in (1) is preserved even if the whole space $\mathbb{R}^n$ is replaced by any open set Ω and u is supposed to be zero on the boundary. The same situation occurs for (2) if moreover Ω contains the origin. Nevertheless if, in particular, Ω is the upper half n-dimensional Euclidean space $\mathbb{R}^n_+ = \{(x,t) \in \mathbb{R}^n : x \in \mathbb{R}^{n-1},\ t > 0\}$, by reflection arguments, we can deduce that both inequalities (1) and (2) still hold (with the same optimal constants) even if u is not necessarily zero on the boundary. Inequalities (1) and (2) then do not give any information about the summability of the trace of u if considered for functions defined in $\mathbb{R}^n_+$. Such summability properties can be deduced from the following result.

Theorem 1 *Let $n \geq 3$ and let u be a real function on $\mathbb{R}^n_+$ vanishing at infinity, such that $|\nabla u| \in L^2(\mathbb{R}^n_+)$. Then the following inequality holds*

$$H_n \int_{\partial\mathbb{R}^n_+} \frac{u^2(x,0)}{|x|}\,dx \leq \int_{\mathbb{R}^n_+} |\nabla u|^2(x,t)\,dx\,dt, \quad u \in W^{1,2}\big(\mathbb{R}^n_+\big). \tag{3}$$

The optimal constant is

$$H_n = 2\frac{\Gamma^2(\frac{n}{4})}{\Gamma^2(\frac{n-2}{4})}. \tag{4}$$

Inequality (3) is known in literature as the Kato inequality and it asserts that $W^{1,2}(\mathbb{R}^n_+)$ is continuously embedded into the Lorentz space $L^{\frac{2(n-1)}{n-2},2}(\partial\mathbb{R}^n_+)$.

The constant (4) is optimal but, similarly to what happens for the standard Hardy inequality (2), it is never attained since functions that are candidates to be extremals, are proportional to the solution of the problem

$$\begin{cases} \Delta\varphi = 0 & \text{in } \mathbb{R}^n_+, \\ \varphi = |x|^{-\frac{n}{2}+1} & \text{on } \partial\mathbb{R}^n_+. \end{cases} \tag{5}$$

Theorem 1 was proved by Herbst in [23] by using dilation analytic techniques. Recently Davila, Dupaigne and Montenegro in [15] gave a different proof based on a suitable change of variables. Moreover, exploiting the lack of the extremals, they improved (3) by adding an extra term on the left hand side, under the additional hypothesis that u is compactly supported.

The aim of this note is to present a different approach in proving Theorem 1 based on the Weierstrass method. Such method has recently been adopted in [3] to find an improvement of the classical Sobolev inequality. It also reveals particularly suitable to prove a class of inequalities that generalize (3) and that provide, at the mean time, the optimal summability both for a function $u \in W^{1,2}(\mathbb{R}^n_+)$ and for its trace on the boundary. Indeed, the following result, obtained in collaboration with A. Alvino and R. Volpicelli in [5], holds.

Theorem 2 *Let $n \geq 3$ and let u be a real function on $\mathbb{R}^n_+$ vanishing at infinity, such that $|\nabla u| \in L^2(\mathbb{R}^n_+)$. Then, for any $2 \leq \beta < n$, there exists a positive constant $H(n,\beta)$ such that*

$$H(n,\beta)\int_{\partial\mathbb{R}^n_+}\frac{u^2}{|x|}\,dx + \frac{(\beta-2)^2}{4}\int_{\mathbb{R}^n_+}\frac{u^2}{|x|^2+t^2}\,dx\,dt \leq \int_{\mathbb{R}^n_+}|\nabla u|^2\,dx\,dt, \tag{6}$$

where

$$H(n,\beta) = 2\frac{\Gamma\left(\frac{n+\beta}{4}-\frac{1}{2}\right)\Gamma\left(\frac{n-\beta}{4}+\frac{1}{2}\right)}{\Gamma\left(\frac{n+\beta}{4}-1\right)\Gamma\left(\frac{n-\beta}{4}\right)}. \tag{7}$$

The constant $H(n,\beta)$ is sharp.

Inequalities (6) can be regarded as a Hardy inequality (or equivalently as a Kato inequality) with a remainder term. Typically, improvements of (2) and (3) consider extra terms on the left-hand side that either involve integrals of $|u|^2$ with weights depending on $|y|$ which are less singular at 0, or integrals of $|\nabla u|^q$ with $q < 2$ (see, for example, [1, 2, 4, 7–11, 14, 16, 18–21, 24, 27, 28]). Nevertheless, this kind of improvements holds only for functions defined in open bounded set, indeed if u is defined in the whole space $\mathbb{R}^n_+$, then no extra terms can be considered [12, 13]. In [5] we show that, starting from the classical Hardy inequality in the half-space, if we replace the optimal constant with a smaller one, then we can add an extra term equal to that one that appears on the left hand side of (3).

It can be easily checked that when $\beta = 2$, then inequality (6) reduces to (3), while when β goes to n, then inequality (6) reduces to (2). The optimal constant $H(n,\beta)$, as expected, is never attained since the candidates for extremals are proportional to the solution of the problems

$$\begin{cases} \Delta\varphi + \dfrac{(\beta-2)^2}{4}\dfrac{\varphi}{|x|^2+t^2} = 0 & \text{in } \mathbb{R}^n_+, \\ \varphi = |x|^{-\frac{n}{2}+1} & \text{on } \partial\mathbb{R}^n_+. \end{cases} \tag{8}$$

The proof of Theorem 2 relies on an adaptation of the proof of Theorem 1: here we will give just a sketch of it and for the details we refer to [5].

The paper is organized as follow: in Sect. 2 we roughly describe the Weierstrass method, in Sect. 3 we prove Theorem 1 and in Sect. 4 we give an outline of the proof of Theorem 2.

2 Preliminary Results: The Weierstrass Method

The aim of this section is to give an idea of a very classical method of Calculus of Variations used to prove that a solution of the Euler Lagrange equation associated to a functional

$$\mathscr{F}(u,\Omega)=\int_{\Omega} f\big(x,u(x),\nabla u(x)\big)\,dx \tag{9}$$

is, in fact, a minimum of the functional. Here Ω is an open bounded domain of $\mathbb{R}^n$, $n\geq 1$, u denotes a function defined in Ω of class C^1 and $f:(x,z,p)\in\mathbb{R}^n\times\mathbb{R}\times\mathbb{R}^n\to\mathbb{R}$ is sufficiently smooth. This method was developed by Weirstrass for one-dimensional integrals and then generalized to multiple integrals by Schwartz, Lichtenstein and Morrey. We refer to the monographs [25] and [22] for the general theory and the references therein.

The basic idea is to consider a whole bundle of extremals, i.e. solutions of the Euler Lagrange equation, instead of a single one. This poses the problem to embed a given extremal in a field of extremals, i.e. in a foliation by extremal surface of codimension one. The surfaces of such functions cover some domain G of $\mathbb{R}^{n+1}$ in the sense that through every point of G there passes exactly one field surface. More precisely, let Σ be an interval of $\mathbb{R}$, G a subset of $\mathbb{R}^{n+1}$ and consider a one parameter family $\{\phi_k\}_{k\in\Sigma}$ of functions defined in Ω satisfying the following conditions:

(I) for every $k\in\Sigma$, ϕ_k is a solution of the Euler-Lagrange equation

$$\sum_{i=1}^{n}\frac{\partial}{\partial x_i}\frac{\partial}{\partial p_i}f(x,\phi,\nabla\phi)=\frac{\partial}{\partial z}f(x,\phi,\nabla\phi),\qquad x\in\Omega \tag{10}$$

associated to (9);

(II) for every point $(x,z)\in G$ there is exactly one value $k=k(x,z)\in\Sigma$ such that $z=\phi_{k(x,z)}(x)$.

As customary in the calculus of variations, we call the field

$$F:(x,k)\in\Omega\times\Sigma\to\big(x,\phi_k(x)\big)\in G\subseteq\mathbb{R}^{n+1}$$

the *Mayer field* on G and the family $\{\mathrm{graph}(\phi_k)\}_{k\in\Sigma}$ the *extremal field* of hypersurface covering G over Ω. The *slope* $\mathscr{P}:(x,z)\in\mathbb{R}^{n+1}\to\mathbb{R}^n$ of the field is defined by

$$\mathscr{P}(x,z)\equiv\nabla\phi_{k(x,z)}(x).$$

Let $\mathscr{S}_0$ be some non parametric hypersurface in $\mathbb{R}^{n+1}$ given by $z=u_0(x)$, $x\in\overline{\Omega}_0$: we say that $\mathscr{S}_0$ is embedded into the field $\{\mathrm{graph}(\phi_k)\}_{k\in\Sigma}$ if there is some $k_0\in\Sigma$ such that $\mathscr{S}_0$ is a part of the leaf $\mathrm{graph}(\phi_{k_0})$, that is $\Omega_0\subseteq\Omega$ and $u_0(x)=\phi_{k_0}(x)$, for all x in Ω_0.

Since the functions $\{\phi_k\}$ solves (10) it is easy to check that the field

$$\begin{aligned}S(x,z)\equiv\Big(&\nabla_p f\big(x,z,\mathscr{P}(x,z)\big),\\ &\nabla_p f\big(x,z,\mathscr{P}(x,z)\big)\cdot\mathscr{P}(x,z)-f\big(x,z,\mathscr{P}(x,z)\big)\Big)\end{aligned} \tag{11}$$

is divergence free (here and in the following $w_1 \cdot w_2$ denotes the scalar product in $\mathbb{R}^m$, $m \geq 2$ of w_1 and w_2). Hence, for any function $u \in C^1(\overline{\Omega})$ with $\operatorname{graph}(u) \subseteq G$, if

$$M(x,z,p) = f\big(x,z,\mathscr{P}(x,z)\big) + \nabla_p f\big(x,z,\mathscr{P}(x,z)\big)\cdot\big(p - \mathscr{P}(x,z)\big),$$

the integral

$$\mathscr{M}(u,\Omega) = \int_\Omega M\big(x,u,\nabla u(x)\big)\,dx \tag{12}$$

is an *invariant integral*, that is

$$\mathscr{M}(u,\Omega) = \mathscr{M}(v,\Omega)$$

for any functions $u, v \in C^1(\overline{\Omega})$, with $\operatorname{graph}(u) \subseteq G$, $\operatorname{graph}(v) \subseteq G$ and $u_{|\partial\Omega} = v_{|\partial\Omega}$. Indeed, by the definition (11) of S, the functional $\mathscr{M}(u,\Omega)$ can be rewritten as

$$\mathscr{M}(u,\Omega) = \int_\Omega S\big(x,u(x)\big)\cdot\big(\nabla u(x), -1\big)\,dx,$$

that is $\mathscr{M}(u,\Omega)$ is the inward flow of the vector field $S(x,u(x))$ through the graph of u. One denotes $\mathscr{M}(u,\Omega)$ as the *Hilbert invariant integral* associated with the Mayer field F.

Using the *Weirstrass excess function* $\mathscr{E}_f$ defined as

$$\mathscr{E}_f\big(x,z,\mathscr{P}(x,z),p\big) = f(x,z,p) - M(x,z,p)$$

we obtain that, for any $v \in C^1(\overline{\Omega})$ with $\operatorname{graph}(v) \subseteq G$,

$$\begin{aligned}\int_\Omega f\big(x,v,\nabla v(x)\big)\,dx = \int_\Omega & M\big(x,v,\nabla v(x)\big)\,dx \\ &+ \int_\Omega \mathscr{E}_f\big(x,v,\mathscr{P}(x,v),\nabla v(x)\big)\,dx,\end{aligned}$$

or, equivalently by definitions (9) and (12)

$$\mathscr{F}(v,\Omega) = \mathscr{M}(v,\Omega) + \int_\Omega \mathscr{E}_f\big(x,v,\mathscr{P}(x,v),\nabla v(x)\big)\,dx \tag{13}$$

that is known as the *Weirstrass representation formula* for (9). Since $\mathscr{M}$ is an invariant integral, we infer that if $u \in C^2(\overline{\Omega})$ is a solution of (10) and u is embedded in a Mayer field F on G with slope $\mathscr{P}$, for any function $v \in C^1(\overline{\Omega})$ with $\operatorname{graph}(v) \subseteq G$ and $v = u$ on $\partial\Omega$ then it follows

$$\mathscr{F}(v,\Omega) = \mathscr{F}(u,\Omega) + \int_\Omega \mathscr{E}_f\big(x,v,\mathscr{P}(x,v),\nabla v(x)\big)\,dx.$$

If in addition, the Mayer field F satisfies the *sufficient Weirstrass condition*

$$\mathscr{E}_f\big(x,z,\mathscr{P}(x,z),p\big) > 0 \quad \text{for } (x,z) \in G \text{ and } p \neq \mathscr{P}(x,z)$$

then it follows that $\mathscr{F}(v,\Omega) > \mathscr{F}(u,\Omega)$ for all function $v \in C^1(\overline{\Omega})$ with $\operatorname{graph}(v) \subseteq G$ and $v = u$ on $\partial\Omega$, that is the extremal u is, in fact, a minimum.

3 Kato's Inequality: An Alternative Proof

In this section we prove Theorem 1. Our aim is to apply the Weirstrass method illustrated in Sect. 2. Nevertheless, the main difficulties in approaching this problem relies on the facts that in this case Ω is an unbounded set and moreover for the functions that are candidates to realize the equality in (3) the functional

$$\mathscr{F}\left(u, \mathbb{R}^n_+\right) = \int_{\mathbb{R}^n_+} |\nabla u|^2(x)\, dx \tag{14}$$

is divergent. Therefore we have to fit the technique illustrated in Sect. 2 to overcome these problems. We, firstly, construct a family of extremals for the functional (14), i.e. a family of harmonic functions on $\mathbb{R}^n_+$, that cover $\mathbb{R}^n_+ \times \mathbb{R}^+$ to define the Mayer field associated. This family is given by $\{k\varphi\}_{k\in\mathbb{R}^+}$ where φ is the solution of (5). To solve problem (5) we start from the family of harmonic functions on $\mathbb{R}^n_+$ that realize the equality in the Trace Sobolev inequality

$$\sqrt{\frac{n-2}{2}}\,\frac{\sqrt[4]{\pi}}{\Gamma^{\frac{1}{2(n-1)}}\left(\frac{n+1}{2}\right)}\|u\|_{L^{\frac{2(n-1)}{n-2}}(\partial\mathbb{R}^n_+)} \le \|\nabla u\|_{L^2(\mathbb{R}^n_+)}.$$

Such family was determined by Escobar in [17] and it is given by

$$\psi_a(x,t) = \left[\frac{a}{(a+t)^2+|x|^2}\right]^{\frac{n-2}{2}}, \quad a\in\mathbb{R}_+.$$

Since the functions ψ_a, $a\in\mathbb{R}_+$, are harmonic, if we integrate such family with respect to the parameter, we still have an harmonic function. If, in particular, we integrate with respect to a suitable weight of the parameter, then we can construct an harmonic function with the required behavior on $\partial\mathbb{R}^n_+$. Indeed, consider the function

$$\psi(x,t) = \int_0^\infty \psi_a(x,t)\frac{da}{a} = \int_0^\infty \frac{a^{\frac{n}{2}-2}}{[(a+t)^2+|x|^2]^{\frac{n}{2}-1}}\, da.$$

By the change of variable $A = \frac{a}{|x|}$ we obtain

$$\begin{aligned}\psi(x,t) &= |x|^{-\frac{n}{2}}\int_0^\infty \left[\frac{a}{|x|}\right]^{\frac{n}{2}-2}\left[\left(\frac{a}{|x|}+\frac{t}{|x|}\right)^2+1\right]^{-\frac{n}{2}+1} da\\ &= |x|^{-\frac{n}{2}+1}\int_0^\infty A^{\frac{n}{2}-2}\left[\left(A+\frac{t}{|x|}\right)^2+1\right]^{-\frac{n}{2}+1} dA.\end{aligned} \tag{15}$$

Since the last integral in (15) is convergent for all $(x,t)\in\mathbb{R}^n_+$, it follows that the function ψ is proportional to the function φ, solution of (5). Moreover, it is easy to check that the family $\{k\psi\}_{k\in\mathbb{R}^n_+}$ covers $\mathbb{R}^n_+\times\mathbb{R}^+$, indeed for any $(x,t,z)\in\mathbb{R}^n_+\times\mathbb{R}^+$ there exists a unique $\overline{k}=\overline{k}(x,t,z)\in\mathbb{R}^n_+$ such that $z=\overline{k}\psi(x,t)$, that is

$$\overline{k}(x,t,z) = \frac{z}{\psi(x,t)}.$$

The Mayer field is then

$$F : (x,t,k) \in \mathbb{R}^n_+ \times \mathbb{R}^+ \to \big(x,t,k\psi(x,t)\big) \in \mathbb{R}^n_+ \times \mathbb{R}^+, \tag{16}$$

the slope is

$$\mathscr{P}(x,t,z) = \frac{z}{\psi(x,t)} \nabla\psi(x,t),$$

and the Weierstrass excess function is

$$\begin{aligned}\mathscr{E}(x,t,z,\mathscr{P},p) &= |p|^2 - \big[\big|\mathscr{P}(x,t,z)\big|^2 + 2\mathscr{P}(x,t,z)\cdot\big(p - \mathscr{P}(x,t,z)\big)\big] \\ &= \big|p - \mathscr{P}(x,t,z)\big|^2.\end{aligned}$$

Therefore, the Mayer field (16) satisfies the sufficient Weierstrass condition. If, here and in the following, $B_\rho(0)$ denotes the ball centered at the origin and of radius ρ, from (13) we deduce

$$\int_{\mathbb{R}^n_+ \setminus B_r(0)} |\nabla u|^2(x)\,dx \geq \mathscr{M}\big(u, \mathbb{R}^n_+ \setminus B_r(0)\big) \tag{17}$$

for every nonnegative function $u \in C^\infty(\mathbb{R}^n_+)$, vanishing outside the ball $B_R(0)$ and for any $0 < r < R$. To evaluate the invariant integral associated $\mathscr{M}(u, \mathbb{R}^n_+ \setminus B_r(0))$, we recall that it can be rewritten as the inward flow of the free divergent vector field

$$\begin{aligned}S(x,t,z) &\equiv \big(2\mathscr{P}(x,t,z), \big|\mathscr{P}(x,t,z)\big|^2\big) \\ &\equiv \left(\frac{2z}{\psi(x,t)}\nabla\psi(x,t), \frac{z^2}{\psi^2(x,t)}|\nabla\psi|^2(x,t)\right)\end{aligned}$$

through the graph of u. Since u is supported on $B_R(0)$, its graphs is a part of the boundary of its undergraph

$$\mathscr{R}(u) = \big\{(x,t,z) \in \big[B_R(0) \setminus B_r(0)\big] \times \mathbb{R}^+ : 0 \leq z \leq u(x,t)\big\}.$$

Since $S \equiv 0$ when $z = 0$, by divergence theorem, it follows that the inward flow through the graph of u equals the sum of the outward flows across the two manifolds

$$\Sigma_1 = \big\{(x,t,v) \in \mathbb{R}^n_+ \times \mathbb{R} : |x| > r,\ t = 0,\ 0 \leq v \leq u(x,0)\big\} \tag{18}$$

$$\Sigma_2 = \big\{(x,t,v) \in \mathbb{R}^n_+ \times \mathbb{R} : |x|^2 + t^2 = r^2,\ 0 \leq v \leq u(x,t)\big\}. \tag{19}$$

Then

$$\mathscr{M}\big(u, \mathbb{R}^n_+ \setminus B_r(0)\big) = \int_{\Sigma_1} S\cdot\nu\,d\mathscr{H}^n + \int_{\Sigma_2} S\cdot\nu\,d\mathscr{H}^n, \tag{20}$$

where, here and in the following, $\mathscr{H}^m$ denotes the m-dimensional Hausdorff measure on $\mathbb{R}^{m+1}$. Let us begin by evaluating the flow of S across Σ_1. Since the only non-zero component of the unit outward normal ν to Σ_1 is $\nu_n = -1$, it follows that

$$\begin{aligned}\int_{\Sigma_1} S\cdot\nu\, d\mathcal{H}^n &= -\int_{|x|>r}\left(\int_0^{u(x,0)}\frac{\psi_t(x,0)}{\psi(x,0)}2z\,dz\right)dx\\ &= -\int_{|x|>r}\left(\frac{\psi_t(x,0)}{\psi(x,0)}\int_0^{u(x,0)}2z\,dz\right)dx\\ &= -\int_{|x|>r}u^2(x,0)\frac{\psi_t(x,0)}{\psi(x,0)}\,dx. \end{aligned}\tag{21}$$

By (15), if we define

$$h(s)=\int_0^\infty A^{\frac{n}{2}-2}\big[(A+s)^2+1\big]^{-\frac{n}{2}+1}\,dA \tag{22}$$

then we can rewrite

$$\psi(x,t)=\frac{1}{|x|^{\frac{n}{2}-1}}h\left(\frac{t}{|x|}\right) \tag{23}$$

from which we easily deduce that

$$\psi_t(x,0)=\frac{1}{|x|^{\frac{n}{2}}}h'(0)=-\frac{n-2}{|x|^{\frac{n}{2}}}\int_0^\infty\frac{A^{\frac{n}{2}-1}}{[A^2+1]^{\frac{n}{2}}}\,dA. \tag{24}$$

Collecting (21)–(24) we get

$$\int_{\Sigma_1} S\cdot\nu\, d\mathcal{H}^n = H_n\int_{|x|>r}\frac{u^2(x,0)}{|x|}\,dx \tag{25}$$

where

$$H_n=(n-2)\left[\int_0^\infty\frac{A^{\frac{n}{2}-1}}{[A^2+1]^{\frac{n}{2}}}\,dA\right]\left[\int_0^\infty\frac{A^{\frac{n}{2}-2}}{[A^2+1]^{\frac{n}{2}-1}}\,dA\right]^{-1}. \tag{26}$$

In order to evaluate explicitly H_n, consider the change of variable $\alpha=\frac{A^2}{A^2+1}$ in both integrals that appear in (26), to obtain

$$\int_0^\infty\frac{A^{\frac{n}{2}-1}}{[A^2+1]^{\frac{n}{2}}}\,dA=\frac{1}{2}\int_0^1\alpha^{\frac{n}{4}-1}(1-\alpha)^{\frac{n}{4}-1}\,d\alpha=\frac{1}{2}\mathfrak{B}\left(\frac{n}{4},\frac{n}{4}\right) \tag{27}$$

and

$$\int_0^\infty\frac{A^{\frac{n}{2}-2}}{[A^2+1]^{\frac{n}{2}-1}}\,dA=\frac{1}{2}\int_0^1\alpha^{\frac{n}{4}-\frac{3}{2}}(1-\alpha)^{\frac{n}{4}-\frac{3}{2}}\,d\alpha=\frac{1}{2}\mathfrak{B}\left(\frac{n-2}{4},\frac{n-2}{4}\right), \tag{28}$$

where $\mathfrak{B}$ denote the Beta function defined as $\mathfrak{B}(\xi,\eta)=\int_0^1\alpha^{\xi-1}(1-\alpha)^{\eta-1}d\alpha$. Recalling that

$$\mathfrak{B}(\xi,\eta)=\frac{\Gamma(\xi)\Gamma(\eta)}{\Gamma(\xi+\eta)}, \tag{29}$$

from (27)–(29) we get

$$H_n = (n-2)\frac{\Gamma^2(\frac{n}{4})}{\Gamma(\frac{n}{2})}\frac{\Gamma(\frac{n}{2}-1)}{\Gamma^2(\frac{n-2}{4})}$$

that jointly with the identity $\Gamma(1+\xi)=\xi\Gamma(\xi)$ gives

$$H_n = 2\frac{\Gamma^2(\frac{n}{4})}{\Gamma^2(\frac{n-2}{4})}.$$

As regards the flow of S across Σ_2, since its outer unit normal is

$$\nu \equiv \left(-\frac{x}{\sqrt{|x|^2+t^2}}, -\frac{t}{\sqrt{|x|^2+t^2}}, 0\right),$$

then

$$\left|\int_{\Sigma_2} S\cdot\nu\, d\mathscr{H}^n\right| \le \int_{\Sigma_2} 2z\left|\frac{\psi_\rho(x,t)}{\psi(x,t)}\right| d\mathscr{H}^n \tag{30}$$

where ψ_ρ denotes the derivative of ψ in the radial direction. Using the representation of ψ given in (23), if we consider the following change of variable

$$|x| = \rho\cos\theta, \qquad t = \rho\sin\theta, \quad \rho \ge 0,\ \theta\in[0,\pi],$$

we easily deduce that

$$\psi(x,t) = \frac{1}{\rho^{\frac{n}{2}-1}}\frac{h(\tan\theta)}{\cos^{\frac{n}{2}-1}\theta},$$

from which we infer

$$\psi_\rho(x,t) = \left(1-\frac{n}{2}\right)\frac{1}{\rho^{\frac{n}{2}}}\frac{h(\tan\theta)}{\cos^{\frac{n}{2}-1}\theta} = \left(1-\frac{n}{2}\right)\frac{\psi(x,t)}{\sqrt{|x|^2+t^2}}. \tag{31}$$

The following inequalities hold

$$\begin{aligned}\left|\int_{\Sigma_2} S\cdot\nu\, d\mathscr{H}^n\right| &\le \frac{n-2}{r}\int_{\partial B_r^+(0)} d\mathscr{H}^{n-1}\int_0^{u(x,t)} z\,dz\\ &= \frac{n-2}{r}\int_{\partial B_r^+(0)} \frac{u^2(x,t)}{2}\, d\mathscr{H}^{n-1}\\ &\le \frac{n-2}{r}\frac{nC_n r^{n-1}}{2}\frac{\|u\|^2_{L^\infty(\partial B_r^+(0))}}{2} = Cr^{n-2},\end{aligned} \tag{32}$$

where the first inequality follows from (30), (31) and from the definition of Σ_2 (here, and in the following, $\partial B_\rho^+(0)$ denotes the boundary of $B_\rho(0)$ contained in $\mathbb{R}^n_+$ and C_n denote the measure of the unit ball of $\mathbb{R}^n$). Finally, from (17), (20), (25) and (32) we deduce that there exists a positive constant C such that

$$\int_{\mathbb{R}^n_+\setminus B_r(0)} |\nabla u|^2(x)\,dx \ge H_n\int_{|x|>r}\frac{u^2(x,0)}{|x|}\,dx - Cr^{n-2},$$

from which (3) follows directly on letting r go to zero and by density arguments.

It remains to show that the constant that appears in (3) is optimal. To do this, let $0 < r < R$ and apply the previous arguments replacing u by ψ. Starting from (13), since $\mathscr{E}(x,t,\psi(x,t),\mathscr{P}(x,t),\nabla\psi(x,t)) = 0$ we get

$$\int_{B_R^+(0)\setminus B_r(0)} |\nabla\psi|^2(x,t)\,dx\,dt = \mathscr{M}\big(\psi, B_R^+(0)\setminus B_r(0)\big).$$

Nevertheless, since ψ is non-zero on $\partial B_R(0)$, instead of (20) we get

$$\int_{B_R^+(0)\setminus B_r(0)} |\nabla\psi|^2(x,t)\,dx\,dt = \int_{\Sigma_1} S\cdot\nu\,d\mathscr{H}^n + \int_{\Sigma_2} S\cdot\nu\,d\mathscr{H}^n + \int_{\mathscr{S}} S\cdot\nu\,d\mathscr{H}^n$$

where Σ_1 and Σ_2 are given in (18) and (19), respectively, by replacing u by ψ, and

$$\mathscr{S} = \big\{(x,t,z)\in\mathbb{R}^n_+\times\mathbb{R} : |x|^2+t^2 = R^2,\ 0\le z\le\psi(x,t)\big\}.$$

By (25) and the definition of S, Σ_1, Σ_2 and $\mathscr{S}$ we infer

$$\begin{aligned}&\int_{B_R^+(0)\setminus B_r(0)} |\nabla\psi|^2(x,t)\,dx\,dt\\ &= H_n\int_{r<|x|<R}\frac{\psi^2(x,0)}{|x|}\,dx + \frac{n-2}{r}\int_{\Sigma_2} z\,d\mathscr{H}^n - \frac{n-2}{R}\int_{\mathscr{S}} z\,d\mathscr{H}^n. \qquad (33)\end{aligned}$$

It is easy to check that the last two integrals in (33) are equal. Indeed, by spherical coordinates and (31), arguing as in (32), we get

$$\begin{aligned}\frac{n-2}{r}\int_{\Sigma_2} z\,d\mathscr{H}^n &= \frac{n-2}{r}\int_{\partial B_1^+(0)}\frac{1}{r^{n-2}}\frac{h^2(\tan\theta)}{2\cos^{n-2}\theta}r^{n-1}\,d\mathscr{H}^{n-1}\\ &= (n-2)\int_{\partial B_1^+(0)}\frac{h^2(\tan\theta)}{2\cos^{n-2}\theta}\,d\mathscr{H}^{n-1}\\ &= \frac{n-2}{R}\int_{\mathscr{S}} z\,d\mathscr{H}^n. \qquad (34)\end{aligned}$$

Collecting (33) and (34) we deduce

$$\lim_{R\to\infty}\lim_{r\to 0}\frac{\int_{B_R^+(0)\setminus B_r(0)}|\nabla\psi|^2(x,t)\,dx\,dt}{\int_{r<|x|<R}\frac{\psi^2(x,0)}{|x|}\,dx} = H_n$$

that shows the optimality of the constant.

4 Sketch of the Proof of Theorem 2

Consider $2\le\beta<n$. The proof contained in Sect. 3 can be adapted to consider the functional

$$\mathscr{F}\big(u,\mathbb{R}^n_+\big) = \int_{\mathbb{R}^n_+}\big|\nabla u(x,t)\big|^2\,dx\,dt - \frac{(\beta-2)^2}{4}\int_{\mathbb{R}^n_+}\frac{u^2(x,t)}{|x|^2+t^2}\,dx\,dt$$

instead of that defined in (14). In this case the family of extremals is given by $\{k\phi\}_{k\in\mathbb{R}^n_+}$, where ϕ is the solution of (8). Writing

$$\phi(x,t) = \rho^{-\frac{n}{2}+1} f(\theta)$$

problem (8) is equivalent to the following limit problem

$$\begin{cases} f''(\theta) - (n-2)\tan\theta f'(\theta) - \left(\dfrac{(n-2)^2}{4} - \dfrac{(\beta-2)^2}{4}\right) f(\theta) = 0, \quad \theta \in \left(0, \dfrac{\pi}{2}\right) \\ f(0) = 1, \quad \lim\limits_{\theta\to\frac{\pi}{2}} f(\theta) \in \mathbb{R}. \end{cases} \tag{35}$$

The solution of (35) lead to the following representation of ϕ

$$\begin{aligned} \phi(x,t) = {} & \frac{\mathfrak{F}\left(\frac{n+\beta}{4} - 1, \frac{n-\beta}{4}, \frac{1}{2}; \frac{t^2}{|x|^2+t^2}\right)}{[|x|^2+t^2]^{\frac{n-2}{4}}} \\ & - t\frac{H(n,\beta)}{[|x|^2+t^2]^{\frac{n}{4}}} \mathfrak{F}\left(\frac{n+\beta}{4} - \frac{1}{2}, \frac{n-\beta}{4} + \frac{1}{2}, \frac{3}{2}; \frac{t^2}{|x|^2+t^2}\right) \end{aligned} \tag{36}$$

where $\mathfrak{F}$ denotes the hypergeometric series

$$\mathfrak{F}(a,b,c;z) = 1 + \frac{\Gamma(c)}{\Gamma(a)\Gamma(b)} \sum_{k=1}^{\infty} \frac{\Gamma(a+k)\Gamma(b+k)}{\Gamma(c+k)} \frac{z^k}{k!}.$$

The result then follows arguing as in Sect. 3, by considering the Mayer field

$$F : (x,t,k) \to \big(x,t,k\phi(x,t)\big) \in \mathbb{R}^n_+.$$

References

1. Adimurthi, Esteban, M.J.: An improved Hardy-Sobolev inequality in $W^{1,p}$ and its application to Schrödinger operators. NoDEA Nonlinear Differ. Equ. Appl. **12**(2), 243–263 (2005)
2. Adimurthi, Chaudhuri, N., Ramaswamy, M.: An improved Hardy-Sobolev inequality and its application. Proc. Am. Math. Soc. **130**(2), 485–505 (2002)
3. Alvino, A.: On a Sobolev-type inequality. Atti Accad. Naz. Lincei Cl. Sci. Fis. Mat. Natur. Rend. Lincei (9) Mat. Appl. **20**(4), 379–386 (2009)
4. Alvino, A., Volpicelli, R., Volzone, B.: On Hardy inequalities with a remainder term. Ric. Mat. **59**(2), 265–280 (2010)
5. Alvino, A., Ferone, A., Volpicelli, R.: Sharp Hardy inequalities in the half space with trace remainder term. Nonlinear Anal. **75**(14), 5466–5472 (2012). doi:10.1016/j.na.2012.04.051
6. Aubin, T.: Problémes isopérimétriques et espaces de Sobolev. J. Differ. Geom. **11**(4), 573–598 (1976)
7. Barbatis, G., Filippas, S., Tertikas, A.: Series expansion for L^p Hardy inequalities. Indiana Univ. Math. J. **52**(1), 171–190 (2003)
8. Bartsch, T., Weth, T., Willem, M.: A Sobolev inequality with remainder term and critical equations on domains with topology for the polyharmonic operator. Calc. Var. Partial Differ. Equ. **18**(3), 253–268 (2003)

9. Brezis, H., Marcus, M.: Hardy's inequalities revisited. Ann. Scuola Norm. Super. Pisa. Cl. Sci. (4) **25**(1–2), 217–237 (1997). 1998
10. Brezis, H., Vazquez, J.L.: Blow-up solutions of some nonlinear elliptic problems. Rev. Mat. Univ. Complut. Madr. **10**(2), 443–469 (1997)
11. Brezis, H., Marcus, M., Shafrir, I.: Extremal functions for Hardy's inequality with weight. J. Funct. Anal. **171**(1), 177–191 (2000)
12. Cianchi, A., Ferone, A.: Best remainder norms in Sobolev-Hardy inequalities. Indiana Univ. Math. J. **58**(3), 1051–1096 (2009)
13. Cianchi, A., Ferone, A.: Improving sharp Sobolev type inequalities by optimal remainder gradient norms. Commun. Pure Appl. Anal. **11**(3), 1363–1386 (2012). doi:10.3934/cpaa.2012.11.1363
14. Dalbeault, J., Esteban, M.J., Loss, M., Vega, L.: An analytical proof of Hardy-like inequalities related to the Dirac operator. J. Funct. Anal. **216**(1), 1–21 (2004)
15. Davila, J., Dupaigne, L., Montenegro, M.: The extremal solution of a boundary reaction problem. Commun. Pure Appl. Anal. **7**(4), 795–817 (2008)
16. Detalla, A., Horiuchi, T., Ando, H.: Missing terms in Hardy-Sobolev inequalities and its application. Far East J. Math. Sci.: FJMS **14**(3), 333–359 (2004)
17. Escobar, J.F.: Sharp constant in a Sobolev trace inequality. Indiana Univ. Math. J. **37**, 687–698 (1988)
18. Filippas, S., Maz'ya, V.G., Tertikas, A.: On a question of Brezis and Marcus. Calc. Var. Partial Differ. Equ. **25**(4), 491–501 (2006)
19. Filippas, S., Maz'ya, V.G., Tertikas, A.: Critical Hardy-Sobolev inequalities. J. Math. Pures Appl. (9) **87**(1), 37–56 (2007)
20. Gazzola, F., Grunau, H.C., Mitidieri, E.: Hardy inequalities with optimal constants and remainder terms. Trans. Am. Math. Soc. **356**(6), 2149–2168 (2004)
21. Ghoussoub, N., Moradifam, A.: On the best possible remaining term in the Hardy inequality. Proc. Natl. Acad. Sci. USA **105**(37), 13746–13751 (2008)
22. Giaquinta, M., Hildebrandt, S.: Calculus of Variations. I. The Lagrangian Formalism. Springer, Berlin (1996)
23. Herbst, I.W.: Spectral theory of the operator $(p^2+m^2)^{1/2} - Ze^2/r$. Commun. Math. Phys. **53**(3), 285–2940 (1977)
24. Hoffmann-Ostenhof, M., Hoffmann-Ostenhof, T., Laptev, A.: A geometrical version of Hardy's inequality. J. Funct. Anal. **189**(2), 539–548 (2002)
25. Sagan, H.: Introduction to the Calculus of Variations. Dover, New York (1992)
26. Talenti, G.: Best constant in Sobolev inequality. Ann. Mat. Pura Appl. **110**, 353–372 (1976)
27. Tidblom, J.: A Hardy inequality in the half-space. J. Funct. Anal. **221**(2), 482–495 (2005)
28. Vazquez, J.L., Zuazua, E.: The Hardy inequality and the asymptotic behaviour of the heat equation with an inverse-square potential. J. Funct. Anal. **173**(1), 103–153 (2000)

Sharp Bounds for the p-Torsion of Convex Planar Domains

Ilaria Fragalà, Filippo Gazzola, and Jimmy Lamboley

Abstract We obtain some sharp estimates for the p-torsion of convex planar domains in terms of their area, perimeter, and inradius. The approach we adopt relies on the use of web functions (*i.e.* functions depending only on the distance from the boundary), and on the behavior of the inner parallel sets of convex polygons. As an application of our isoperimetric inequalities, we consider the shape optimization problem which consists in maximizing the p-torsion among polygons having a given number of vertices and a given area. A long-standing conjecture by Pólya-Szegö states that the solution is the regular polygon. We show that such conjecture is true within the subclass of polygons for which a suitable notion of "asymmetry measure" exceeds a critical threshold.

Keywords Isoperimetric inequalities · Shape optimization · Web functions · Convex shapes

1 Introduction

Let $\Omega \subset \mathbb{R}^2$ be an open bounded domain and let $p \in (1, +\infty)$. Consider the boundary value problem

$$\begin{cases} -\Delta_p u = 1 & \text{in } \Omega \\ u = 0 & \text{on } \partial\Omega, \end{cases} \tag{1}$$

I. Fragalà (✉) · F. Gazzola
Politecnico di Milano, Piazza Leonardo da Vinci 32, 20133 Milano, Italy
e-mail: ilaria.fragala@polimi.it

F. Gazzola
e-mail: filippo.gazzola@polimi.it

J. Lamboley
Ceremade, Université Paris-Dauphine, Place du Maréchal de Lattre de Tassigny, 75775 Paris, France
e-mail: lamboley@ceremade.dauphine.fr

R. Magnanini et al. (eds.), *Geometric Properties for Parabolic and Elliptic PDE's*, Springer INdAM Series 2, DOI 10.1007/978-88-470-2841-8_7,

where $\Delta_p u = \mathrm{div}(|\nabla u|^{p-2}\nabla u)$ denotes the p-Laplacian. The *p-torsion* of Ω is defined by

$$\tau_p(\Omega) := \int_\Omega |\nabla u_p|^p = \int_\Omega u_p, \tag{2}$$

being u_p the unique solution to (1) in $W_0^{1,p}(\Omega)$. Notice that the second equality in (2) is obtained by testing (1) by u_p and integrating by parts. Since (1) is the Euler-Lagrange equation of the variational problem

$$\min_{u \in W_0^{1,p}(\Omega)} J_p(u), \quad \text{where } J_p(u) = \int_\Omega \left(\frac{1}{p}|\nabla u|^p - u \right), \tag{3}$$

there holds

$$\tau_p(\Omega) = \frac{p}{1-p} \min_{u \in W_0^{1,p}(\Omega)} J_p(u).$$

A further characterization of the p-torsion is provided by the equality $\tau_p(\Omega) = S(\Omega)^{1/(p-1)}$, where $S(\Omega)$ is the best constant for the Sobolev inequality $\|u\|_{L^1(\Omega)}^p \leq S(\Omega)\|\nabla u\|_{L^p(\Omega)}^p$ on $W_0^{1,p}(\Omega)$.

The purpose of this paper is to provide some sharp bounds for $\tau_p(\Omega)$, holding for a convex planar domain Ω, in terms of its area, perimeter, and inradius (in the sequel denoted respectively by $|\Omega|$, $|\partial\Omega|$, and R_Ω). The original motivation for studying this kind of shape optimization problem draws its origins in the following long-standing conjecture by Pólya and Szegö:

Among polygons with a given area and N vertices,

the regular N-gon maximizes τ_p. (4)

A similar conjecture is stated by the same Authors also for the principal frequency and for the logarithmic capacity, see [15]. For $N = 3$ and $N = 4$ these conjectures were proved by Pólya and Szegö themselves [15, p. 158]. For $N \geq 5$, to the best of our knowledge, the unique solved case is the one of logarithmic capacity, see the beautiful paper [16] by Solynin and Zalgaller; the cases of torsion and principal frequency are currently open. In fact let us remind that, for $N \geq 5$, the classical tool of Steiner symmetrization fails because it may increase the number of sides, see [10, Sect. 3.3].

The approach we adopt in order to provide upper and lower bounds for the p-torsion in terms of geometric quantities, is based on the idea of considering a proper subspace $\mathscr{W}_p(\Omega)$ of $W_0^{1,p}(\Omega)$ and to address the minimization problem for the functional J_p on $\mathscr{W}_p(\Omega)$. More precisely, we consider the subspace of functions depending only on the distance $d(x) = \mathrm{dist}(x, \partial\Omega)$ from the boundary:

$$\mathscr{W}_p(\Omega) = \{u \in W_0^{1,p}(\Omega) : u(x) = u(d(x))\}.$$

Functions in $\mathscr{W}_p(\Omega)$ have the same level lines as d, namely the boundaries of the so-called *inner parallel sets*, $\Omega_t := \{x \in \Omega : d(x) > t\}$, which were first used in variational problems by Pólya and Szegö [15, Sect. 1.29], see also [13]. Later, in

[9], the elements of $\mathscr{W}_p(\Omega)$ were called *web functions*, because in case of planar polygons the level lines of d recall the pattern of a spider web. We refer to [4, 5] for some estimates on the minimizing properties of web-functions, and to the subsequent papers [2, 6, 7] for applications of these functions in different frameworks. In particular, the papers [6, 7] deal with the problem of estimating how efficiently $\tau_p(\Omega)$ can be approximated by the *web p-torsion*, defined as

$$w_p(\Omega) := \frac{p}{1-p} \min_{u \in \mathscr{W}_p(\Omega)} J_p(u).$$

While the value of $\tau_p(\Omega)$ is in general not known (because the solution to problem (1) cannot be determined except for some special geometries of Ω), the value of $w_p(\Omega)$ admits the following explicit expression in terms of the parallel sets Ω_t:

$$w_p(\Omega) = \int_0^{R_\Omega} \frac{|\Omega_t|^q}{|\partial\Omega_t|^{q-1}}\, dt, \tag{5}$$

where $q = \frac{p}{p-1}$ is the conjugate exponent of p, and R_Ω is the inradius of Ω (see [7]).

Clearly, since $\mathscr{W}_p(\Omega) \subset W_0^{1,p}(\Omega)$, $w_p(\Omega)$ bounds $\tau_p(\Omega)$ from below. On the other hand, when Ω is convex, $\tau_p(\Omega)$ can be bounded from above by a constant multiple of $w_p(\Omega)$, for some constant which tends to 1 as $p \to +\infty$. In fact, in [7] it is proved that, for any $p \in (1, +\infty)$, the following estimates hold and are sharp:

$$\forall \Omega \in \mathscr{C}, \quad \frac{q+1}{2^q} < \frac{w_p(\Omega)}{\tau_p(\Omega)} \le 1 \tag{6}$$

where $\mathscr{C}$ denotes the class of planar bounded convex domains; moreover the right inequality holds as an equality if and only if Ω is a disk. Note that, if $p \to +\infty$, then $q \to 1$ and the constant in the left hand side of (6) tends to 1.

In this paper, we prove some geometric estimates for $\tau_p(\Omega)$ in the class $\mathscr{C}$, which have some implications in the conjecture (4). More precisely, we consider the following shape functionals:

$$\Omega \mapsto \frac{\tau_p(\Omega)|\partial\Omega|^q}{|\Omega|^{q+1}} \quad \text{and} \quad \Omega \mapsto \frac{\tau_p(\Omega)}{R_\Omega^q |\Omega|}. \tag{7}$$

Let us remark that the above quotients are invariant under dilations and that convex sets which agree up to rigid motions (translations and rotations) are systematically identified throughout the paper.

Our main results are Theorems 1 and 3, which give sharp bounds for the functionals (7) when Ω varies in $\mathscr{C}$. We also exhibit minimizing and maximizing sequences. These bounds are obtained by combining sharp bounds for the web p-torsion (see Theorem 2 and the second part of Theorem 3) with (6). As a consequence of our results we obtain the validity of some weak forms of Pólya-Szegö conjecture (4). On the class $\mathscr{P}$ of convex polygons we introduce a sort of "asymmetry measure" such as

$$\forall \Omega \in \mathscr{P}, \quad \gamma(\Omega) := \frac{|\partial\Omega|}{|\partial\Omega^{\circledast}|} \in [1, +\infty),$$

where $\Omega^{\circledast}$ denotes the regular polygon with the same area and the same number of vertices as Ω. Then, if the p-torsion $\tau_p(\Omega)$ is replaced by the web p-torsion $w_p(\Omega)$, (4) holds in the following refined form:

$$\forall \Omega \in \mathscr{P}, \quad w_p(\Omega) \leq \gamma(\Omega)^{-q} w_p\big(\Omega^{\circledast}\big). \tag{8}$$

Consequently, on the class $\mathscr{P}_N$ of convex polygons with N vertices, conjecture (4) holds true for those Ω which are sufficiently "far" from $\Omega^{\circledast}$, meaning that $\gamma(\Omega)$ exceeds a threshold depending on N and p:

$$\forall \Omega \in \mathscr{P}_N\colon \quad \gamma(\Omega) \geq \Gamma_{N,p}, \quad \tau_p(\Omega) < \tau_p\big(\Omega^{\circledast}\big). \tag{9}$$

The value of the threshold $\Gamma_{N,p}$ can be explicitly characterized (see Corollary 2) and tends to 1 as $p \to +\infty$.

The paper is organized as follows. Section 2 contains the statement of our results, which are proved in Sect. 4 after giving in Sect. 3 some preliminary material of geometric nature. Section 5 is devoted to some related open questions and perspectives.

2 Results

We introduce the following classes of convex planar domains:

$\mathscr{C}$ = the class of bounded convex domains in $\mathbb{R}^2$;
$\mathscr{C}_o$ = the subclass of $\mathscr{C}$ given by tangential bodies to a disk;
$\mathscr{P}$ = the class of convex polygons;
$\mathscr{P}_N$ = the class of convex polygons having N vertices ($N \geq 3$).

Tangential bodies to a disk are domains $\Omega \in \mathscr{C}$ such that, for some disk D, through each point of $\partial\Omega$ there exists a tangent line to Ω which is also tangent to D. Domains in $\mathscr{P} \cap \mathscr{C}_o$ are circumscribed polygons, whereas domains in $\mathscr{C}_o \setminus \mathscr{P}$ can be obtained by removing from a circumscribed polygon some connected components of the complement (in the polygon itself) of the inscribed disk. In particular, the disk itself belongs to $\mathscr{C}_o$.

Our first results are the following sharp bounds for the p-torsion of convex planar domains. We recall that, for any given $p \in (1, +\infty)$, $q := \frac{p}{p-1}$ denotes its conjugate exponent.

Theorem 1 *For any $p \in (1, +\infty)$, it holds*

$$\forall \Omega \in \mathscr{C}, \quad \frac{1}{q+1} < \frac{\tau_p(\Omega)|\partial\Omega|^q}{|\Omega|^{q+1}} < \frac{2^{q+1}}{(q+2)(q+1)}. \tag{10}$$

Moreover,

- *the left inequality holds asymptotically with equality sign for any sequence of thinning rectangles;*
- *the right inequality holds asymptotically with equality sign for any sequence of thinning isosceles triangles.*

By sequence of thinning rectangles or triangles, we mean that the ratio between their minimal width and diameter tends to 0. We point out that, in the particular case when $p = 2$, the statement of Theorem 1 is already known. Indeed, the left inequality in (10) holds true for any simply connected set in $\mathbb{R}^2$ as discovered by Pólya [14]; the right inequality in (10) for convex sets is due to Makai [12], though its method of proof, which is different from ours, does not allow to obtain the *strict* inequality.

Our approach to prove Theorem 1 employs as a major ingredient the following sharp estimates for the web p-torsion of convex domains, which may have their own interest.

Theorem 2 *For any* $p \in (1, +\infty)$, *it holds*

$$\forall \Omega \in \mathscr{C}, \quad \frac{1}{q+1} < \frac{w_p(\Omega)|\partial\Omega|^q}{|\Omega|^{q+1}} \leq \frac{2}{q+2}. \tag{11}$$

Moreover,

- *the left inequality holds asymptotically with equality sign for any sequence of thinning rectangles*;
- *the right inequality holds with equality sign for* $\Omega \in \mathscr{C}_o$.

Let us now discuss the implications of the above results in the shape optimization problem which consists in maximizing τ_p in the class of convex polygons with a given area and a given number of vertices:

$$\max\{\tau_p(\Omega) : \Omega \in \mathscr{P}_N, \ |\Omega| = m\}. \tag{12}$$

We recall that, for any $\Omega \in \mathscr{P}$, $\Omega^\circledast$ denotes the regular polygon with the same area and the same number of vertices as Ω. Moreover, we set

$$\forall \Omega \in \mathscr{P}, \quad \gamma(\Omega) := \frac{|\partial\Omega|}{|\partial\Omega^\circledast|};$$

notice that by the isoperimetric inequality for polygons (see Proposition 1), $\gamma(\Omega) \in [1, +\infty)$ and $\gamma(\Omega) > 1$ if $\Omega \neq \Omega^\circledast$. With this notation, it is straightforward to deduce from Theorem 2 the following

Corollary 1 *The regular polygon is the unique maximizer of* w_p *over polygons in* $\mathscr{P}$ *with a given area and a given number of vertices. More precisely, the following refined isoperimetric inequality holds*:

$$\forall \Omega \in \mathscr{P}, \quad w_p(\Omega) \leq \gamma(\Omega)^{-q} w_p(\Omega^\circledast). \tag{13}$$

As a consequence, using (6), we obtain some information on the shape optimization problem (12):

Corollary 2 *Let* $\Gamma_{N,p} := (\frac{w_p(\Omega^\circledast)}{\tau_p(\Omega^\circledast)})^{1/q} \frac{2}{(q+1)^{1/q}}$. *Then,*

$$\forall \Omega \in \mathscr{P}_N, \quad \tau_p(\Omega) < \Gamma_{N,p}^q \gamma(\Omega)^{-q} \tau_p(\Omega^\circledast).$$

In particular, the p-torsion of the regular N-gon is larger than the p-torsion of any polygon in $\mathscr{P}_N$ having the same area and an asymmetry measure larger than the threshold $\Gamma_{N,p}$:

$$\forall \Omega \in \mathscr{P}_N, \quad \gamma(\Omega) \geq \Gamma_{N,p} \quad \Rightarrow \quad \tau_p(\Omega) < \tau_p(\Omega^{\circledast}). \tag{14}$$

Some comments on Corollary 2 are gathered in the next remark.

Remark 1 (i) Using again (6) we infer

$$1 < \Gamma_{N,p} < \frac{2}{(q+1)^{1/q}} < 2 \quad \forall N, p, \qquad \lim_{p\to+\infty} \Gamma_{N,p} = 1.$$

Hence, asymptotically with respect to p, the condition $\gamma(\Omega) \geq \Gamma_{N,p}$ appearing in (14) becomes not restrictive. Moreover, if $p = 2$, we have $\Gamma_{N,2} \leq 2/\sqrt{3} \approx 1.15$ and the dependence on N of $\Gamma_{N,2}$ can be enlightened by using the numerical values given in [4]:

N	3	4	5	6	7	8	9	10	20
$\Gamma_{N,2} \approx$	1.054	1.089	1.108	1.121	1.129	1.135	1.138	1.141	1.149

(ii) Though the validity of (4) is known for triangles, in order to give an idea of the efficiency of Corollary 2, consider the case $N = 3$ and $p = 2$. The equilateral triangle

$$T^{\circledast} := \left\{(x, y) \in \mathbb{R}^2;\ y > 0,\ -\frac{1}{2} + \frac{y}{\sqrt{3}} < x < \frac{1}{2} - \frac{y}{\sqrt{3}}\right\}$$

satisfies $|T^{\circledast}| = \frac{\sqrt{3}}{4}$ and $|\partial T^{\circledast}| = 3$. The solution to (1) is explicitly given by

$$u(x, y) = \frac{\sqrt{3}}{8}\left(y - \frac{4}{\sqrt{3}}y^2 + \frac{4}{3}y^3 - 4x^2 y\right)$$

so that $\tau_2(T^{\circledast}) = \sqrt{3}/640$. Moreover, by (27) below we find $w_2(T^{\circledast}) = \sqrt{3}/768$ and, in turn, that $\Gamma_{3,2} = \sqrt{10}/3 \approx 1.054$.

Consider now the isosceles triangles T_k having the basis of length $k > 0$ and the two equal sides of length

$$\ell_k = \sqrt{\frac{3}{4k^2} + \frac{k^2}{4}} \quad \text{so that} \quad |\partial T_k| = k + \sqrt{\frac{3}{k^2} + k^2} \quad \text{and} \quad |T_k| = \frac{\sqrt{3}}{4} = \left|T^{\circledast}\right|,$$

(notice that $T_1 = T^{\circledast}$). Therefore,

$$\gamma(T_k) = \frac{k + \sqrt{\frac{3}{k^2} + k^2}}{3}$$

and $\gamma(T_k) \geq \Gamma_{3,2}$ if and only if $2\sqrt{10}k^3 - 10k^2 + 3 \geq 0$, which approximatively corresponds to $k \notin (0.760, 1.301)$.

We conclude this section with a variant of Theorems 1 and 2.

Theorem 3 *For every $p \in (1, +\infty)$, it holds*

$$\forall \Omega \in \mathscr{C}, \quad \frac{1}{(q+2)2^{q-1}} \leq \frac{\tau_p(\Omega)}{R_\Omega^q |\Omega|} < \frac{2^q}{(q+1)^2} \tag{15}$$

$$\forall \Omega \in \mathscr{C}, \quad \frac{1}{(q+2)2^{q-1}} \leq \frac{w_p(\Omega)}{R_\Omega^q |\Omega|} < \frac{1}{q+1}. \tag{16}$$

Moreover,

- *the left inequality in* (15) *holds with equality sign for balls;*
- *the left inequality in* (16) *holds with equality sign for $\Omega \in \mathscr{C}_o$;*
- *the right inequality in* (16) *holds asymptotically with equality sign for a sequence of thinning rectangles.*

The right inequality in (15) *is not* sharp. In fact, for $p = 2$, one has the sharp inequalities

$$\forall \Omega \in \mathscr{C}, \quad \frac{1}{8} \leq \frac{\tau_2(\Omega)}{R_\Omega^2 |\Omega|} \leq \frac{1}{3},$$

see [15, p. 100] for the left one, and [12] for the right one.

Using the isoperimetric inequalities (15) and (16), one can also derive statements similar to Corollaries 1 and 2, where $\gamma(\Omega)$ is replaced by another "asymmetry measure" given by

$$\widetilde{\gamma}(\Omega) = \frac{R_{\Omega^{\circledast}}}{R_\Omega}.$$

3 Geometric Preliminaries

In this section we present some useful geometric properties of convex polygons, which will be exploited to prove Theorem 2. First, we recall an improved form of the isoperimetric inequality in the class $\mathscr{P}$, whose proof can be found for instance in [6, Theorem 2]. For any $\Omega \in \mathscr{P}$, we set

$$C_\Omega := \sum_i \operatorname{cotan} \frac{\theta_i}{2}, \quad \text{being } \theta_i \text{ the inner angles of } \Omega. \tag{17}$$

Proposition 1 *For every $\Omega \in \mathscr{P}$, it holds*

$$|\Omega| \leq \frac{|\partial \Omega|^2}{4C_\Omega}, \tag{18}$$

with equality sign if and only if $\Omega \in \mathscr{P} \cap \mathscr{C}_o$, namely when Ω is a circumscribed polygon.

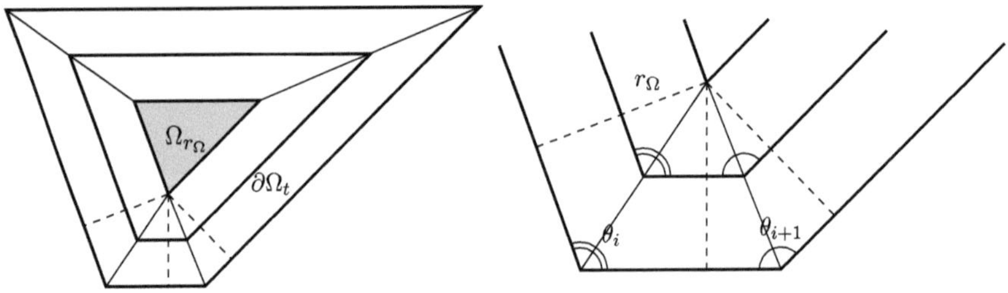

Fig. 1 Intersection of bisectors

Next, we recall that, denoting by R_Ω the inradius of any $\Omega \in \mathscr{P}$, for every $t \in [0, R_\Omega]$, the *inner parallel* sets of Ω are defined by

$$\Omega_t := \{x \in \Omega : \mathrm{dist}(x, \partial\Omega) > t\}$$

(notice in particular that $\Omega_{R_\Omega} = \emptyset$). Then we focus our attention on the behavior of the map $t \mapsto C_{\Omega_t}$ on the interval $[0, R_\Omega]$, and on the related expression of Steiner formulae. For every $\Omega \in \mathscr{P}$, we set

$$r_\Omega := \sup\{t \in [0, R_\Omega] : \Omega_t \text{ has the same number of vertices as } \Omega\}.$$

Clearly, if $r_\Omega < R_\Omega$, the number of vertices of Ω_t is strictly less than the number of vertices of Ω for every $t \in [r_\Omega, R_\Omega)$.

Proposition 2 *For every $\Omega \in \mathscr{P}$ and $t \geq 0$, $\Omega_t \in \mathscr{P}$ and the map $t \mapsto C_{\Omega_t}$ is piecewise constant on $[0, R_\Omega)$. Moreover, for every $t \in [0, r_\Omega]$, it holds*

$$|\Omega_t| = |\Omega| - |\partial\Omega| t + C_\Omega t^2 \quad \text{and} \quad |\partial\Omega_t| = |\partial\Omega| - 2C_\Omega t. \tag{19}$$

Finally, for every $t \in [0, R_\Omega]$, it holds

$$|\partial\Omega_t| \leq |\partial\Omega| - 2\pi t. \tag{20}$$

Proof For t small enough, the sides of Ω_t are parallel and at distance t from the sides of Ω, and the corners of Ω_t are located on the bisectors of the angles of Ω. r_Ω is actually the first time when two of these bisectors intersect at a point having distance t from at least two sides, see Fig. 1.

Therefore, for $t < r_\Omega$, Ω_t has the same angles as Ω (so $C_{\Omega_t} = C_\Omega$ by (17)), and we notice that the perimeter of grey areas in Fig. 2 is $2t\,\mathrm{cotan}(\theta_i/2)$, and their areas are $t^2\,\mathrm{cotan}(\theta_i/2)$, which gives (19) (still valid for $t = r_\Omega$ by continuity).

Let us now show that the map $t \mapsto C_{\Omega_t}$ is piecewise constant on $[0, R_\Omega)$, assuming that $r_\Omega < R_\Omega$. Once $t = r_\Omega$, Ω_t still has sides parallel to the ones of Ω but loses at least one of them. Again, C_{Ω_t} is constant for $t \geq r_\Omega$ until the next value of t such that another intersection of bisectors appears (we now consider bisectors of Ω_{r_Ω}). The number of discontinuities of $t \mapsto C_{\Omega_t}$ is finite since Ω has a finite number of sides, and therefore iterating the previous argument, we get that $t \mapsto C_{\Omega_t}$ is piecewise constant.

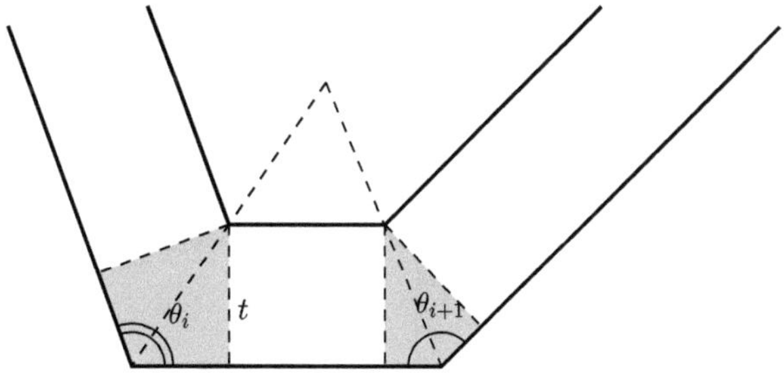

Fig. 2 How to derive Steiner formulae

Finally, from (17) we infer that $C_\Omega \geq \pi$ for any $\Omega \in \mathscr{P}$, so that (20) follows from the concavity of the map $t \mapsto |\partial\Omega_t|$ on $[0, R_\Omega]$ (see [1, Sects. 24 and 55]). □

A special role is played by polygons $\Omega \in \mathscr{P}$ such that $r_\Omega = R_\Omega$, namely polygons Ω whose inner parallel sets all have the same number of vertices as Ω itself. These are *polygonal stadiums*, characterized by the following

Definition 1 We call $\mathscr{S}$ the class of *polygonal stadiums*, namely polygons $P^\ell \in \mathscr{P}$ such that there exist a circumscribed polygon $P \in \mathscr{P} \cap \mathscr{C}_o$ having two parallel sides, and a nonnegative number ℓ such that, by choosing a coordinate system with origin in the center of the disk inscribed in P and the x-axis directed as two parallel sides of P, P^ℓ can be written as

$$P^\ell := \left(P_- - \left(\frac{\ell}{2}, 0\right)\right) \bigcup \left(\left[-\frac{\ell}{2}, \frac{\ell}{2}\right] \times (-R_P, R_P)\right) \bigcup \left(P_+ + \left(\frac{\ell}{2}, 0\right)\right), \quad (21)$$

where P_- (resp. P_+) denotes the set of points $(x, y) \in P$ with $x < 0$ (resp. $x > 0$), and R_P is the inradius of P, see Fig. 3.

Proposition 3 *Let $\Omega \in \mathscr{P}$. There holds $r_\Omega = R_\Omega$ if and only if $\Omega \in \mathscr{S}$.*

Proof We use the same notation as in Definition 1. Assume that $\Omega = P^\ell \in \mathscr{S}$. Then the bisectors of the angles of Ω intersect either at $(-\frac{\ell}{2}, 0)$ or at $(\frac{\ell}{2}, 0)$, which are at distance R_Ω from the boundary, see Fig. 4. In particular, if Ω is circumscribed to a disk, namely if $\ell = 0$, then the bisectors of the angles of Ω all intersect at the center of the disk. Therefore, Ω_t has the same number of sides as Ω if $t < R_\Omega$.

Conversely, assume that $R_\Omega = r_\Omega$. The set $\{x \in \Omega : d(x) = R_\Omega\}$ is convex with empty interior, so either it is a point, or a segment. If it is a point, then its distance to each side is the same, and therefore the disk having this point as a center and radius R_Ω is tangent to every side of Ω, so that Ω is circumscribed to a disk. If it is a segment, we choose coordinates such that this segment is $[(-\frac{\ell}{2}, 0); (\frac{\ell}{2}, 0)]$ for some positive number ℓ. Every point of this segment is at distance R_Ω from the boundary, so Ω contains the rectangle $(-\frac{\ell}{2}, \frac{\ell}{2}) \times (-R_\Omega, R_\Omega)$. Considering

$$P := \left(\Omega \cap \left\{x \leq -\frac{\ell}{2}\right\} + \left(\frac{\ell}{2}, 0\right)\right) \bigcup \left(\Omega \cap \left\{x \geq \frac{\ell}{2}\right\} + \left(-\frac{\ell}{2}, 0\right)\right),$$

we have that P is circumscribed and $\Omega = P^\ell$. □

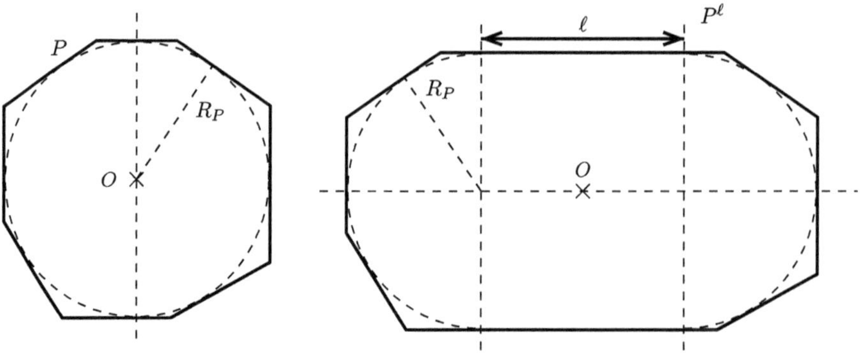

Fig. 3 A circumscribed polygon P and a polygonal stadium P^ℓ

Remark 2 Thanks to Proposition 3, for any polygonal stadium P^ℓ, the validity of the Steiner formulae (19) extends for t ranging over the whole interval $[0, R_{P^\ell}]$. Moreover, the value of the coefficients $|P^\ell|$, $|\partial P^\ell|$ and C_{P^ℓ} appearing therein, can be expressed only in terms of $|P|$, R_P, and ℓ (see Sect. 4). It is enough to use the following elementary equalities deriving from decomposition (21)

$$|P^\ell| = |P| + 2\ell R_P, \qquad |\partial P^\ell| = |\partial P| + 2\ell, \qquad C_{P^\ell} = C_P, \qquad R_{P^\ell} = R_P,$$

and the following identities holding for every $P \in \mathscr{P} \cap \mathscr{C}_o$

$$C_P = \frac{|P|}{R_P^2}, \qquad |\partial P| = \frac{2|P|}{R_P}. \tag{22}$$

Finally, we show that the parallel sets of any convex polygon Ω are polygonal stadiums for t sufficiently close to R_Ω:

Proposition 4 *For every $\Omega \in \mathscr{P}$, there exists $\bar{t} \in [0, R_\Omega)$ such that the parallel sets Ω_t belong to $\mathscr{S}$ for every $t \in [\bar{t}, R_\Omega)$.*

Proof We define $\bar{t}$ as the last time $t < R_\Omega$ such that Ω loses a side (we may have $\bar{t} = 0$). Therefore $\forall t \in [\bar{t}, R_\Omega]$, Ω_t has a constant number of sides, and so is in the class $\mathscr{S}$ by Proposition 3. □

4 Proofs

4.1 Proof of Theorem 2

We first prove Theorem 2 for $\Omega \in \mathscr{P}$, then we prove it for all $\Omega \in \mathscr{C}$.

Step 1: comparison with inner parallel sets. For a given $\Omega \in \mathscr{P}$, we wish to compare the value of the energy $\frac{w_p(\Omega)|\partial\Omega|^q}{|\Omega|^{q+1}}$ with the one of its parallel set Ω_ε for

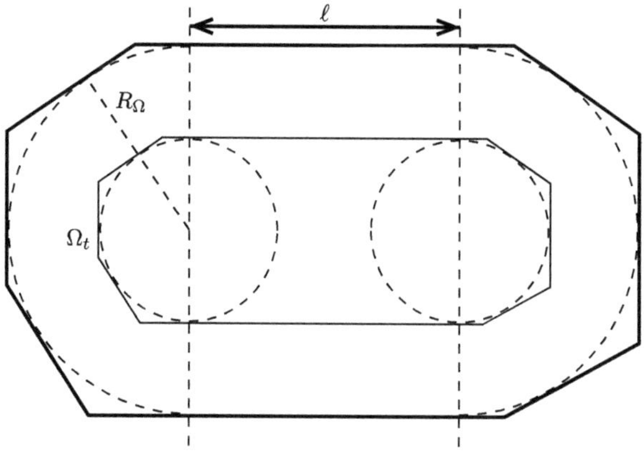

Fig. 4 Parallel sets of a polygonal stadium P^ℓ

small ε. To that aim, we use the representation formula (5) for $w_p(\Omega)$, and Steiner's formulae (19). In applying them we recall that, by Proposition 2 the map $t \mapsto C_{\Omega_t}$ is piecewise constant for $t \in [0, R_\Omega)$, and in particular it equals C_Ω on $[0, r_\Omega]$. Taking also into account that $(\Omega_\varepsilon)_t = \Omega_{\varepsilon+t}$, as $\varepsilon \to 0$ we have

$$\begin{aligned}
\frac{w_p(\Omega_\varepsilon)|\partial\Omega_\varepsilon|^q}{|\Omega_\varepsilon|^{q+1}} &= \frac{\int_0^{R_\Omega-\varepsilon} \frac{|(\Omega_\varepsilon)_t|^q}{|\partial(\Omega_\varepsilon)_t|^{q-1}}\, dt\, |\partial\Omega_\varepsilon|^q}{|\Omega_\varepsilon|^{q+1}} \\
&= \frac{[w_p(\Omega) - \int_0^\varepsilon \frac{|\Omega_t|^q}{|\partial\Omega_t|^{q-1}}\, dt][|\partial\Omega| - 2C_\Omega\, \varepsilon]^q}{[|\Omega| - |\partial\Omega|\varepsilon]^{q+1}} + o(\varepsilon), \\
&= \frac{|\partial\Omega|^q}{|\Omega|^{q+1}}\left[w_p(\Omega) - \frac{|\Omega|^q}{|\partial\Omega|^{q-1}}\varepsilon\right]\left[1 - 2q\frac{C_\Omega}{|\partial\Omega|}\varepsilon\right]\left[1 + (q+1)\frac{|\partial\Omega|}{|\Omega|}\varepsilon\right] + o(\varepsilon), \\
&= \frac{w_p(\Omega)|\partial\Omega|^q}{|\Omega|^{q+1}} \\
&\quad + \left[(q+1)\frac{|\partial\Omega|^{q+1}}{|\Omega|^{q+2}} w_p(\Omega) - \frac{|\partial\Omega|}{|\Omega|} - 2q\frac{C_\Omega w_p(\Omega)|\partial\Omega|^{q-1}}{|\Omega|^{q+1}}\right]\varepsilon + o(\varepsilon),
\end{aligned} \tag{23}$$

so that

$$\begin{aligned}
&\frac{w_p(\Omega_\varepsilon)|\partial\Omega_\varepsilon|^q}{|\Omega_\varepsilon|^{q+1}} - \frac{w_p(\Omega)|\partial\Omega|^q}{|\Omega|^{q+1}} \\
&= \left[(q+1)\frac{|\partial\Omega|^{q+1}}{|\Omega|^{q+2}} w_p(\Omega) - \frac{|\partial\Omega|}{|\Omega|} - 2q\frac{C_\Omega w_p(\Omega)|\partial\Omega|^{q-1}}{|\Omega|^{q+1}}\right]\varepsilon + o(\varepsilon).
\end{aligned} \tag{24}$$

As we shall see in the next steps, formula (24) will enable us to reach a contradiction if (11) fails.

Step 2: if (11) *fails for some convex polygon then it also fails for a polygonal stadium.* Let $\Omega \in \mathscr{P} \setminus \mathscr{S}$, and assume that (11) fails. We have to distinguish two cases.

First case: Assume that

$$\frac{w_p(\Omega)|\partial\Omega|^q}{|\Omega|^{q+1}} > \frac{2}{q+2}. \tag{25}$$

Using the isoperimetric inequality (18) and (25), one gets

$$\left[(q+1)\frac{|\partial\Omega|^{q+1}}{|\Omega|^{q+2}}w_p(\Omega) - \frac{|\partial\Omega|}{|\Omega|} - 2q\frac{C_\Omega w_p(\Omega)|\partial\Omega|^{q-1}}{|\Omega|^{q+1}}\right]$$
$$\geq \frac{q+2}{2}\frac{|\partial\Omega|}{|\Omega|}\left[\frac{w_p(\Omega)|\partial\Omega|^q}{|\Omega|^{q+1}} - \frac{2}{q+2}\right] > 0.$$

Inserting this information into (24) shows that

$$\frac{w_p(\Omega_\varepsilon)|\partial\Omega_\varepsilon|^q}{|\Omega_\varepsilon|^{q+1}} - \frac{w_p(\Omega)|\partial\Omega|^q}{|\Omega|^{q+1}} > 0$$

for sufficiently small ε. In fact, more can be said. By Proposition 2 we know that $C_{\Omega_t} = C_\Omega$ for all $t \in [0, r_\Omega)$. By extending the above argument to all such t, we obtain that, if (25) holds, then the map $t \mapsto \frac{w_p(\Omega_t)|\partial\Omega_t|^q}{|\Omega_t|^{q+1}}$ is strictly increasing for $t \in [0, r_\Omega)$. In particular, by (25),

$$\frac{w_p(\Omega_\varepsilon)|\partial\Omega_\varepsilon|^q}{|\Omega_\varepsilon|^{q+1}} > \frac{w_p(\Omega)|\partial\Omega|^q}{|\Omega|^{q+1}} > \frac{2}{q+2} \quad \forall\varepsilon \in (0, r_\Omega].$$

So, if $\Omega_{r_\Omega} \in \mathscr{S}$, we are done since it violates (11). At $t = r_\Omega$ the number of sides of Ω_t varies. If $\Omega_{r_\Omega} \notin \mathscr{S}$, we repeat the previous argument to the next interval where C_{Ω_t} remains constant. Again, the map $t \mapsto \frac{w_p(\Omega_t)|\partial\Omega_t|^q}{|\Omega_t|^{q+1}}$ is strictly increasing on such interval. In view of Proposition 4, this procedure enables us to obtain some polygonal stadium such that (25) holds.

Second case: Assume that

$$\frac{w_p(\Omega)|\partial\Omega|^q}{|\Omega|^{q+1}} \leq \frac{1}{q+1}. \tag{26}$$

Hence,

$$\left[(q+1)\frac{|\partial\Omega|^{q+1}}{|\Omega|^{q+2}}w_p(\Omega) - \frac{|\partial\Omega|}{|\Omega|} - 2q\frac{C_\Omega w_p(\Omega)|\partial\Omega|^{q-1}}{|\Omega|^{q+1}}\right]$$
$$= (q+1)\frac{|\partial\Omega|}{|\Omega|}\left[\frac{|\partial\Omega|^q}{|\Omega|^{q+1}}w_p(\Omega) - \frac{1}{q+1} - \frac{2q}{q+1}\frac{C_\Omega w_p(\Omega)|\partial\Omega|^{q-2}}{|\Omega|^q}\right] < 0.$$

Inserting this into (24) and arguing as in the previous case, we see that the map $t \mapsto \frac{w_p(\Omega_t)|\partial\Omega_t|^q}{|\Omega_t|^{q+1}}$ is strictly decreasing for $t \in [0, R_\Omega)$. In view of Proposition 4, this proves that there exists some polygonal stadium such that (26) holds.

Step 3: explicit computation for a polygonal stadium. Let $\Omega = P^\ell \in \mathscr{S}$ be a polygonal stadium. We are going to derive an explicit expression for the function

$$F(\ell) := \frac{w_p(P^\ell)|\partial P^\ell|^q}{|P^\ell|^{q+1}} \qquad \forall \ell \geq 0.$$

We point out that, in the special case $\ell = 0$, $\Omega \in \mathscr{P} \cap \mathscr{C}_o$ (namely Ω is a circumscribed polygon), and it is proven in [7, Proposition 2] that

$$\forall \Omega \in \mathscr{C}_o, \qquad \frac{w_p(\Omega)|\partial \Omega|^q}{|\Omega|^{q+1}} = \frac{2}{q+2}. \tag{27}$$

In particular, formula (27) shows that the upper bound in (11) is achieved when $\Omega \in \mathscr{C}_o$.

We now show that the above formula can be suitably extended also to the case $\ell > 0$. Our starting point is the representation formula (5). Therein, we use the Steiner formulae (19); in particular, by Propositions 2 and 3, we know that $C_{\Omega_t} \equiv C_\Omega$ for every $t \in [0, R_\Omega)$. Moreover, since $P \in \mathscr{P} \cap \mathscr{C}_o$, we can exploit identities (22). Setting for brevity

$$A := |P|, \qquad R := R_P, \qquad x := \frac{2R\ell}{A},$$

we obtain

$$\begin{aligned} F(\ell) &= \frac{(2\frac{A}{R} + 2\ell)^q}{(A + 2R\ell)^{q+1}} \int_0^R \frac{(A + 2R\ell - 2\ell t - 2\frac{A}{R}t + \frac{A}{R^2}t^2)^q}{(2\ell + 2\frac{A}{R} - 2\frac{A}{R^2}t)^{q-1}}\, dt \\ &= \frac{(x+2)^q}{(x+1)^{q+1}} \int_0^1 \frac{(1 + x - xt - 2t + t^2)^q}{(x + 2 - 2t)^{q-1}}\, dt \\ &= \frac{(x+2)^q}{(x+1)^{q+1}} \int_0^1 \frac{t^q (x+t)^q}{(x+2t)^{q-1}}\, dt. \end{aligned} \tag{28}$$

Of course, taking $x = 0$ in (28) gives again (27); on the other hand, taking $x \to \infty$ gives the asymptotic behavior for thinning polygonal stadiums.

Step 4. In view of equality (28) obtained in Step 3, the estimate (11) will be proved for any polygonal stadium, provided we show that for all $q \in (1, +\infty)$ one has

$$\frac{1}{q+1} < \frac{(x+2)^q}{(x+1)^{q+1}} \int_0^1 \frac{t^q (x+t)^q}{(x+2t)^{q-1}}\, dt < \frac{2}{q+2} \qquad \forall x \in (0, +\infty). \tag{29}$$

With the change of variables $t = xs$, the inequalities in (29) become

$$\begin{aligned} &\frac{1}{q+1} \frac{(x+1)^{q+1}}{x^{q+2}(x+2)^q} \\ &< \int_0^{1/x} \frac{s^q (1+s)^q}{(1+2s)^{q-1}}\, ds < \frac{2}{q+2} \frac{(x+1)^{q+1}}{x^{q+2}(x+2)^q} \qquad \forall x \in (0, +\infty). \end{aligned}$$

In turn, by putting $y = 1/x$, the latter inequalities become

$$\frac{1}{q+1}\frac{y^{q+1}(1+y)^{q+1}}{(1+2y)^q} < \int_0^y \frac{s^q(1+s)^q}{(1+2s)^{q-1}}\,ds < \frac{2}{q+2}\frac{y^{q+1}(1+y)^{q+1}}{(1+2y)^q} \quad \forall y\in(0,+\infty). \tag{30}$$

In order to prove the right inequality in (30), consider the function

$$\Phi(y) := \int_0^y \frac{s^q(1+s)^q}{(1+2s)^{q-1}}\,ds - \frac{2}{q+2}\frac{y^{q+1}(1+y)^{q+1}}{(1+2y)^q} \qquad y\in(0,+\infty)$$

and we need to prove that $\Phi(y)<0$ for all $y>0$. This is a consequence of the two following facts:

$$\Phi(0)=0, \qquad \Phi'(y) = -\frac{q}{q+2}\frac{y^q(1+y)^q}{(1+2y)^{q+1}} < 0.$$

In order to prove the left inequality in (30), consider the function

$$\Psi(y) := \int_0^y \frac{s^q(1+s)^q}{(1+2s)^{q-1}}\,ds - \frac{1}{q+1}\frac{y^{q+1}(1+y)^{q+1}}{(1+2y)^q} \qquad y\in(0,+\infty)$$

and we need to prove that $\Psi(y)>0$ for all $y>0$. This is a consequence of the two following facts:

$$\Psi(0)=0, \qquad \Psi'(y) = \frac{2q}{q+1}\frac{y^{q+1}(1+y)^{q+1}}{(1+2y)^{q+1}} > 0.$$

Both inequalities in (30) are proved and (29) follows.

We point out that, in the case $q=2$, some explicit computations give the stronger result that the map

$$x \mapsto \frac{(x+2)^q}{(x+1)^{q+1}}\int_0^1 \frac{t^q(x+t)^q}{(x+2t)^{q-1}}\,dt$$

is decreasing. We believe that this is true for any q, but we do not have a simple proof of this property.

Step 5: conclusion. Let $\Omega\in\mathscr{P}$ and assume for contradiction that Ω violates (11). Then by Step 2 we know that there exists a polygonal stadium which also violates (11). This contradicts Step 4, see (29). We have so far proved that (11) holds for all $\Omega\in\mathscr{P}$. By a density argument we then infer that

$$\forall\Omega\in\mathscr{C}, \quad \frac{1}{q+1} \le \frac{w_p(\Omega)|\partial\Omega|^q}{|\Omega|^{q+1}} \le \frac{2}{q+2}. \tag{31}$$

Therefore, in order to complete the proof we need to show that the left inequality in (31) is strict. Assume for contradiction that there exists $\Omega\in\mathscr{C}$ such that

$$\frac{w_p(\Omega)|\partial\Omega|^q}{|\Omega|^{q+1}} = \frac{1}{q+1}. \tag{32}$$

Take any sequence $\Omega^k\in\mathscr{P}$ such that $\Omega^k\supset\Omega$ and $\Omega^k\to\Omega$ in the Hausdorff topology. Similar computations as in (23), combined with (20), enable us to obtain

$$\frac{w_p(\Omega_\varepsilon^k)|\partial\Omega_\varepsilon^k|^q}{|\Omega_\varepsilon^k|^{q+1}} \leq \frac{[w_p(\Omega^k) - \int_0^\varepsilon \frac{|\Omega_t^k|^q}{|\partial\Omega_t^k|^{q-1}}dt][|\partial\Omega^k| - 2\pi\varepsilon]^q}{[|\Omega^k| - |\partial\Omega^k|\varepsilon]^{q+1}}$$
$$\leq \frac{w_p(\Omega^k)|\partial\Omega^k|^q}{|\Omega^k|^{q+1}}$$
$$+\left[(q+1)\frac{|\partial\Omega^k|^{q+1}}{|\Omega^k|^{q+2}}w_p(\Omega^k) - \frac{|\partial\Omega^k|}{|\Omega^k|} - 2q\frac{\pi w_p(\Omega^k)|\partial\Omega^k|^{q-1}}{|\Omega^k|^{q+1}}\right]\varepsilon + \alpha\varepsilon^2,$$

where α is some positive constant, depending on Ω but not on k. Therefore, since $\Omega_t^k \to \Omega_t$ for all $t \in [0, R_\Omega]$, we have

$$\frac{w_p(\Omega_\varepsilon)|\partial\Omega_\varepsilon|^q}{|\Omega_\varepsilon|^{q+1}} - \frac{w_p(\Omega)|\partial\Omega|^q}{|\Omega|^{q+1}}$$
$$= \frac{w_p(\Omega_\varepsilon^k)|\partial\Omega_\varepsilon^k|^q}{|\Omega_\varepsilon^k|^{q+1}} - \frac{w_p(\Omega^k)|\partial\Omega^k|^q}{|\Omega^k|^{q+1}} + o(1)$$
$$\leq \left[o(1) - 2q\frac{\pi w_p(\Omega^k)|\partial\Omega^k|^{q-1}}{|\Omega^k|^{q+1}}\right]\varepsilon + \alpha\varepsilon^2 + o(1)$$

where $o(1)$ are infinitesimals (independent of ε) as $k \to \infty$. Hence, by letting $k \to \infty$ and taking ε sufficiently small, we obtain $\frac{w_p(\Omega_\varepsilon)|\partial\Omega_\varepsilon|^q}{|\Omega_\varepsilon|^{q+1}} < \frac{1}{q+1}$, which contradicts (31).

4.2 *Proof of Theorem 1*

The inequalities (10) follow directly from (11) and (6) so we just need to show that they are sharp.

For the right inequality, take a sequence of thinning isosceles triangles T_k. Then, by Theorem 2 we have

$$\frac{w_p(T_k)|\partial T_k|^q}{|T_k|^{q+1}} = \frac{2}{q+2} \quad \text{for all } k.$$

On the other hand, by [7, Proposition 3] and (6) we know that

$$\lim_{k\to\infty}\frac{w_p(T_k)}{\tau_p(T_k)} = \frac{q+1}{2^q}$$

and therefore

$$\lim_{k\to\infty}\frac{\tau_p(T_k)|\partial T_k|^q}{|T_k|^{q+1}} = \frac{2^{q+1}}{(q+2)(q+1)}.$$

For the left inequality, we seek an upper bound for $\tau_p(\Omega)$ by using the maximum principle. For all $\ell \in (0, +\infty)$ let $\Omega^\ell = (-\frac{\ell}{2}, \frac{\ell}{2}) \times (-1, 1)$ and let u_ℓ be the unique solution to

$$-\Delta_p u_\ell = 1 \quad \text{in } \Omega^\ell, \qquad u_\ell = 0 \quad \text{on } \partial\Omega^\ell.$$

Let $u_\infty(x,y)=\frac{p-1}{p}(1-|y|^{p/(p-1)})$ so that

$$-\Delta_p u_\infty = 1 \quad \text{in } \Omega^\ell, \qquad u_\infty \geq 0 \quad \text{on } \partial\Omega^\ell.$$

By the maximum principle, we infer that $u_\infty \geq u_\ell$ in Ω^ℓ so that

$$\begin{aligned}\tau_p(\Omega^\ell) &= \int_{\Omega^\ell} u_\ell \leq \int_{\Omega^\ell} u_\infty = \frac{2(p-1)}{p}\ell \int_0^1 \left(1-y^{p/(p-1)}\right) dy \\ &= \frac{2(p-1)}{2p-1}\ell = \frac{2\ell}{q+1}.\end{aligned}$$

Hence,

$$1 \geq \liminf_{\ell\to\infty} \frac{w_p(\Omega^\ell)}{\tau_p(\Omega^\ell)} \geq \liminf_{\ell\to\infty} \frac{(q+1)w_p(\Omega^\ell)}{2\ell} = 1$$

where the last equality follows from Theorem 2. Combined with Theorem 2, this proves that

$$\lim_{\ell\to\infty} \frac{\tau_p(\Omega^\ell)|\partial\Omega^\ell|^q}{|\Omega^\ell|^{q+1}} = \frac{1}{q+1}.$$

4.3 Proof of Theorem 3

Since it follows closely the proof of Theorem 1, we just sketch it. We first prove the counterpart of Theorem 2 and we follow the same steps.

Step 1. Given $\Omega \in \mathscr{P}$ and using $R_{\Omega_\varepsilon} = R_\Omega - \varepsilon + o(\varepsilon)$ we prove:

$$\frac{w_p(\Omega_\varepsilon)}{R^q_{\Omega_\varepsilon}|\Omega_\varepsilon|} - \frac{w_p(\Omega)}{R^q_\Omega|\Omega|} = \frac{\varepsilon}{R^q_\Omega|\Omega|}\left(w_p(\Omega)\left[\frac{q}{R_\Omega} + \frac{|\partial\Omega|}{|\Omega|}\right] - \frac{|\Omega|^q}{|\partial\Omega|^{q-1}}\right) + o(\varepsilon). \tag{33}$$

Step 2. We prove that, if (16) fails for some $\Omega \in \mathscr{P}$, then it also fails for a polygonal stadium. To that end, we estimate the sign in (33) with the help of the following classical geometric inequalities (see [1])

$$\forall \Omega \in \mathscr{C}, \qquad \frac{|\Omega|}{R_\Omega} < |\partial\Omega| \leq \frac{2|\Omega|}{R_\Omega}.$$

Step 3. Again, explicit computations can be done for a polygonal stadium, and with the same notation as in the proof of Theorem 2, we get:

$$\frac{w_p(P^\ell)}{R^q_{P^\ell}|P^\ell|} = \frac{1}{x+1}\int_0^1 \frac{t^q(x+t)^q}{(x+2t)^{q-1}}\,dt \quad \forall P^\ell \in \mathscr{S}.$$

Step 4. In view of Step 3, estimate (16) is proved for any polygonal stadium, provided for all $q \in (1,+\infty)$ one has

$$\frac{1}{(q+2)2^{q-1}} < \frac{1}{x+1}\int_0^1 \frac{t^q(x+t)^q}{(x+2t)^{q-1}}\,dt < \frac{1}{q+1} \qquad \forall x \in (0,+\infty). \tag{34}$$

With the change of variables $t = xs$ and putting $y = 1/x$, the inequalities in (34) become

$$\frac{y^{q+2}+y^{q+1}}{(q+2)2^{q-1}} < \int_0^y \frac{s^q(1+s)^q}{(1+2s)^{q-1}}\,ds < \frac{y^{q+2}+y^{q+1}}{q+1} \qquad \forall y \in (0,+\infty). \tag{35}$$

Some tedious but straightforward computations show that

$$\frac{y^{q+1}}{2^{q-1}} + \frac{q+1}{(q+2)2^{q-1}}y^q < \frac{y^q(1+y)^q}{(1+2y)^{q-1}} < \frac{q+2}{q+1}y^{q+1} + y^q \quad \forall y \in (0,+\infty)$$

and (35) follows after integration over $(0, y)$.

Step 5. The previous steps leads to (16) for polygons and by density for convex domains. The strict right inequality in (16) can be obtained by reproducing carefully the computations in Step 1, similarly as done in Step 5 of Sect. 4.1.

Now the counterpart of Theorem 2 is proved, and we may use (6) in order to get (15) from (16). Balls realize equality in the left inequality of (15) because they are at the same time circumscribed and maximal for the quotient w_p/τ_p.

5 Some Open Problems

We briefly suggest here some perspectives which might be considered, in the light of our results.

Sharp bounds for the p-torsion in higher dimensions. In higher dimensions the shape functionals τ_p and w_p can be defined in the analogous way as for $n = 2$. In [3], Crasta proved the following sharp bounds:

$$\forall \Omega \text{ bounded convex set } \subset \mathbb{R}^n, \qquad \frac{n+1}{2n} < \frac{w_2(\Omega)}{\tau_2(\Omega)} \le 1.$$

Therefore it seems natural to ask: what kind of isoperimetric inequality can be proved for w_p and τ_p among convex sets in $\mathbb{R}^n$? In this direction, let us quote an inequality proven in [8], obtained by a strategy similar to our approach, that is by looking at the level sets of the support function:

$$\forall \Omega \text{ bounded convex set } \subset \mathbb{R}^n, \quad \frac{\tau_2(\Omega)|\partial\Omega|}{R_\Omega|\Omega|^2} \ge \frac{\tau_2(B)|\partial B|}{R_B|B|^2} \quad \big(B \text{ is a ball of } \mathbb{R}^n\big).$$

Sharp bounds for the principal frequency. A notion of "web principal frequency" can be defined (in any space dimension) similarly as done for the web torsion, that is

$$\lambda_1^+(\Omega) := \inf\left\{\frac{\int_\Omega |\nabla u|^2}{\int_\Omega u^2} : u \in \mathscr{W}_2(\Omega)\right\}.$$

Writing the optimality condition in the space $\mathscr{W}_2(\Omega)$, one can express $\lambda_1^+(\Omega)$ as

$$\lambda_1^+(\Omega) = \inf\left\{\frac{\int_0^{R_\Omega} \alpha\rho'^2}{\int_0^{R_\Omega} \alpha\rho^2} : \rho \in H^1(0,R_\Omega),\ \rho(0)=0\right\}, \qquad \text{where } \alpha(t) = |\partial\Omega_t|.$$

It is clear that $\lambda_1^+(\Omega) \geq \lambda_1(\Omega)$, with equality sign when Ω is a ball. On the other hand, the following questions can be addressed:

- Find a sharp bound from above for the ratio $\lambda_1^+(\Omega)/\lambda_1(\Omega)$ among bounded convex subsets of $\mathbb{R}^n$.
- Is it possible to apply successfully the same strategy of this paper, that is find sharp bounds for $\lambda_1^+(\Omega)$ and then use the estimates on the ratio $\lambda_1^+(\Omega)/\lambda_1(\Omega)$, to deduce sharp bounds for $\lambda_1(\Omega)$? In particular, this approach might allow to retrieve the following known inequalities holding for any bounded convex domain $\Omega \subset \mathbb{R}^2$ (see [11, 12, 14]):

$$\frac{\pi^2}{16} \leq \lambda_1(\Omega)\frac{|\Omega|^2}{|\partial\Omega|^2} \leq \frac{\pi^2}{4} \quad \text{and} \quad \frac{\pi^2}{4} \leq \lambda_1(\Omega)R_\Omega^2 \leq j_0^2.$$

Acknowledgements This work was realized thanks to the financial support of the Italian Gruppo Nazionale per l'Analisi Matematica, la Probabilità e le loro Applicazioni, as part of the project "Problemi di ottimizzazione e disuguaglianze geometrico-funzionali", and by the French Agence Nationale de la Recherche, as part of the project ANR-09-BLAN-0037 "Geometric analysis of optimal shapes".

References

1. Bonnesen, T., Fenchel, W.: Theory of Convex Bodies. BSC Associates, Moscow (1987)
2. Brandolini, B., Nitsch, C., Trombetti, C.: An upper bound for nonlinear eigenvalues on convex domains by means of the isoperimetric deficit. Arch. Math. **94**, 391–400 (2010)
3. Crasta, G.: Estimates for the energy of the solutions to elliptic Dirichlet problems on convex domains. Proc. R. Soc. Edinb., Sect. A **134**, 89–107 (2004)
4. Crasta, G., Gazzola, F.: Web functions: survey of results and perspectives. Rend. Ist. Mat. Univ. Trieste **33**, 313–326 (2001)
5. Crasta, G., Gazzola, F.: Some estimates of the minimizing properties of web functions. Calc. Var. **15**, 45–66 (2002)
6. Crasta, G., Fragalà, I., Gazzola, F.: A sharp upper bound for the torsional rigidity of rods by means of web functions. Arch. Ration. Mech. Anal. **164**, 189–211 (2002)
7. Crasta, G., Fragalà, I., Gazzola, F.: On the role of energy convexity in the web function approximation. Nonlinear Differ. Equ. Appl. **12**, 93–109 (2005)
8. Freitas, P., Krejčiřík, D.: A sharp upper bound for the first Dirichlet eigenvalue and the growth of the isoperimetric constant of convex domains. Proc. Am. Math. Soc. **136**, 2997–3006 (2008)
9. Gazzola, F.: Existence of minima for nonconvex functionals in spaces of functions depending on the distance from the boundary. Arch. Ration. Mech. Anal. **150**, 57–76 (1999)
10. Henrot, A.: Extremum Problems for Eigenvalues of Elliptic Operators. Birkhäuser, Basel (2006)
11. Hersch, J.: Sur la fréquence fondamentale d'une membrane vibrante: évaluations par défaut et principe de maximum. Z. Angew. Math. Phys. **11**, 387–413 (1960)
12. Makai, E.: On the Principal Frequency of a Membrane and the Torsional Rigidity of a Beam. Studies in Mathematical Analysis and Related Topics, pp. 227–231. Stanford University Press, Stanford (1962)
13. Payne, L.E., Weinberger, H.F.: Some isoperimetric inequalities for membrane frequencies and torsional rigidity. J. Math. Anal. Appl. **2**, 210–216 (1961)

14. Pólya, G.: Two more inequalities between physical and geometrical quantities. J. Indian Math. Soc. **24**, 413–419 (1960)
15. Pólya, G., Szegö, G.: Isoperimetric Inequalities in Mathematical Physics. Princeton University Press, Princeton (1951)
16. Solynin, A.Y., Zalgaller, V.A.: An isoperimetric inequality for logarithmic capacity of polygons. Ann. of Math. (2) **159**, 277–303 (2004)

Geometric Analysis of Fractional Phase Transition Interfaces

Giovanni Franzina and Enrico Valdinoci

Abstract We discuss some recent results on phase transition models driven by non-local operators, also in relation with their limit (either local or nonlocal) interfaces.

Keywords Fractional Laplacian · Nonlocal minimal surfaces · Asymptotics · Rigidity and regularity theory

1 The Fractional Laplacian Operator

This note is devoted to report some recent advances concerning the fractional powers of the Laplace operator and some related problems arising in pde's and geometric measure theory. Namely, the s-Laplacian of a (sufficiently regular) function u can be defined as an integral in the principal value sense by the formula

$$\begin{aligned} -(-\Delta)^s u(x) &:= C_{n,s}\,\text{P.V.} \int_{\mathbb{R}^n} \frac{u(x+y)-u(x)}{|y|^{n+2s}}\,dy \\ &:= C_{n,s} \lim_{\varepsilon\to 0^+} \int_{\mathbb{R}^n\setminus B_\varepsilon} \frac{u(x+y)-u(x)}{|y|^{n+2s}}\,dy, \end{aligned} \tag{1}$$

where $s \in (0,1)$, and

$$C_{n,s} = \pi^{-2s+n/2} \frac{\Gamma(n/2+s)}{\Gamma(-s)}$$

is a normalization constant (blowing up as $s \to 1^-$ and $s \to 0^+$, because of the singularities of the Euler Γ-function). Note that the integral here is singular in the

G. Franzina (✉)
Dipartimento di Matematica, Università degli Studi di Trento, Via Sommarive 14, 38050 Povo, Trento, Italy
e-mail: giovanni.franzina@unitn.it

E. Valdinoci
Dipartimento di Matematica, Università degli Studi di Milano, Via Cesare Saldini 50, 20133 Milano, Italy
e-mail: enrico.valdinoci@unimi.it

R. Magnanini et al. (eds.), *Geometric Properties for Parabolic and Elliptic PDE's*, Springer INdAM Series 2, DOI 10.1007/978-88-470-2841-8_8,

case $s \geq 1/2$, but it converges if $s < 1/2$ as it can be estimated via an elementary argument by splitting the domain of integration.

We point out that an equivalent definition may be given by integrating against a singular kernel, which suitably averages a second-order incremental quotient. Indeed, thanks to the symmetry of the kernel under the map $y \mapsto -y$, by performing a standard change of variables, one obtains

$$\int_{\mathbb{R}^n \setminus B_\varepsilon} \frac{u(x+y)-u(x)}{|y|^{n+2s}}\,dy = \frac{1}{2}\int_{\mathbb{R}^n \setminus B_\varepsilon} \frac{u(x+y)+u(x-y)-2u(x)}{|y|^{n+2s}}\,dy,$$

for all $\varepsilon > 0$; thus (1) becomes

$$-(-\Delta)^s u(x) = \frac{C_{n,s}}{2}\int_{\mathbb{R}^n} \frac{u(x+y)+u(x-y)-2u(x)}{|y|^{n+2s}}\,dy. \tag{2}$$

It is often convenient to use the expression in (2), that deals with a convergent Lebesgue integral, rather than the one in (1), that needs a principal value to be well-posed.

The fractional Laplacian may be equivalently defined by means of the Fourier symbol $|\xi|^{2s}$ by simply setting, for every $s \in (0, 1)$,

$$\mathscr{F}\big((-\Delta)^s u\big) = \big(|\xi|^{2s}(\mathscr{F}u)\big), \tag{3}$$

for all $u \in \mathscr{S}'(\mathbb{R}^n)$, and this occurs in analogy with the limit case $s = 1$, when (3) is consistent with the well-known behavior of the (distributional) Fourier transform $\mathscr{F}$ on Laplacians.

Note that this operator is invariant under the action of the orthogonal transformations in $\mathbb{R}^n$ and that the following scaling property holds:

$$(-\Delta)^s u_\lambda(x) = \lambda^{2s}\big((-\Delta)^s u\big)(\lambda x), \quad \text{for all } x \in \mathbb{R}^n, \tag{4}$$

were we denoted $u_\lambda(x) = u(\lambda x)$. For the basics of the fractional Laplace operators and related functional settings see, for instance, [30] and references therein.

After being studied for a long time in potential theory and harmonic analysis, fractional operators defined via singular integrals are riveting attention due to the pliant use that can be made of their nonlocal nature and their applications to models of concrete interest. In particular, equations involving the fractional Laplacian or similar nonlocal operators naturally surface in several applications. For instance, the s-Laplacian is an interesting specific example of infinitesimal generator for rotationally invariant $2s$-stable Lévy processes, taking the "hydrodynamic limit" of the discrete random walk with possibly long jumps: in this case the probability density is described by a fractional heat equation where the classical Laplace operator is replaced by the s-Laplacian, see e.g. [60] for details. An analogous fractional diffusion arises in the asymptotic analysis of the distribution associated with some collision operators in kinetic theory, see [45]. Also, lower dimensional obstacle problems and the fractional Laplacian were intensively studied, see e.g. [57]; for the regularity of the solutions and of that of the free boundary in the obstacle problem and in the thin obstacle problem, we refer to [20]. Nonlocal operators arise also in elasticity problems [56], and in several phenomena, such as water waves [24, 25], flame propagation [22], stability of matter [36], quasi-geostrophic flows [43], crystal dislocation [59], soft thin films [42], stratified materials [51] and others.

The fractional Laplacian presents several technical and conceptual difficulties. First of all, the operator is nonlocal, hence, during the computations, one needs to estimate also the contribution coming from far. Also, since integrating is usually harder than differentiating, constructing barriers and checking the existence of sub/supersolutions is often much harder than in the case of the Laplacian. Furthermore, a psychological unease may arise from the fact that an "integral" operator behaves in fact as a "differential" one. The counterpart of this difficulties lies in the nice averaging properties of the fractional Laplacian, that makes a function revert to its nearby mean, and this has somewhat to control the oscillations.

2 Back to the Laplacian Case

2.1 The Allen-Cahn Equation

In our opinion, an interesting topic of research involving the s-Laplacian, $s \in (0,1)$, concerns the analysis of the geometric properties of the solutions of the equation

$$(-\Delta)^s u = u - u^3. \tag{5}$$

The formal limit of Eq. (5) for $s \to 1^-$ is the well-studied Allen-Cahn (or scalar Ginzburg-Landau) equation

$$-\Delta u = u - u^3 \tag{6}$$

which describes (among other things) a two-phase model, where, roughly speaking, the pure phases correspond to the states $u \sim +1$ and $u \sim -1$ and the level set $\{u = 0\}$ is the interface which separates the pure phases.

2.2 A Conjecture of De Giorgi

In 1978 De Giorgi [26] conjectured that if u is a smooth, bounded solution of equation (6) in the whole of $\mathbb{R}^n$ and it is monotone in the variable x_n (i.e. $\partial_{x_n} u(x) > 0$ for any $x \in \mathbb{R}^n$), then u depends in fact only on one Euclidean variable and its level sets are hyperplanes—at least if $n \leq 8$.

This conjecture seems to have a strong analogy with the celebrated Bernstein problem, which claims that all entire minimal graphs in $\mathbb{R}^n$ are hyperplanes: indeed, it is well-known that this property holds true for all $n \leq 8$ but it is false for $n \geq 9$.

Similarly, it is known that the conjecture of De Giorgi is true if $n \leq 3$: in the case $n = 2$ this was proved by N. Ghoussoub and C. Gui [40] (see also the paper of Berestycki, Caffarelli and Nirenberg [9]), whereas the result for $n = 3$ is due to Ambrosio and Cabré [5] (see also the paper of Alberti, Ambrosio and Cabré [4]). In his PhD Thesis [50], Savin proved that the claim holds true also for $4 \le n \le 8$ under the additional assumption

$$\lim_{x_n \to \pm\infty} u\big(x', x_n\big) = \pm 1, \quad \text{for all } x' \in \mathbb{R}^{n-1}. \tag{7}$$

On the other hand the same conjecture turns out to be false in general if $n \ge 9$: the final answer was given by Del Pino, Kowalczyk and Wei, see [27–29] by manufacturing a solution to the Allen-Cahn equation, monotone in the direction of x_9, whose zero level set lies closely to the same graph exhibited as a counterexample to the Bernstein problem in $\mathbb{R}^9$ by Bombieri, De Giorgi and Giusti in [11].

Nevertheless, if one assumes the limit (7) to hold uniformly for $x' \in \mathbb{R}^{n-1}$, the conjecture (which follows under the name of Gibbons conjecture in this case and it has some importance in cosmology) is true for all dimension $n \in \mathbb{N}$, see [7, 10, 32]. We point out that the problem is still open in dimension $4 \le n \le 8$ if the extra assumptions are dropped. For more information, see [33].

3 Some Research Lines for $(-\Delta)^s$

3.1 The Symmetry Problem for $(-\Delta)^s$

One can ask a question similar to the one posed by De Giorgi for the fractional Laplacian: that is, given $s \in (0, 1)$ and u a smooth, bounded solution of

$$(-\Delta)^s u = u - u^3$$

in the whole of $\mathbb{R}^n$, with the monotonicity condition

$$\partial_{x_n} u > 0,$$

we may wonder whether or not u depends only on one Euclidean variable and its level sets are hyperplanes—at least in small dimension.

In this framework, the first positive answer was given in the pioneering work of Cabré and Solà-Morales [16] when $n = 2$ and $s = 1/2$.

The answer to this question is also positive when $n = 2$ for any $s \in (0, 1)$, see [15, 58], and when $n = 3$ and $s \in [1/2, 1)$, see [13, 14].

Moreover, as it happens in the classical case when $s = 1$, the answer is positive for any $n \in \mathbb{N}$ and any $s \in (0, 1)$ if one assumes the limit condition in (7) to hold uniformly for $x' \in \mathbb{R}^{n-1}$ (this is a byproduct of the results in [15, 34]).

The problem is open for $n \ge 4$ and any $s \in (0, 1)$, and even for $n = 3$ and $s \in (0, 1/2)$. No counterexample is known, in any dimension.

It is worth pointing out that monotone solutions of fractional Allen Cahn equation are local minimizers of a fractional energy, see [49].

3.2 Γ-Convergence for $(-\Delta)^s$

In this section we discuss the asymptotics of a variational problem related to the fractional Laplacian, and specifically to the fractional Allen-Cahn equation (5). Namely, we will consider the free energy defined by

$$\mathfrak{I}_s^{\varepsilon}(u,\Omega) = \varepsilon^{2s}\mathscr{K}_s(u,\Omega) + \int_{\Omega} W(u)\,dx, \tag{8}$$

where, at the right-hand side, the dislocation energy of a suitable double-well potential W vanishing at ± 1 (e.g., $W(u) = (1-u^2)^2/4$) is penalized by a small contribution given by the nonlocal functional

$$\mathscr{K}_s(u,\Omega) = \frac{1}{2}\int_{\Omega}\int_{\Omega}\frac{|u(x)-u(y)|^2}{|x-y|^{n+2s}}\,dx\,dy + \int_{\Omega}\int_{\mathscr{C}\Omega}\frac{|u(x)-u(y)|^2}{|x-y|^{n+2s}}\,dx\,dy, \tag{9}$$

where $s \in (0,1)$. The presence of a nonlocal contribution here, due to the term $\mathscr{K}_s$, constitutes the main difference of $\mathfrak{I}_s^{\varepsilon}$ from the perturbed energies arising in the standard theory of phase transitions, such as the functional

$$\varepsilon^2\int_{\Omega}|\nabla u|^2\,dx + \int_{\Omega} W(u)\,dx,$$

where the potential energy, which is minimized by the system at the equilibrium, is compensated by a term proportional (up to a small factor representing the surface tension coefficient) to the perimeter of the interface between the two phases. In the nonlocal model in (8), this gradient term is replaced by the fractional Sobolev seminorm of u, which is responsible for the effects of the long-range particle interactions and affects the interface between the two phases, which can have a *fractional dimension* (see [61, 62] for instance). Such models—or some variation of them obtained after replacing the singular kernel in (8) by a suitable anisotropic Kac potential—under suitable boundary conditions arise in the study of surface tension effects, and they have been investigated, jointly with the limit properties of functionals in the sense of Γ-convergence, by many authors: we mention [1–3, 37–39, 41] among the others.

Moreover, as a robust Γ-convergence theory is available, it is customary to discuss the asymptotic behavior of the latter model problems as $\varepsilon \to 0^+$: for instance, in the classical case of $s = 1$, i.e. of the Allen-Cahn equation (6), the Γ-limit is strictly related to the perimeter of a set in the sense of De Giorgi (see [46–48]). On the other hand, when dealing with the nonlocal functional in (8), the Γ-convergence to the classical perimeter functional holds true provided that $s \in [1/2, 1)$, while for all $s \in (0, 1/2)$ the Γ-limit is related to a suitable nonlocal perimeter that will be introduced and discussed in the incoming Sect. 3.4. From a physical point of view, these results locate at $s = 1/2$ a critical threshold[1] for the size of the range

[1]The reader has noticed that this is the same threshold for the known results discussed in Sect. 3.1 when $n = 3$. On the other hand, while the threshold $s = 1/2$ is optimal here, the optimality for the results of Sect. 3.1 is completely open. Further regularity results for the limit interface when $s \to (1/2)^-$ will be discussed in Sect. 3.5.

of all possible interactions among particles contributing to affect the limit interface. We recall that in the nonlocal case, in analogy with the classical case $s = 1$, by the scaling property (4), the solution obtained passing to the microscopic variables $v(x) = u(x/\varepsilon)$ satisfies

$$\varepsilon^{2s}(-\Delta)^s v = v - v^3 \quad \text{in } \mathbb{R}^n,$$

and the latter can be regarded to as the Euler-Lagrange equation associated with the variational problem of minimizing the scaled energy $\mathfrak{I}_s^\varepsilon(\cdot, \mathbb{R}^n)$ where the potential is $W(u) = (1 - u^2)^2/4$.

To make all these statements precise, first of all one has to specify the metric space where the functionals involved in the Γ-limit are defined: we will denote by X the metric space given by the set

$$\{u \in L^\infty(\mathbb{R}^n) : \|u\|_\infty \le 1\},$$

equipped with the convergence in $L^1_{loc}(\mathbb{R}^n)$. For all $\varepsilon > 0$ we define a functional $\mathscr{F}_s^\varepsilon : X \to \mathbb{R} \cup \{+\infty\}$ by setting

$$\mathscr{F}_s^\varepsilon(u, \Omega) = \begin{cases} \varepsilon^{-2s}\mathfrak{I}_s^\varepsilon(u, \Omega), & \text{if } s \in (0, 1/2), \\ |\varepsilon \log \varepsilon|^{-1}\mathfrak{I}_s^\varepsilon(u, \Omega), & \text{if } s = 1/2, \\ \varepsilon^{-1}\mathfrak{I}_s^\varepsilon(u, \Omega), & \text{if } s \in (1/2, 1). \end{cases}$$

The proof of the following Theorems 1 and 2 can then be found in [53].

Theorem 1 *Let Ω be a bounded open set in $\mathbb{R}^n$, $\varepsilon \in (0, 1]$ and $s \in [1/2, 1)$. Then, there exists a constant $c_* > 0$, possibly depending on W and s, such that the functional $\mathscr{F}_s^\varepsilon$ of X to $\mathbb{R} \cup \{+\infty\}$ defined by (8) Γ-converges in X to the functional $\mathscr{F}_L$ of X to $\mathbb{R} \cup \{+\infty\}$ defined by*

$$\mathscr{F}_L(u, \Omega) = \begin{cases} c_* \operatorname{Per}(E, \Omega), & \text{if } u_{|\Omega} = \chi_E - \chi_{\mathscr{C}E}, \text{ for some } E \subset \Omega, \\ +\infty, & \text{otherwise}, \end{cases} \tag{10}$$

for[2] all $u \in X$.

Theorem 2 *Let Ω be a bounded open set in $\mathbb{R}^n$, $\varepsilon \in (0, 1]$ and $s \in (0, 1/2)$. Then, the functional $\mathscr{F}_s^\varepsilon$ of X to $\mathbb{R} \cup \{+\infty\}$ defined by (8) Γ-converges in X to the functional $\mathscr{F}_L$ of X to $\mathbb{R} \cup \{+\infty\}$ defined by*

$$\mathscr{F}_{NL}^\varepsilon(u, \Omega) = \begin{cases} \mathscr{K}_s(u, \Omega), & \text{if } u_{|\Omega} = \chi_E - \chi_{\mathscr{C}E}, \text{ for some } E \subset \Omega, \\ +\infty, & \text{otherwise}, \end{cases} \tag{11}$$

for all $u \in X$.

The proof of the Γ-convergence stated in Theorem 2 for the case $s \in (0, 1/2)$ turns out to be quite direct, while a much finer analysis is needed in order to deal with the case $s \in [1/2, 1)$. The claim of Theorem 1, besides giving an explicit Γ-limit, says that a localization property occurs as ε goes to zero when $s \in [1/2, 1)$.

Moreover, the following pre-compactness criterion was proved in [53]:

[2] As usual, χ_E denotes the characteristic function of the set E.

Theorem 3 *Let Ω be a bounded open set in $\mathbb{R}^n$, and $s \in (0,1)$. For all sequences $\{u_\varepsilon\}_{\varepsilon>0}$ such that*

$$\sup_{\varepsilon\in(0,1]} \mathscr{F}_s^\varepsilon(u_\varepsilon, \Omega) < \infty,$$

by possibly passing to a subsequence we have that

$$u_\varepsilon \to u_* = \chi_E - \chi_{\mathscr{C}E}, \qquad \text{in } L^1(\Omega), \tag{12}$$

for some $E \subset \Omega$.

Actually, when dealing with minimizers, the convergence in (12) can be enhanced, thanks to the optimal uniform density estimates of [54]. Indeed, it turns out that the level sets of the ε-minimizers u_ε of the scaled functional $\mathscr{F}_s^\varepsilon$ converge to ∂E locally uniformly. The content of such density estimates will be briefly described in the next section.

3.3 Density Estimates for the Level Sets

One of the interesting feature of the solutions to the fractional Allen-Cahn equation is that it is possible to gain informations on the measure occupied by their level sets in a given ball, and this may be interpreted as the probability of finding a phase in a fixed region. In the classical case $s = 1$, corresponding to the Allen-Cahn equation in (6), these density estimates were proved by Caffarelli and Cordoba [17]. A nonlocal counterpart indeed holds, as the following result states, in the case of the fractional Laplacian. The proof, for which we refer to the paper of the second author and Savin [54], makes use of a fine analysis on a weighted double integral and the measure theoretical properties of the minimizers; an alternative proof based on a fractional Sobolev inequality is contained in [52]. In order to obtain the density estimates, the results from the paper [49] are helpful.

Theorem 4 *Let $R > 0$ and u be a minimizer of the functional $\mathfrak{I}_s^1(\cdot, B_R)$. Then, there exist a function $\bar{R} : (-1,1) \times (-1,1) \to (0,+\infty)$ and a positive constant $\bar{c} > 0$, depending on n, s, W, such that if*

$$u(0) > \theta_1,$$

then

$$\left|\{u > \theta_2\} \cap B_R\right| \geq \bar{c}R^n, \tag{13}$$

provided $R \geq \bar{R}(\theta_1, \theta_2)$.

By a scaling argument it follows that if u_ε minimizes the functional $\mathscr{F}_s^\varepsilon(\cdot, B_r)$ and $u_\varepsilon(0) > \theta_1$, then

$$\left|\{u_\varepsilon > \theta_2\} \cap B_r\right| \geq \bar{c}\, r^n,$$

provided that $\bar{R}\varepsilon \leq r$.

3.4 Few Words on the Nonlocal Perimeter

Let $s \in (0, 1/2)$ and E be a measurable subset of $\mathbb{R}^n$. The s-perimeter of a set E in Ω is defined by

$$\begin{aligned}\mathrm{Per}_s(E,\Omega) := &\int_{E\cap\Omega}\int_{(\mathscr{C}E)\cap\Omega}\frac{1}{|x-y|^{n+2s}}\,dy\,dx\\ &+\int_{E\cap\Omega}\int_{(\mathscr{C}E)\cap(\mathscr{C}\Omega)}\frac{1}{|x-y|^{n+2s}}\,dy\,dx\\ &+\int_{E\cap(\mathscr{C}\Omega)}\int_{(\mathscr{C}E)\cap\Omega}\frac{1}{|x-y|^{n+2s}}\,dy\,dx,\end{aligned} \tag{14}$$

for all bounded and connected open set Ω in $\mathbb{R}^n$.

We point out that, in the notation of Sect. 3.2, we have

$$\mathrm{Per}_s(E,\Omega) = \mathscr{K}_s(\chi_E,\Omega).$$

The properties of the above fractional perimeter were studied in [61, 62] where a generalized co-area formula was established and some nonlocal functionals defined similarly as in (14) were used to define a suitable concept of *fractal dimension*.

As customary in calculus of variations, it worths considering the problem of finding local minimizers of the functional $\mathrm{Per}_s(\cdot,\Omega)$, i.e. measurable sets E in $\mathbb{R}^n$ such that

$$\mathrm{Per}_s(E,\Omega) \le \mathrm{Per}_s(F,\Omega), \quad \text{for all } F\subset\mathbb{R}^n \text{ such that } E\cap\mathscr{C}\Omega = F\cap\mathscr{C}\Omega. \tag{15}$$

If E satisfies (15) (with $\mathrm{Per}_s(E,\Omega) < +\infty$), we say that E is s-minimal in Ω. Since the functional $\mathrm{Per}_s(\cdot,\Omega)$ is lower semi-continuous thanks to Fatou's Lemma, the existence of a minimizer E such as in (15) follows by the compact fractional Sobolev embedding. These nonlocal minimal surfaces were introduced by Caffarelli, Roquejoffre and Savin in [21], were the regularity issues were studied; among the several results therein obtained, we may mention the following

Theorem 5 *Let $\Omega = B_1$ be the ball of radius 1 centered at the origin in $\mathbb{R}^n$. If E is a measurable subset of $\mathbb{R}^n$ such that* (15) *holds, then $\partial E\cap B_{1/2}$ is locally a $\mathscr{C}^{1,\alpha}$ hypersurface out of a closed a set $N\subset\partial E$ having finite $(n-2)$-dimensional Hausdorff measure.*

In order to prove Theorem 5, the strategy in [21] is to obtain for all the s-minimizers $E\subset\mathbb{R}^n$ some density estimates and a geometric condition such as the one displayed in Fig. 1. Namely, there is a universal constant $c\in(0,1)$ such that for all points $\xi\in\partial E$ and all $r\in(0,c)$ one can find balls $B_{cr}(a)\subset E\cap B_r(\xi)$, and $B_{cr}(b)\subset(\mathscr{C}E)\cap B_r(\xi)$.

Several improvements of this regularity results have been obtained in the recent years, see [6, 18, 19, 23, 55]. For instance, Caffarelli and the second author have shown in [18] that the constant c of the clean ball condition is uniform as $s\to(1/2)^-$ and they deduced several limit properties of nonlocal minimal surfaces that can be summarized in the following

Fig. 1 Clean ball condition for the s-minimizers

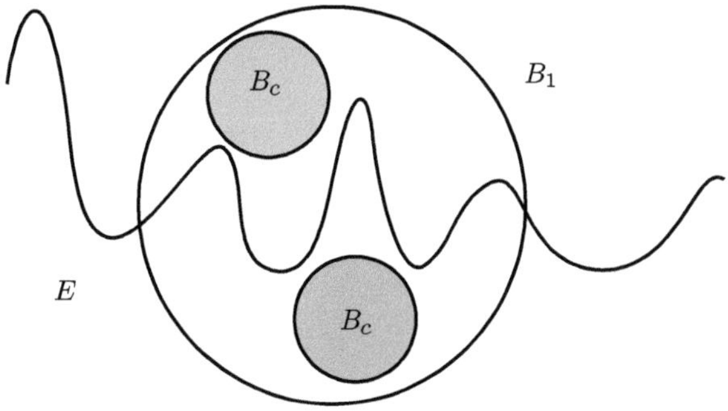

Theorem 6 *Let $R' > R > 0$, $\{s_k\}_{k\in\mathbb{N}} \subset (0, 1/2)$ be a sequence converging to $1/2$ and let us denote by E_k an s_k-minimal set in $B_{R'}$, for all $k \in \mathbb{N}$. Then the following hold:*

- *by possibly passing to a subsequence, the set $\{E_k\}_{k\in\mathbb{N}}$ converge to some limit set E uniformly in B_R;*
- *E is a set of minimal classical perimeter in B_R.*

The convergence of minimizers stated in Theorem 6, was obtained under mild assumptions via dual convergence techniques in the paper by Ambrosio, de Philippis and Martinazzi [6], where the authors proved the equi-coercivity and the Γ-convergence of the fractional perimeter (up to the scaling factor $\omega_{n-1}^{-1}(1/2 - s)$) to the classical perimeter in the sense of De Giorgi, whence they also deduced a local convergence result for minimizers.

In the case of a general (smooth) subset $E \subset \mathbb{R}^n$—possibly being not s-minimal—the s-perimeter can however be related to the classical perimeter in a ball, as the following pointwise convergence result of [18] states:

Theorem 7 *Let $R > 0$ and $E \subset \mathbb{R}^n$ be such that $\partial E \cap B_R$ is $\mathscr{C}^{1,\alpha}$ for some $\alpha \in (0, 1)$. Then, there exists a countable subset $N \subset (0, R)$ such that*

$$\lim_{s\to(1/2)^-} (1/2 - s)\,\mathrm{Per}_s(E, B_r) = \omega_{n-1}\,\mathrm{Per}(E, \Omega), \tag{16}$$

for all $r \in (0, 1) \setminus N$.

The presence of the normalizing factor $(1/2 - s)$, vanishing as $s \to (1/2)^-$, in front of the fractional perimeter here is consistent with the fact that the first integral in (14) diverges when $s = 1/2$ unless either $E \cap \Omega$ or $\mathscr{C}E \cap \Omega$ is empty (see [12], and the remark after Theorem 1 in [6]). In order to prove Theorem 7 one has to estimate the integral contribution to the s-perimeter coming from the smooth transition surface, whereas the boundary contributions are responsible for the possible presence of the negligible set N of radii where (16) may fail. Namely, it turns out that the $(n - s)$-dimensional fractional perimeter of the set E inside a ball B_r, i.e.

$$\mathrm{Per}_s^L(E, B_r) = \int_{E\cap B_r}\int_{(\mathscr{C}E)\cap B_r} \frac{1}{|x-y|^{n+2s}}\,dy\,dx,$$

up to constants, does indeed approach the classical perimeter, whereas the integral contributions

$$\mathrm{Per}_s^{NL}(E,B_r)=\int_{E\cap B_r}\int_{(\mathscr{C}E)\cap(\mathscr{C}B_r)}\frac{1}{|x-y|^{n+2s}}\,dy\,dx+\int_{E\cap(\mathscr{C}B_r)}\int_{(\mathscr{C}E)\cap B_r}\frac{1}{|x-y|^{n+2s}}\,dy\,dx$$

can be estimated by the $(n-1)$-dimensional Hausdorff measure of $\partial E\cap\partial B_r$, which vanishes for all radii r out of a countable subset of $\mathbb{R}$.

Some words are also in order about the asymptotic behavior of the s-perimeter as $s\to 0^+$. The results proved in the paper [44] imply that

$$\lim_{s\to 0^+} s\mathrm{Per}_s\big(E,\mathbb{R}^n\big)=n\omega_n|E|, \tag{17}$$

where $|\cdot|$ here stands for the n-dimensional Lebesgue measure. If Ω is any bounded open set in $\mathbb{R}^n$, it is easily seen that the subadditivity property of $\mathrm{Per}_s(\cdot,\Omega)$, is preserved by taking the limit and thus

$$\mu(E)=\lim_{s\to 0^+} s\,\mathrm{Per}_s(E,\Omega) \tag{18}$$

defines a subadditive set function on the family $\mathscr{E}$ of sets E such that the limit (18) exists. For instance, all the bounded measurable sets E such that $\mathrm{Per}_s(E,\Omega)<\infty$ for some $s\in(0,1/2)$ turn out to be included in $\mathscr{E}$. Unfortunately, μ is not a measure, as it was shown in the very recent work [31]. However, in the same paper it is proved that μ is finitely additive on the bounded and separated subsets or $\mathbb{R}^n$ belonging to $\mathscr{E}$ and, in turn, it coincides with a rescaled Lebesgue measure of the intersection $E\cap\Omega$, namely

$$\mu(E)=n\omega_n|E\cap\Omega|,$$

provided $E\in\mathscr{E}$ is bounded. Furthermore, it can be proved that whenever $E\in\mathscr{E}$ and the integrals

$$\int_{E\cap(\mathscr{C}B_1)}\frac{1}{|y|^{n+2s}}\,dy$$

do converge to some limit $\alpha(E)$ as s tends to zero, then

$$\mu(E)=\big(n\omega_n-\alpha(E)\big)|E\cap\Omega|+\alpha(E)|\Omega\cap\mathscr{C}E|.$$

We end this section with the following result.

Theorem 8 *Let $s_o\in(0,1/2)$, $s\in[s_o,1/2)$ and E be s-minimal. There exists a universal $\varepsilon_*>0$, possibly depending on s_o but independent of s and E, such that if*

$$\partial E\cap B_1\subset\big\{x:|x_n|\le\varepsilon_*\big\} \tag{19}$$

then $\partial E\cap B_{1/2}$ is a $\mathscr{C}^\infty$-graph in the n-th Euclidean direction.

Such a uniform flatness property was shown in [19], with a proof of the $\mathscr{C}^{1,\alpha}$ regularity, and the $\mathscr{C}^{\infty}$ smoothness for all s-minimal $\mathscr{C}^{1,\alpha}$-graphs was proved in the recent paper [8] by the second author in collaboration with Barrios Barrera and Figalli.

This result improves some previous work in [21] where a similar condition was also obtained with $\varepsilon_* > 0$ possibly depending on s. The independence of s will be crucial for the subsequent Theorems 10, 11 and 12.

3.5 Singularities of Nonlocal Minimal Surfaces

As it was discussed in the previous Sect. 3.4, the s-perimeter Γ-converges, up to scaling factors, to the classical perimeter and the s-minimal sets converge to the classical minimal surfaces. Thus, the regularity of the classical minimal surfaces in suitably low dimension naturally drops hints that the regularity results available for nonlocal minimal surfaces might not be sharp. Namely, by Theorem 5, the boundary of any s-minimal set E in Ω is a $\mathscr{C}^{1,\alpha}$ manifold out of a set N, which is somehow negligible. The question is whether or not

$$N = \varnothing. \tag{20}$$

As far as we know, the only occurrence in which (20) can be proved for any $s \in (0, 1/2)$ is in dimension 2:

Theorem 9 *Let $s \in (0, 1/2)$. If $E \subset \mathbb{R}^2$ is an s-minimal cone in $\mathbb{R}^2$ then E is a half-plane.*

We refer to the recent paper by the second author and O. Savin [55] for the details of the proof, in which a central role is played by some estimates related to a compactly supported domain-perturbation of the cone which is almost linear about the origin. Thus, through a classical dimension reduction argument due to Federer [35]—that was adapted to the fractional case in [21, Theorem 10.3]—it is possible to give an improved estimate for the size of the singular set by replacing $n-2$ with $n-3$ in Theorem 5.

The question of the possible existence of s-minimal cones different from half-space is open in higher dimension, for any $s \in (0, 1/2)$. On the other hand, when s is sufficiently close to $1/2$ the s-minimal sets inherit some of the regularity properties of the classical minimal surfaces. The full regularity results summarized in the following Theorems follow by combining the rigidity results of [19, Theorems 3, 4, 5], where the $\mathscr{C}^{1,\alpha}$ regularity is proved, and the Schauder-type estimates recently provided in the paper [8].

Theorem 10 *Let $n \le 7$. There exists $\varepsilon > 0$ such that, if $2s \in (1-\varepsilon, 1)$, then any s-minimal surface in $\mathbb{R}^n$ is locally a $\mathscr{C}^{\infty}$-hypersurface for some $\alpha \in (0, 1)$.*

Theorem 11 *There exists $\varepsilon > 0$ such that, if $2s \in (1-\varepsilon, 1)$, then any s-minimal surface in $\mathbb{R}^8$ is locally a $\mathscr{C}^\infty$-hypersurface out of a countable set.*

Theorem 12 *Let $n \geq 9$. There exists $\varepsilon > 0$ such that, if $2s \in (1-\varepsilon, 1)$, then any s-minimal surface in $\mathbb{R}^n$ is locally a $\mathscr{C}^\infty$-hypersurface out of a set whose Hausdorff measure is at most $n-8$.*

Of course, we think that it would be desirable to determine sharply the above ε, to better analyze the regularity theory of nonlocal minimal surfaces for other ranges of $s \in (0, 1/2)$ and $n \in \mathbb{N}$, and to understand the possible relation between this regularity theory and the symmetry property of minimal or monotone solutions of the fractional Allen-Cahn equation.

Acknowledgements The second author is supported by the ERC grant ε *Elliptic Pde's and Symmetry of Interfaces and Layers for Odd Nonlinearities* and the FIRB project A&B *Analysis and Beyond.*

References

1. Alberti, G., Bellettini, G.: A nonlocal anisotropic model for phase transitions, I. The optimal profile problem. Math. Ann. **310**(3), 527–560 (1998)
2. Alberti, G., Bouchitté, G., Seppecher, P.: Un résultat de perturbations singulières avec la norme $H^{1/2}$. C. R. Acad. Sci. Paris **319**(I), 333–338 (1994)
3. Alberti, G., Bellettini, G., Cassandro, M., Presutti, E.: Surface tension in Ising systems with Kac potentials. J. Stat. Phys. **82**, 743–796 (1996)
4. Alberti, G., Ambrosio, L., Cabré, X.: On a long-standing conjecture of E. De Giorgi: symmetry in 3D for general nonlinearities and a local minimality property. Acta Appl. Math. **65**(1–3), 9–33 (2001)
5. Ambrosio, L., Cabré, X.: Entire solutions of semilinear elliptic equations in R3 and a conjecture of De Giorgi. J. Am. Math. Soc. **13**, 725–739 (2000)
6. Ambrosio, L., de Philippis, G., Martinazzi, L.: Γ-convergence of nonlocal perimeter functionals. Manuscr. Math. **134**, 377–403 (2011)
7. Barlow, M.T., Bass, R.F., Gui, C.: The Liouville property and a conjecture of De Giorgi. Commun. Pure Appl. Math. **53**(8), 1007–1038 (2000)
8. Barrios Barrera, B., Figalli, A., Valdinoci, E.: Bootstrap regularity for integro-differential operators and its application to nonlocal minimal surfaces. Ann. Sc. Norm. Super. Pisa, Cl. Sci. (2012, accepted for publication)
9. Berestycki, H., Caffarelli, L., Nirenberg, L.: Further qualitative properties for elliptic equations in unbounded domains. Ann. Sc. Norm. Super. Pisa. Cl. Sci. (4) **25**(1–2), 69–94 (1997–1998) (Dedicated to Ennio De Giorgi)
10. Berestycki, H., Hamel, F., Monneau, R.: One-dimensional symmetry of bounded entire solutions of some elliptic equations. Duke Math. J. **103**(3), 375–396 (2000)
11. Bombieri, E., De Giorgi, E., Giusti, E.: Minimal cones and the Bernstein problem. Invent. Math. **7**, 243–268 (1969)
12. Brezis, H.: How to recognize constant functions. Connections with Sobolev spaces. Usp. Mat. Nauk **57**(346), 59–74 (2002)
13. Cabré, X., Cinti, E.: Energy estimates and 1-D symmetry for nonlinear equations involving the half-Laplacian. Discrete Contin. Dyn. Syst. **28**(3), 1179–1206 (2010)

14. Cabré, X., Cinti, E.: Sharp energy estimates for nonlinear fractional diffusion equations. Calc. Var. Partial Differ. Equ. (2012, accepted for publication)
15. Cabré, X., Sire, Y.: Nonlinear equations for fractional Laplacians II: existence, uniqueness, and qualitative properties of solutions. Preprint. http://arxiv.org/abs/1111.0796
16. Cabré, X., Solà-Morales, J.: Layer solutions in a half-space for boundary reactions. Commun. Pure Appl. Math. **58**(12), 1678–1732 (2005)
17. Caffarelli, L., Córdoba, A.: Uniform convergence of a singular perturbation problem. Commun. Pure Appl. Math. **48**, 1–12 (1995)
18. Caffarelli, L., Valdinoci, E.: Uniform estimates and limiting arguments for nonlocal minimal surfaces. Calc. Var. Partial Differ. Equ. **41**(1–2), 203–240 (2011)
19. Caffarelli, L., Valdinoci, E.: Regularity properties of nonlocal minimal surfaces via limiting arguments. Preprint, http://www.ma.utexas.edu/mp_arc/c/11/11-69.pdf
20. Caffarelli, L., Salsa, S., Silvestre, L.: Regularity estimates for the solution and the free boundary of the obstacle problem for the fractional Laplacian. Invent. Math. **171**(2), 425–461 (2008)
21. Caffarelli, L., Roquejoffre, J.M., Savin, O.: Nonlocal minimal surfaces. Commun. Pure Appl. Math. **63**(9), 1111–1144 (2010)
22. Caffarelli, L., Roquejoffre, J.M., Sire, Y.: Variational problems for free boundaries for the fractional Laplacian. J. Eur. Math. Soc. **12**(5), 1151–1179 (2010)
23. Caputo, M.C., Guillen, N.: Regularity for non-local almost minimal boundaries and applications. Preprint. http://arxiv.org/pdf/1003.2470.pdf
24. Craig, W., Groves, M.D.: Hamiltonian long-wave approximations to the water-wave problem. Wave Motion **19**(4), 367–389 (1994)
25. Craig, W., Schanz, U., Sulem, C.: The modulational regime of three-dimensional water waves and the Davey-Stewartson system. Ann. Inst. Henri Poincaré, Anal. Non Linéaire **5**(14), 615–667 (1997)
26. De Giorgi, E.: Convergence problems for functionals and operators. In: Proc. Int. Meeting on Recent Methods in Nonlinear Analysis, Rome, 1978, pp. 131–188. Pitagora, Bologna (1979)
27. del Pino, M., Kowalczyk, M., Wei, J.: A counterexample to a conjecture by De Giorgi in large dimensions. C. R. Math. **346**, 23–24 (2008)
28. del Pino, M., Kowalczyk, M., Wei, J.: Entire solutions of the Allen-Cahn equation and complete embedded minimal surfaces. Rev. Unión Mat. Argent. **50**(2), 95–107 (2009)
29. del Pino, M., Kowalczyk, M., Wei, J.: On De Giorgi's conjecture in dimension $N \geq 9$. Ann. Math. **174**, 1485–1569 (2011)
30. Di Nezza, E., Palatucci, G., Valdinoci, E.: Hitchhiker.s guide to the fractional Sobolev spaces. Bull. Sci. Math. **136**(5), 521–573 (2012). doi:10.1016/j.bulsci.2011.12.004. http://www.sciencedirect.com/science/article/pii/S0007449711001254
31. Dipierro, S., Figalli, A., Palatucci, G., Valdinoci, E.: Asymptotics of the s-perimeter as $s \searrow 0$. Discrete Contin. Dyn. Syst., Ser. A (2012, accepted for publication)
32. Farina, A.: Symmetry for solutions of semilinear elliptic equations in $\mathbb{R}^N$ and related conjectures. Ric. Mat. **48**, 129–154 (1999)
33. Farina, A., Valdinoci, E.: The state of art for a conjecture of De Giorgi and related questions. In: Reaction-Diffusion Systems and Viscosity Solutions. World Scientific, Singapore (2008)
34. Farina, A., Valdinoci, E.: Rigidity results for elliptic PDEs with uniform limits: an abstract framework with applications. Indiana Univ. Math. J. **60**, 121–142 (2011)
35. Federer, H.: The singular sets of area minimizing rectifiable currents with codimension one and of area minimizing flat chains modulo two with arbitrary codimension. Bull. Am. Math. Soc. **76**, 767–771 (1970)
36. Fefferman, C., de la Llave, R.: Relativistic stability of matter, I. Rev. Mat. Iberoam. **2**(1–2), 119–213 (1986)
37. Garroni, A., Müller, S.: Γ-limit of a phase-field model of dislocations. SIAM J. Math. Anal. **36**, 1943–1964 (2005)
38. Garroni, A., Müller, S.: A variational model for dislocations in the line tension limit. Arch. Ration. Mech. Anal. **181**, 535–578 (2006)

39. Garroni, A., Palatucci, G.: A singular perturbation result with a fractional norm. In: Variational Problems in Material Science. Progr. Nonlinear Differential Equations Appl., vol. 68, pp. 111–126. Birkhäuser, Basel (2006)
40. Ghoussoub, N., Gui, C.: On a conjecture of De Giorgi and some related problems. Math. Ann. **311**, 481–491 (1998)
41. González, M.D.M.: Γ-convergence of an energy functional related to the fractional Laplacian. Calc. Var. Partial Differ. Equ. **36**, 173–210 (2009)
42. Kurzke, M.: A nonlocal singular perturbation problem with periodic well potential. ESAIM Control Optim. Calc. Var. **12**(1), 52–63 (2006) (electronic)
43. Majda, A.J., Tabak, E.G.: A 2D model for quasigeostrophic flow: comparison with the two-dimensional Euler flow. Physica D **98**(2–4), 515–522 (1996). Nonlinear phenomena in ocean dynamics, Los Alamos, NM (1995)
44. Maz'ya, V., Shaposhnikova, T.: On the Bourgain, Brezis, and Mironescu theorem concerning limiting embeddings of fractional Sobolev spaces. J. Funct. Anal. **195**(2), 230–238 (2002)
45. Mellet, A., Mischler, S., Mouhot, C.: Fractional diffusion limit for collisional kinetic equations. Preprint. http://arxiv.org/abs/0809.2455
46. Modica, L.: Convergence to minimal surfaces problem and global solutions of $\Delta u = 2(u^3 - u)$. In: Proceedings of the International Meeting on Recent Methods in Nonlinear Analysis, Rome, 1978, pp. 223–244. Pitagora, Bologna (1979)
47. Modica, L.: The gradient theory of phase transitions and the minimal interface criterion. Arch. Ration. Mech. Anal. **98**, 23–142 (1987)
48. Modica, L., Mortola, S.: Un esempio di Γ^--convergenza. Boll. Unione Mat. Ital, B **14**, 285–299 (1977)
49. Palatucci, G., Savin, O., Valdinoci, E.: Local and global minimizers for a variational energy involving a fractional norm. Ann. Mat. Pura Appl. (2012). doi:10.1007/s10231-011-0243-9
50. Savin, O.: Regularity of flat level sets in phase transitions. Ann. Math. **169**(1), 41–78 (2009)
51. Savin, O., Valdinoci, E.: Elliptic PDEs with fibered nonlinearities. J. Geom. Anal. **19**(2), 420–432 (2009)
52. Savin, O., Valdinoci, E.: Density estimates for a nonlocal variational model via the Sobolev inequality. SIAM J. Math. Anal. **43**, 2675–2687 (2011)
53. Savin, O., Valdinoci, E.: Γ-convergence for nonlocal phase transitions. Ann. Inst. H. Poincaré, Anal. Non Linéaire (2012). doi:10.1016/j.anihpc.2012.01.006
54. Savin, O., Valdinoci, E.: Density estimates for a variational model driven by the Gagliardo norm. Preprint. http://www.math.columbia.edu/~savin/s-density-estimates.pdf
55. Savin, O., Valdinoci, E.: Regularity of nonlocal minimal cones in dimension 2. Preprint. http://math.columbia.edu/~savin/cone_jan16.pdf
56. Signorini, A.: Questioni di elasticità non linearizzata e semilinearizzata. Rend. Mat. Appl. **18**, 95–139 (1959). http://www.scienzadellecostruzioni.co.uk/Documenti/Signorini-Prima.pdf
57. Silvestre, L.: Regularity of the obstacle problem for a fractional power of the Laplace operator. Commun. Pure Appl. Math. **60**(1), 67–112 (2007)
58. Sire, Y., Valdinoci, E.: Fractional Laplacian phase transitions and boundary reactions: a geometric inequality and a symmetry result. J. Funct. Anal. **256**(6), 1842–1864 (2009)
59. Toland, J.F.: The Peierls-Nabarro and Benjamin-Ono equations. J. Funct. Anal. **145**(1), 136–150 (1997)
60. Valdinoci, E.: From the long jump random walk to the fractional Laplacian. Bol. Soc. Esp. Mat. Apl. SeMA **49**, 33–34 (2009)
61. Visintin, A.: Generalized coarea formula and fractal sets. Jpn. J. Ind. Appl. Math. **8**, 175–201 (1991)
62. Visintin, A.: Non convex functionals related to multi-phase systems. In: Proceedings of the Colloquium on Free Boundary Problems held in Irsee in June, 1987

Existence of Solutions to Some Classical Variational Problems

Antonio Greco

Abstract Some non-coercive variational integrals are considered, including the classical time-of-transit functional arising in the problem of the brachistochrone, and the area functional in the problem of the minimal surface of revolution. A minimizer is constructed by means of the direct method. More precisely, each admissible curve is replaced by its convex envelope, and the functional is shown to decrease. Hence, there exists a minimizing sequence made up of convex curves, which in turn possesses a locally uniformly converging subsequence. The limiting curve is a minimizer because the functionals under consideration are continuous under such a convergence.

Keywords Brachistochrone · Convex envelope · Direct method

1 Introduction

Galileo compared the time of descent along an arc of circle with that of a corresponding chord, and also with a piecewise-smooth curve made up of two chords, concluding that *brevissimo sopra tutti i tempi sarà quello della caduta per l'arco* (shortest among all times will be that of the fall along the arc) [14, Fourth Day]. However, there exist curves whose time of transit is smaller.

The classical problem of the brachistochrone consists in finding a curve γ_0 joining two given points A, B in a vertical plane, so that a point mass placed at rest in A, constrained to γ_0 and subject to a uniform gravity field reaches B in the shortest time. The effect of friction is neglected. More generally, the point mass may be given an initial velocity whose modulus $v_0 \geq 0$ is prescribed.

The solution γ_0 is called *brachistochrone* from the Greek $\beta\rho\alpha\chi\iota\varsigma\tau o\sigma$ (shortest) and $\chi\rho o\nu o\sigma$ (time) [3]. Together with Newton's problem of minimal resistance ([17, 18], [20, pp. 1–3]), the problem of the brachistochrone is considered as one of the starting problems of the calculus of variations. It was posed by Johann Bernoulli

A. Greco (✉)
Dipartimento di Matematica e Informatica, Università degli Studi di Cagliari, via Ospedale 72, 09124 Cagliari, Italy
e-mail: greco@unica.it

R. Magnanini et al. (eds.), *Geometric Properties for Parabolic and Elliptic PDE's*, Springer INdAM Series 2, DOI 10.1007/978-88-470-2841-8_9,

in 1696 as a mathematical defy, and then solved by himself and some of his contemporaries. From a modern point of view, they characterized any possible solution, but underestimated the importance of an existence proof. Details are found in several textbooks [1, 3, 11, 13, 22].

Proving existence of the brachistochrone is more delicate: the result is often achieved by means of the concept of *field of extremals* and related theory, which has its roots in the work of Weierstrass (cf. [3, 12, 21]). Alternative approaches are found in [4, 15] and in the present paper.

Let us recall a well-known mathematical model of the problem. Let $I=(a,b)$ be a nonempty, bounded, open interval on the real line, and denote by X the class of all functions u in the Sobolev space $W^{1,1}(I)$ attaining prescribed boundary values $u(a)=0$ and $u(b)=u_b\le 0$. Here and in the sequel, each function in $W^{1,1}(I)$ is identified with its continuous representative, which is in turn extended by continuity up to the endpoints. The graphs of the functions $u\in X$ provide the admissible curves connecting the points $A=(a,0)$ and $B=(b,u_b)$ in the plane xy.

Conservation of energy (in case of vanishing initial speed) reads as $\frac{1}{2}v^2+gy=0$, where v is the velocity of the point mass, and g is the gravity acceleration. Hence $v=\sqrt{2g|y|}$. Since $v=ds/dt$, and since the element of arc ds along the graph of u is given by $ds=\sqrt{1+(u'(x))^2}\,dx$, we get $dt=\sqrt{1+(u'(x))^2}\,dx/\sqrt{2g|u(x)|}$ and we are led to minimize the (possibly infinite) time of transit $T[u]$ given by

$$T[u]=\frac{1}{\sqrt{2g}}\int_a^b\frac{\sqrt{1+(u'(x))^2}}{\sqrt{|u(x)|}}\,dx. \tag{1}$$

The core of the existence proof found in the next sections is the direct method, and therefore consists of a compactness and continuity argument. Before proceeding further, note that a minimizing sequence of T may fail to be weakly precompact in $W^{1,p}(I)$: indeed, the Lagrangian $\mathcal{L}(x,u,u')=\sqrt{1+(u')^2}/\sqrt{|u|}$ is asymptotic to $|u'|^p$ with $p=1$ as $|u'|\to+\infty$, and the weak compactness property of bounded sequences in the Sobolev space $W^{1,p}(I)$ requires $p>1$.

Here, as in [15], compactness is recovered by converting non-convex functions into convex ones. However, the transformation is different. Instead of using rearrangements, we replace each admissible function u with its *convex envelope*, i.e., the largest convex function that does not exceed u (pointwise). Since this reduces the time of transit (Theorem 3), we may consider a minimizing sequence made up of convex functions, although the space of competing functions contains both convex and non-convex elements. By the way, the convex envelope has been applied to the theory of partial differential equations for different purposes (see, for instance, [2, 8]).

Convexity (and boundedness) of the functions in the minimizing sequence allows us to extract a subsequence converging to a limit function u_0 locally uniformly. The limit function u_0 belongs, in principle, to the slightly wider class X^- of functions possibly having jumps at the endpoints. The time-of-transit functional admits a natural extension (see (3) below) to such kind of functions, which is continuous with respect to locally uniform convergence of convex functions [15]. Consequently, the limit function u_0 is a minimizer in the extended class. This in turn implies further regularity of u_0, including boundary continuity, and we may finally conclude that u_0 belongs to the class X defined at the beginning. The method also applies to functionals of the form

$$F[u] = \int_a^b f\big(u(x)\big)\sqrt{1 + \big(u'(x)\big)^2}\, dx \tag{2}$$

where f is any positive, non-decreasing function of class C^1 blowing up at some finite $y_0 \in \mathbb{R}$. More precisely, we have:

Theorem 1 (Brachistochrone) *Let u_a, u_b be prescribed real numbers, and define $y_0 = \max\{u_a, u_b\}$. Let $f \in C^1((-\infty, y_0))$ be a positive, non-decreasing function such that $\lim_{y \to y_0^-} f(y) = +\infty$ and satisfying*

$$\int_{-\infty}^{y_0 - 1} f(y)\, dy = +\infty \quad \text{and} \quad \int_{y_0-1}^{y_0} f(y)\, dy < +\infty.$$

Then, the functional F in (2) has a minimizer u_0 in the class X of all functions $u \in W^{1,1}(I)$ such that $u(x) < y_0$ for $x \in I$, $u(a) = u_a$, $u(b) = u_b$. Furthermore, every minimizer in X is a convex function belonging to the class $C^2(I)$ and satisfying the associated Euler equation.

Details on the special case when $u_a = 0$, $u_b \le 0$ and $f(y) = |y|^{-1/2}$ are given in Sect. 3. A further application of the procedure is the following. Recall that if $f(y) = 2\pi y$ then the minimization problem of the functional (2) can be interpreted as the problem of minimizing the area of the surface of revolution generated by the graph of $u > 0$ around the x-axis. The boundary of the surface is prescribed, and it is composed of the two circles lying on the planes $x = a$ and $x = b$, respectively, centered on the x-axis and whose radii are u_a and u_b. It is well known that if such radii are small enough in comparison with $b - a$, then the solution is the disconnected surface made up of the two discs bounded by those circles. In order to take such a case into account, we proceed as follows.

As mentioned before, each function in $W^{1,1}(I)$ is extended by continuity up to the endpoints. Thus, we make a distinction between u_a, u_b (the prescribed boundary values) and the values $u(a), u(b)$ (defined by continuity). Let X^- be the class of all functions $u \in W^{1,1}(I)$ satisfying $u \ge 0$ in I, $u(a) \le u_a$, $u(b) \le u_b$. A rotational surface is obtained from a function $u \in X^-$ by rotating around the x-axis the portion of the graph of u lying in the open half-plane $y > 0$, and then by adding to it two annuli: the annulus lying in the plane $x = a$ whose radii are $u(a), u_a$, and the one lying in the plane $x = b$ whose radii are $u(b), u_b$. Finally, we take the closure in $\mathbb{R}^3$ of the set constructed that way (this is to recover any points belonging to the surface but lying on the x-axis). In particular, the couple of discs mentioned before is associated to the function $u_0 \equiv 0$. Consequently, we are led to extend the functional F as follows: for every $u \in X^-$ define

$$\overline{F}[u] = \int_{u(a)}^{u_a} f(y)\, dy + F[u] + \int_{u(b)}^{u_b} f(y)\, dy. \tag{3}$$

It is readily seen that if $f(y) = 2\pi y$, and if u_0 vanishes identically, then $\overline{F}[u_0]$ is the sum of the areas of the two mentioned discs. It is also apparent that if, instead, $u(a) = u_a$ and $u(b) = u_b$, then $\overline{F}[u] = F[u]$. The same extension enters in the proof of Theorem 1 (given in Sect. 3) and has also been used in [15] in connection with the convex rearrangement. Concerning the minimization problem of the functional $\overline{F}$, we have:

Theorem 2 (Minimal surface of revolution) *Let $u_a, u_b > 0$ be prescribed real numbers, and let $f \in C^1((0,+\infty)) \cap C^0([0,+\infty))$ be a non-decreasing function such that $f(0) = 0$ and $f(y) > 0$ for $y > 0$. Then, the extended functional $\overline{F}$ in* (3) *has a minimizer u_0 in the class X^- of all non-negative functions $u \in W^{1,1}(I)$ such that $u(a) \le u_a$ and $u(b) \le u_b$. Furthermore, every minimizer either vanishes identically, or it is a positive, convex function belonging to the class $C^2([a,b])$, attaining the boundary values $u(a) = u_a$ and $u(b) = u_b$, and satisfying the Euler equation associated to the functional F in* (2).

The proof of the stated theorems is found in Sect. 3. In the next section we prove that the convex envelope reduces the functionals under consideration. The definition and some properties of the cycloid are recalled in the Appendix. Further results on the existence of minimizers of non-coercive functionals are found, for instance, in [6, 7, 9, 10].

2 Convex Envelope

Let u be a (continuous) function in $W^{1,1}(I)$, extended by continuity up to the endpoints. Following [2], let us denote by u_{**} the convex envelope (also called the convex hull) of u, i.e., the largest convex function on $[a,b]$ not exceeding u at any point. Assuming monotonicity of the function f in (2), this section is devoted to the proof of the fundamental inequality (5) below, which allows us to transform any minimizing sequence into a sequence of *convex* functions that still minimizes the functional F. Some characterizations of u_{**} are found in [2, 8, 19]. Here, it suffices to recall that u_{**} coincides with u at a, b, and for every $x_0 \in I$ such that $u_{**}(x_0) < u(x_0)$ there exist suitable endpoints $a \le \alpha(x_0) < x_0 < \beta(x_0) \le b$ such that the restriction of u_{**} to the closed interval $[\alpha(x_0), \beta(x_0)]$ is the affine function

$$u_{**}(x) = u\big(\alpha(x_0)\big) + \frac{u(\beta(x_0)) - u(\alpha(x_0))}{\beta(x_0) - \alpha(x_0)}\big(x - \alpha(x_0)\big). \tag{4}$$

In particular, u_{**} is continuous in the closed interval $[a,b]$ and satisfies $\min_{[a,b]} u_{**} = \min_{[a,b]} u$.

Theorem 3 *Take $u \in W^{1,1}(I)$ and define $y_0 = \max_{[a,b]} u$ and $\mu = \min_{[a,b]} u$. Let $f : [\mu, y_0) \to [0,+\infty)$ be a continuous, non-negative, non-decreasing function (whose limit $f(y_0^-) = \lim_{y \to y_0^-} f(y)$ is allowed to be $+\infty$). Then*

$$F[u_{**}] \le F[u] \tag{5}$$

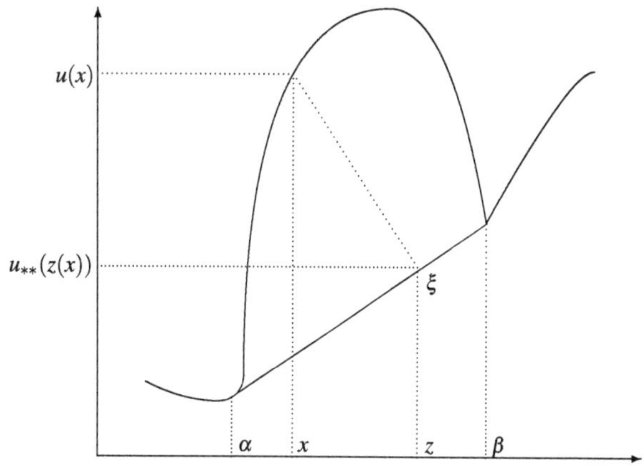

Fig. 1 The change of variables

where one or both sides may be infinite.

Proof By definition we have $u_{**} \le u$ in $[a,b]$, therefore for almost every x in the contact set $K = \{x \in [a,b] \mid u_{**}(x) = u(x)\}$ the derivatives u', u'_{**} (exist and) coincide. Here and in the sequel, u'_{**} stands for $(u_{**})'$, for shortness. Hence the function under the sign of integral in $F[u]$ coincides with that in $F[u_{**}]$ almost everywhere in K, and may restrict our attention to the contribution of the complement $K^c = [a,b] \setminus K = \{x \in I \mid u_{**}(x) < u(x)\}$, assuming that it is nonempty. In this case, K^c is the union of a finite, or at most countable, number of open intervals (α_n, β_n), in which u_{**} has the form (4), and it is enough to show that in every interval (α_n, β_n) we have

$$\int_\alpha^\beta f\big(u_{**}(x)\big)\sqrt{1+m^2}\,dx \le \int_\alpha^\beta f\big(u(x)\big)\sqrt{1+\big(u'(x)\big)^2}\,dx \tag{6}$$

where we have written α, β in place of α_n, β_n for simplicity, and the constant m is given by $m = u'_{**}(x)$ for $x \in (\alpha, \beta)$. The proof of (6) is based on some suitable change of variables, depicted in Fig. 1. In order to include in the argument the case when $f(y_0^-) = +\infty$, observe that if $u_{**} = y_0$ at both α and β, then by convexity we have $u_{**} \equiv u \equiv y_0$ in all of I, and (5) trivially holds with equality. Thus, we may assume that $u_{**} < y_0$ in the interval (α, β), and possibly $u_{**} = y_0$ at most at one of the two endpoints (in which case, that endpoint coincides with the corresponding endpoint of I). Without loss of generality, let us assume $u(\alpha) < y_0$, and let us give the first integral in (6) a more convenient expression. First of all, we denote the variable of integration by z for reasons that will be clear in a moment. Secondly, by considering the abscissa $\xi = z\sqrt{1+m^2}$ running on the straight line ℓ containing the graph of u_{**}, we may transform that integral as follows:

$$\int_\alpha^\beta f\big(u_{**}(z)\big)\sqrt{1+m^2}\,dz = \int_{\alpha\sqrt{1+m^2}}^{\beta\sqrt{1+m^2}} f\big(u_{**}(\xi/\sqrt{1+m^2})\big)\,d\xi. \tag{7}$$

Denoting by q the value $u(\alpha) - m\alpha = u(\beta) - m\beta$, the equation of the line ℓ is $y = mz + q$, and we may rewrite (7) in the following form:

$$\int_{\alpha}^{\beta} f\big(u_{**}(z)\big)\sqrt{1+m^2}\,dz = \int_{\alpha\sqrt{1+m^2}}^{\beta\sqrt{1+m^2}} f\big(m\xi/\sqrt{1+m^2}+q\big)\,d\xi. \tag{8}$$

Finally, we interpret the point ξ on the line ℓ as the orthogonal projection onto ℓ of some point (there may be many) belonging to the graph of u. More precisely, we intend to perform the change of variable

$$\xi(x) = \frac{x + mu(x) - mq}{\sqrt{1+m^2}} \quad \text{for } x \in [\alpha, \beta]. \tag{9}$$

The change of variable above requires some care for two reasons: firstly, the function $\xi(x)$ may be non-invertible. Secondly, although the endpoints α, β are transformed into $\xi(\alpha) = \alpha\sqrt{1+m^2}$ and $\xi(\beta) = \beta\sqrt{1+m^2}$, it is not told that the interval $[\alpha, \beta]$ is transformed into $\sqrt{1+m^2}\,[\alpha, \beta]$. Indeed, if $m \neq 0$, and if not only $u(\alpha) < y_0$, but also $u(\beta) < y_0$, it may well happen that the image $\xi([\alpha, \beta])$ contains the interval $\sqrt{1+m^2}\,[\alpha, \beta]$ strictly. This is also seen in Fig. 1 letting x approach β from below.

These difficulties are overcome because in dimension 1 the change-of-variable formula is a consequence of the chain rule and the fundamental theorem of calculus. First of all, we need to ensure that when x varies in $[\alpha, \beta]$, the values of $m\xi(x)/\sqrt{1+m^2} + q$, to be inserted in (8), keep inside the domain of the function f. To ensure this, recall that $mx + q \le u(x)$ for all $x \in [\alpha, \beta]$. Hence by (9) we get

$$m\xi(x)/\sqrt{1+m^2} + q \le u(x) < y_0 \qquad \text{for all } x \in [\alpha, \beta), \tag{10}$$

and therefore the quantity $m\xi(x)/\sqrt{1+m^2} + q$ can be inserted as argument of f in (8), as claimed. Next, observe that since f is continuous in $[\mu, y_0)$ the integral function

$$\Phi(\xi) = \int_{\alpha\sqrt{1+m^2}}^{\xi} f\big(mt/\sqrt{1+m^2}+q\big)\,dt$$

is well defined over the interval $J = \xi([\alpha, \beta))$ and belongs to the class $C^1(J)$. Furthermore, the composite function $\Phi(\xi(x))$ is of class $W^{1,1}_{\mathrm{loc}}([\alpha, \beta))$ and for almost every $x \in [\alpha, \beta)$ we have

$$\big(\Phi\big(\xi(x)\big)\big)' = f\big(mz(x)+q\big)\xi'(x),$$

where we have written $z(x) = \xi(x)/\sqrt{1+m^2}$, for shortness. By integrating both sides in dx over the interval (α, β) we obtain:

$$\int_{\alpha\sqrt{1+m^2}}^{\beta\sqrt{1+m^2}} f\big(m\xi/\sqrt{1+m^2}+q\big)\,d\xi = \int_{\alpha}^{\beta} f\big(mz(x)+q\big)\frac{1+mu'(x)}{\sqrt{1+m^2}}\,dx.$$

It remains to check that the last integral does not exceed the right-hand side of (6). Observe that the inequality

$$\frac{1+mu'(x)}{\sqrt{1+m^2}} \le \sqrt{1+\big(u'(x)\big)^2} \tag{11}$$

follows immediately from the Cauchy inequality in $\mathbb{R}^2$. In view of the next remark, observe, further, that the inequality in (11) becomes strict whenever $u'(x) \neq m$, and the set $\{x \in (\alpha, \beta) \mid u'(x) \neq m\}$ has positive Lebesgue measure. Using (11), since f is non-negative and non-decreasing, and by (10), we arrive at (6) and the proof is complete. □

Remark 1 (The equality case) A glance at the last part of the proof shows that if both sides in (5) are finite, and if $f(y) > 0$ for $y \in (\mu, y_0)$, then equality holds in (5) if and only if $u = u_{**}$, i.e., if and only if u is convex.

3 Proofs of the Main Theorems

Proof of Theorem 1 The scheme of the proof is the following. Firstly, we prove that the extended functional $\overline{F}$ in (3) admits a minimizer in the class X^- of all functions $u \in W^{1,1}(I)$ such that $u(x) < y_0$ for $x \in I$, $u(a) \leq u_a$, $u(b) \leq u_b$. Secondly, we prove that every minimizer attains the prescribed boundary values and complete the proof of the theorem. Before proceeding into details, observe that $\inf \overline{F} < +\infty$ because there are admissible functions u such that $\overline{F}[u] < +\infty$: indeed, since $\int_{y_0-1}^{y_0} f(y)\,dy < +\infty$ we may let u be any constant less than $\min\{u_a, u_b\}$. Now let $(u_n)_{n\geq 1}$ be a minimizing sequence in X^-, and proceed as follows. By Theorem 3 we may assume that each u_n is a convex function.

Step 1. Uniform boundedness from below. Since $f > 0$ we have $\int_{u_n(a)}^{u_a} f(y)dy < \overline{F}[u_n]$ for all n, and the last term is bounded from above because $(u_n)_{n\geq 1}$ is a minimizing sequence. Consequently, since $\int_{-\infty}^{y_0-1} f(y)dy = +\infty$, we get that $u_n(a)$ is bounded from below. A uniform bound on $u_n(x)$ follows from the following estimate, which is obtained by the change of variable $y = u_n(x)$:

$$\left| \int_{u_n(x)}^{u_n(a)} f(y)\,dy \right| \leq \int_a^x f\big(u_n(t)\big)\big|u_n'(t)\big|\,dt \leq \overline{F}[u_n].$$

Step 2. There exists a subsequence converging to a convex function u_0 locally uniformly, and the functional $\overline{F}$ is continuous under such a convergence. The first claim is a known result in convex analysis [19, Theorem 10.9] whose proof is also recalled in [15, Sect. 3]. Continuity was proved in Theorem 4.2 of the last paper. The same theorem implies that the subsequence cannot converge to the constant function y_0 (nor can converge to y_0 at any interior point) because $\overline{F}$ would tend to $+\infty$. Hence u_0 is a minimizer belonging to the class X^-.

Step 3. Regularity. Let u_0 be *any* minimizer of the extended functional $\overline{F}$ in the class X^-. Clearly, the same u_0 also minimizes the original functional F in the subclass of X^- composed of all functions coinciding with u_0 at the endpoints, and therefore u_0 satisfies the Euler equation in integral form associated to F:

$$f\big(u(x)\big)\frac{u'(x)}{\sqrt{1+(u'(x))^2}} - \int_{x_0}^{x} f'\big(u(t)\big)\sqrt{1+\big(u'(t)\big)^2}\,dt = \text{const.} \qquad (12)$$

The formula above is obtained following the usual procedure to derive the Euler equation of a functional, apart from the fact that Du Bois-Reymond's lemma [5, Lemma 1.8] replaces the fundamental lemma of the calculus of variations (which requires more regularity of the minimizer).

Of course, any minimizer of the functional F in X also satisfies (12). Since f is of class C^1 and satisfies $f(u_0(x)) \in (0, +\infty)$ for all $x \in I$, and since the function $t \mapsto t/\sqrt{1+t^2}$ is invertible, it follows that u_0 is of class $C^2(I)$ and satisfies the Euler equation in strong form

$$f(u)u'' = f'(u)\big(1 + (u')^2\big) \tag{13}$$

which in turn has the prime integral

$$\sqrt{1 + (u')^2} = Cf(u) \tag{14}$$

where C is a positive constant. This is relevant in order to prove that any minimizer u_0 of $\overline{F}$ in X^- attains the prescribed boundary values. Indeed, if we assume $u_0(a) < u_a$ then we have $u_0 \in C^2([a, b))$ and we reach the contradiction described in the next step.

Step 4. Boundary continuity. Let u_0 be any minimizer in the extended class X^-, and let us prove that $u_0(a) = u_a$. Supposing $u_0(a) < u_a$ we now construct a function $u_\varepsilon \in X^-$ such that $\overline{F}[u_\varepsilon] < \overline{F}[u_0]$, a contradiction. The argument is taken from the proof of Theorem 1.2 in [15], where the more restrictive condition $f' > 0$ was required (and the sharper conclusion $u_0'' > 0$ was obtained). By Step 3, the assumption $u_0(a) < u_a$ implies that $u_0'(a)$ is finite. Fix $y_1 \in (u_0(a), u_a)$, and for $\varepsilon > 0$ so small that $u_0(a+\varepsilon) < y_1$ define

$$u_\varepsilon(x) = \begin{cases} y_1 + (u_0(a+\varepsilon) - y_1)(x-a)/\varepsilon, & x \in [a, a+\varepsilon), \\ u_0(x), & x \in [a+\varepsilon, b]. \end{cases}$$

By the (affine) change of variable $y = u_\varepsilon(x)$ in the integral

$$\int_a^{a+\varepsilon} f\big(u_\varepsilon(x)\big)\sqrt{1 + \big(u_\varepsilon'(x)\big)^2}\,dx$$

we arrive at

$$\begin{aligned}\overline{F}[u_\varepsilon] = &\int_{y_1}^{u_a} f(y)\,dy + \sqrt{1 + \left(\frac{\varepsilon}{u_0(a+\varepsilon) - y_1}\right)^2}\int_{u_0(a+\varepsilon)}^{y_1} f(y)\,dy \\ &+ \int_{a+\varepsilon}^{b} f\big(u_0(x)\big)\sqrt{1 + \big(u_0'(x)\big)^2}\,dx + \int_{u_0(b)}^{u_b} f(y)\,dy.\end{aligned}$$

Observe that the preceding equality extends continuously to $\varepsilon = 0$. We may also differentiate at $\varepsilon = 0$ and find

$$\left(\frac{d}{d\varepsilon}\overline{F}[u_\varepsilon]\right)_{\varepsilon=0} = -f\big(u_0(a)\big)\Big(u_0'(a) + \sqrt{1 + \big(u_0'(a)\big)^2}\Big).$$

Since $f > 0$, it follows that $\overline{F}[u_\varepsilon] < \overline{F}[u_0]$ for small $\varepsilon > 0$, as claimed. This contradiction proves that $u_0(a) = u_a$, and by a similar argument $u_0(b) = u_b$. Consequently, every minimizer in the extended class X^- does indeed belong to the smaller class X, and the proof is complete. □

Proof of Theorem 2 Let us denote by $(u_n)_{n\geq 1}$ a minimizing sequence of the functional $\overline{F}$ in the class X^-, and let us follow the same steps as before. By Theorem 3 we may assume that each u_n is a convex function. Step 1 is now trivial because all functions in X^- are non-negative. Step 2 leads to the existence of a minimizer. Since $f(y) > 0$ for $y > 0$, by Theorem 3 and Remark 1 it follows that *every* minimizer u_0 is a convex function. Suppose, now, that some minimizer u_0 is positive at an interior point $x_0 \in (a,b)$, and let $(c,d) \subset (a,b)$ be the largest open interval containing x_0 and such that $u_0 > 0$ in all of (c,d). Then, u_0 satisfies the prime integral (14) in (c,d), and therefore $f(u_0(x)) \geq 1/C$ for every $x \in (c,d)$. Since both f and u_0 are continuous, and since $f(0) = 0$ by assumption, this prevents u_0 from vanishing at the endpoints c, d. But since the interval (c,d) is maximal, we deduce $(c,d) = (a,b)$. Therefore, every minimizer that does not vanish identically must be positive in the whole interval $[a,b]$. In the last case, the claimed properties of the minimizer follow by Step 3 and Step 4 in the proof of Theorem 1. In addition, from (13) we see that every positive minimizer belongs to the class $C^2([a,b])$ because f does not blow up at a finite point. □

Acknowledgements I wish to thank the two referees, whose careful reports have helped me to improve the paper.

Appendix: The Cycloid

For the reader's convenience, a parametric representation and some basic properties of this celebrated curve are recalled here. We also prove the following

Proposition 1 *When $u_a = 0$, $u_b \leq 0$ and $f(y) = |y|^{-1/2}$, the minimizer whose existence is assured by Theorem 1 is unique and it is given by the unique cycloid connecting the points $(a,0)$ and (b,u_b) and having the parametric representation* (20)–(21) *below.*

The proof is found at the end of the section. To construct a cycloid, consider the circumference of radius R in the plane xy whose center $C = (x_C, y_C)$ moves along the line $y = -R$ according to the law

$$x_C(t) = \lambda t, \qquad y_C(t) = -R \tag{15}$$

where λ is a positive constant and t the time variable. If the circumference rolls without slipping under the x-axis, then each of its points describes a cycloid. In particular, let P be the point of the circumference that coincides with O at $t = 0$. Then, the coordinates (x, y) of P for all $t \in \mathbb{R}$ are found as follows:

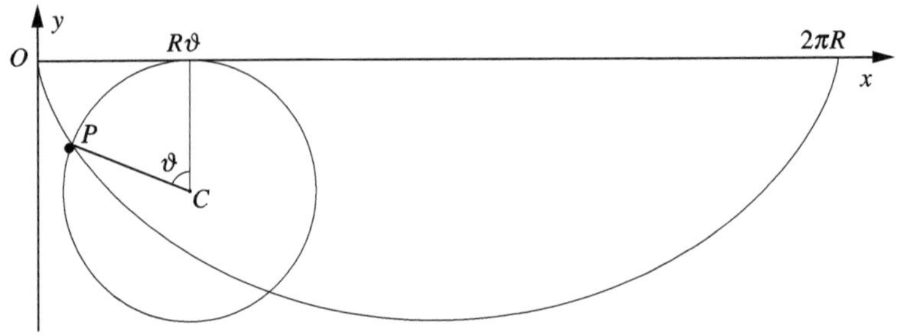

Fig. 2 Construction of a cycloid

$$x(t) = x_C(t) - R\sin\vartheta(t) \tag{16}$$

$$y(t) = y_C(t) + R\cos\vartheta(t) \tag{17}$$

where $\vartheta(t)$ is the angle between the ray $\overrightarrow{CP}$ and the unit vector $e_2 = (0, 1)$. Since the circumference rolls frictionless on the x-axis, the abscissa λt of the contact point equals the arc $R\vartheta$. Hence from (15) and (16)–(17) we finally get a parametric representation of the path of P:

$$x = R(\vartheta - \sin\vartheta) \tag{18}$$

$$y = R(-1 + \cos\vartheta). \tag{19}$$

The construction above is represented in Fig. 2. Animated pictures are found over the Internet, and in some cases the user may perform interactive experiments (see, for instance, [16]).

From the kinematical point of view it is interesting to observe that if the circumference rotates uniformly with $\lambda = \sqrt{gR}$ in (15), then the velocity v of P, obtained by differentiation of $(x(t), y(t))$, satisfies $\frac{1}{2}v^2 + gy = 0$, hence P moves just like a point mass constrained to the cycloid and put initially at O with speed $v_0 = 0$. In particular, we may easily compute the period p, i.e., the time needed by the particle to reach the point $(2\pi R, 0)$ and then to come back to the origin: $p = 4\pi R/\sqrt{gR} = 4\pi\sqrt{R/g}$.

The curve given by (18)–(19) for $\vartheta \in [0, 2\pi]$ is strictly convex in the sense that the angle α between the tangent vector $(dx/d\vartheta, dy/d\vartheta)$ and the unit vector $e_1 = (1, 0)$ is a strictly increasing function of ϑ: more precisely, by giving α the same sign as $dy/d\vartheta$ we have $\alpha = (\vartheta - \pi)/2$. This also shows that the cycloid is orthogonal to the x-axis at its endpoints.

Note that two cycloids given by (18)–(19) for two values of R are homothetic: this is apparent since R may be interpreted as a scale factor [1, pp. 181–182].

Strict convexity and homotheticity, together with the orthogonality of the cycloid to the x-axis at the origin, imply that for every $x > 0$ and $y \le 0$ there exist a unique $R = R(x, y) > 0$ and a unique $\vartheta \in (0, 2\pi]$ such that (18)–(19) hold, i.e. there exists a unique cycloid of the given form passing through the point (x, y). The value of R as a function of (x, y) may be interpreted as the gauge function, also called the Minkowski functional of the convex set bounded by the x-axis and the cycloid determined by letting $R = 1$ in (18)–(19). More generally, by admitting translations in the direction of the x-axis, we come to the following known proposition:

Proposition 2 *For every couple of points $A = (x_A, y_A)$ and $B = (x_B, y_B)$ such that $x_A < x_B$ and $y_A, y_B \leq 0$ there exist a unique $x_0 \leq x_A$ and a unique $R > 0$ such that the cycloid $\gamma(x_0, R)$ whose parametric representation is*

$$x = x_0 + R(\vartheta - \sin\vartheta) \tag{20}$$

$$y = R(-1 + \cos\vartheta) \tag{21}$$

connects them.

Proof By a suitable translation, we may assume $x_A = 0$. The case when also $y_A = 0$ has been just discussed: in this case, $x_0 = 0$ and $R = R(x_B, y_B)$ (see above). It remains to consider $y_A < 0$. By homotheticity, we may assume $y_A = 2$. Consider the cycloid $\gamma(-\pi, 1)$, which has its lowest point at A. From the mentioned discussion it follows that for every $x_0 < 0$ there exists a unique R, which we now denote by $R = R(x_0)$, such that the cycloid $\gamma(x_0, R(x_0))$ passes through A. In particular, we have $R(-\pi) = 1$. Now we move x_0 passing from $x_0 = -\pi$ to $-\infty$, and then from $x_0 = -\pi$ to 0. When $x_0 \searrow -\infty$ ranging in the interval $(-\infty, -\pi]$, the continuous, strictly decreasing function $R(x_0)$ tends to $+\infty$, and the arc of $\gamma(x_0, R(x_0))$ lying in the half-plane $x > 0$ tends to the vertical line-segment from the origin to A. If, instead, we let x_0 vary in $[-\pi, 0)$ and tend to 0, then $R(x_0)$ tends again to $+\infty$, and the plane region bounded by $\gamma(x_0, R(x_0))$ and the x-axis invades continuously and monotonically the quadrant $x > 0$, $y \leq 0$. Since $x_B > 0 = x_A$ and $y_B \leq 0$ by assumption, the lemma follows. □

Now we can prove the statement made at the beginning.

Proof of Proposition 1 The uniqueness of the minimizer follows from the fact that any minimizer is a classical solution, negative in the interval I, to the Euler equation (13). Indeed, if the difference $w = u - v$ of two solutions coinciding at the endpoints does not vanish identically, then it has at least an interior, positive maximum, or an interior negative minimum. Without loss of generality we may assume we are in the last case, and there exists $x_0 \in I$ such that $w(x_0) = \min_{x \in \bar{I}} w(x) < 0$. But then $u'(x_0) = v'(x_0)$ and $u(x_0) < v(x_0) < 0$. Since $f(y) = |y|^{-1/2}$ and $f'(y) = -\frac{1}{2}\operatorname{sgn}(y)|y|^{-3/2}$, from (13) we get $w''(x_0) < 0$, which is impossible at a minimum. Therefore the minimizer is unique, as claimed.

To complete the proof, observe that by Proposition 2 there exists a unique $R > 0$ such that the cycloid $\gamma(a, R)$, whose parametric representation is (20)–(21) and which starts from the point $(a, 0)$, also passes through (b, u_b). By computation, we find that $\gamma(a, R)$ is the graph of a function $y(x)$ which satisfies $y'(x) = -(1 - \cos\vartheta)^{-1}\sin\vartheta$ and $y''(x) = (1 - \cos\vartheta)^{-2}R^{-1}$, where x is related to ϑ by (20), and $\vartheta \in (0, 2\pi)$. Hence such a function satisfies the Euler equation (13), and the conclusion follows. □

References

1. Akhiezer, N.I.: The Calculus of Variations. A Blaisdell Book in the Pure and Applied Sciences. Blaisdell, New York (1962)
2. Alvarez, O., Lasry, J.-M., Lions, P.-L.: Convex viscosity solutions and state constraints. J. Math. Pures Appl. **76**, 265–288 (1997)
3. Bliss, G.A.: Calculus of Variations. The Carus Mathematical Monographs, vol. 1. The Mathematical Association of America/The Open Court Publishing Company, Washington/La Salle (1925)
4. Botteron, B., Marcellini, P.: A general approach to the existence of minimizers of one-dimensional non-coercive integrals of the calculus of variations. Ann. Inst. Henri Poincaré, Anal. Non Linéaire **8**, 197–223 (1991)
5. Buttazzo, G., Giaquinta, M., Hildebrandt, S.: One-Dimensional Variational Problems. An Introduction. Oxford University Press, New York (1998)
6. Cellina, A., Ferriero, A., Marchini, E.M.: On the existence of solutions to a class of minimum time control problems and applications to Fermat's principle and to the brachistochrone. Syst. Control Lett. **55**, 119–123 (2006)
7. Clarke, F.H.: An indirect method in the calculus of variations. Trans. Am. Math. Soc. **336**, 655–673 (1993)
8. Colesanti, A., Salani, P., Francini, E.: Convexity and asymptotic estimates for large solutions of Hessian equations. Differ. Integral Equ. **13**, 1459–1472 (2000)
9. Cupini, G., Marcelli, C.: Monotonicity properties of minimizers and relaxation for autonomous variational problems. ESAIM Control Optim. Calc. Var. **17**, 222–242 (2011)
10. Cupini, G., Guidorzi, M., Marcelli, C.: Existence of minimizers of free autonomous variational problems via solvability of constrained ones. Ann. Inst. Henri Poincaré, Anal. Non Linéaire **26**, 1183–1205 (2009)
11. Elsgolts, L.: Differential Equations and the Calculus of Variations. MIR, Moscow (1970)
12. Giaquinta, M., Hildebrandt, S.: Calculus of Variations. I. Grundlehren der Mathematischen Wissenschaften, vol. 310. Springer, Berlin (1996)
13. Giaquinta, M., Hildebrandt, S.: Calculus of Variations. II. Grundlehren der Mathematischen Wissenschaften, vol. 311. Springer, Berlin (1996)
14. Galilei, G.: Dialogo sopra i due massimi sistemi del mondo. Opere di G. Galilei. II. UTET, Torino (1999)
15. Greco, A.: Minimization of non-coercive integrals by means of convex rearrangement. Adv. Calc. Var. **5**, 231–249 (2012)
16. http://web.math.unifi.it/users/magnanin/Online1/DiFuria/tesi.html
17. Kline, M.: Mathematical Thought from Ancient to Modern Times. Oxford University Press, New York (1972)
18. Marcellini, P.: Non convex integrals of the calculus of variations. In: Methods of Nonconvex Analysis, Varenna, 1989. Lecture Notes in Math., vol. 1446, pp. 16–57. Springer, Berlin (1990)
19. Rockafellar, R.T.: Convex Analysis. Princeton Mathematical Series, vol. 28. Princeton University Press, Princeton (1970)
20. Tonelli, L.: Fondamenti di Calcolo delle Variazioni. I. Nicola Zanichelli, Bologna (1922)
21. Troutman, J.L.: Variational Calculus with Elementary Convexity. Springer, New York (1983)
22. Weinstock, R.: Calculus of Variations with Applications to Physics and Engineering. McGraw-Hill, New York (1952)

Existence of Minimizers for Some Coupled Nonlinear Schrödinger Equations

Norihisa Ikoma

Abstract The existence and nonexistence of minimizers for two minimizing problems are considered. These problems are related to three coupled nonlinear Schrödinger equations which appear in the field of nonlinear optics. We observe the relationship between two minimizing problems and also study the asymptotic behavior when the coefficient standing for the nonlocal nonlinearity tend to 0.

Keywords Minimization problem · Two constraints conditions · Nonlocal nonlinearity

1 Introduction

In this paper, we are interested in minimizing problems (see (3) and (4) below) related to the following coupled nonlinear Schrödinger equations:

$$\begin{cases} i\partial_t\psi_1 + \Delta\psi_1 + \mu_1\psi_1\psi_3 = 0 & \text{in } \mathbf{R}\times\mathbf{R}^N, \\ i\partial_t\psi_2 + \Delta\psi_2 + \mu_2\psi_2\psi_3 = 0 & \text{in } \mathbf{R}\times\mathbf{R}^N, \\ -\varepsilon^2\Delta\psi_3 + \omega\psi_3 = \mu_1|\psi_1|^2 + \mu_2|\psi_2|^2 & \text{in } \mathbf{R}\times\mathbf{R}^N \end{cases} \tag{1}$$

where $\partial_t = \partial/\partial t$, $\mu_1, \mu_2, \varepsilon, \omega > 0$ are constants, $\psi_1, \psi_2 : \mathbf{R}\times\mathbf{R}^N \to \mathbf{C}$ and $\psi_3 : \mathbf{R}\times\mathbf{R}^N \to \mathbf{R}$ are unknown functions. Since (1) is symmetric with respect to the indices 1 and 2, we may suppose $0 < \mu_2 \le \mu_1$ without loss of generality.

Equation (1) appears in the filed of nonlinear optics to describe the propagation of two mutually incoherent laser beams in the media which has a nonlocal nonlinearity [15]. The functions ψ_1 and ψ_2 stand for light field amplitudes, and the term ψ_3 the effect of nonlocal nonlinearity. For details, see [15] and references therein.

In [15], the authors studied the existence of standing wave solution of (1) and its profiles by the numerical method. Here the standing wave solution of (1) is

N. Ikoma (✉)
Mathematical Institute, Tohoku University, 6-3, Aoba, Aramaki, Aoba-ku Sendai-Shi, Miyagi 980-8578, Japan
e-mail: ikoma@math.tohoku.ac.jp

R. Magnanini et al. (eds.), *Geometric Properties for Parabolic and Elliptic PDE's*, Springer INdAM Series 2, DOI 10.1007/978-88-470-2841-8_10,

a solution of form $\psi_1(t,x) = e^{i\lambda_1 t}u_1(x)$, $\psi_2(t,x) = e^{i\lambda_2 t}u_2(x)$, $\psi_3(t,x) = u_3(x)$ where $\lambda_1, \lambda_2 \in \mathbf{R}$ and u_1, u_2, u_3 are solutions of

$$\begin{cases} -\Delta u_1 + \lambda_1 u_1 = \mu_1 u_1 u_3 & \text{in } \mathbf{R}^N, \\ -\Delta u_2 + \lambda_2 u_2 = \mu_2 u_2 u_3 & \text{in } \mathbf{R}^N, \\ -\varepsilon^2 \Delta u_3 + \omega u_3 = \mu_1 |u_1|^2 + \mu_2 |u_2|^2 & \text{in } \mathbf{R}^N. \end{cases} \tag{2}$$

In the present paper we have two aims. The first one is to give a rigorous proof of the existence of standing wave solutions of (1). Secondly, we observe asymptotic behaviors of standing wave solutions as $\varepsilon \to 0$.

In order to find a solution of (2), we note that (1) has the following three conserved quantities: $\|\psi_1(t)\|_{L^2} = \|\psi_1(0)\|_{L^2}$, $\|\psi_2(t)\|_{L^2} = \|\psi_2(0)\|_{L^2}$ and $E(\psi(t)) = E(\psi(0))$. Here $\psi(t,x) = (\psi_1(t,x), \psi_2(t,x), \psi_3(t,x))$ is a solution of (1) and for $u = (u_1, u_2, u_3)$ the functional $E(u)$ is defined by

$$\begin{aligned} E(u) := &\frac{1}{2}\int_{\mathbf{R}^N} \left(|\nabla u_1|^2 + |\nabla u_2|^2\right) dx + \frac{1}{4}\int_{\mathbf{R}^N} \left(\varepsilon^2 |\nabla u_3|^2 + \omega u_3^2\right) dx \\ &- \frac{1}{2}\int_{\mathbf{R}^N} \left(\mu_1 |u_1|^2 + \mu_2 |u_2|^2\right) u_3 \, dx. \end{aligned}$$

We consider the following minimizing problems: For each $\alpha_1, \alpha_2 > 0$,

$$d_1(\varepsilon, \alpha_1, \alpha_2) := \inf\left\{ E(u) : u \in H,\ \|u_1\|_{L^2}^2 = \alpha_1,\ \|u_2\|_{L^2}^2 = \alpha_2 \right\} \tag{3}$$

where $H := H^1 \times H^1 \times H^1_{\mathbf{R}}$ and $H^1 := H^1(\mathbf{R}^N, \mathbf{C})$, $H^1_{\mathbf{R}} := H^1(\mathbf{R}^N, \mathbf{R})$. Remark that any minimizer of (3) is a solution of (2) with some $\lambda_1, \lambda_2 \in \mathbf{R}$, i.e., standing wave solution of (1).

Henceforth, we consider the solvability of (3). We define a set of minimizers:

$$\mathscr{M}_1(\varepsilon, \alpha_1, \alpha_2) := \left\{ u \in H : \|u_1\|_{L^2}^2 = \alpha_1,\ \|u_2\|_{L^2}^2 = \alpha_2,\ E(u) = d_1(\varepsilon, \alpha_1, \alpha_2) \right\}.$$

Our first result is concerned with the existence and nonexistence result of minimizing problem (3).

Theorem 1 *Let* $\varepsilon, \omega > 0$ *and* $\mu_1 \ge \mu_2 > 0$. *Then*

(i) *When* $N = 1$, $\mathscr{M}_1(\varepsilon, \alpha_1, \alpha_2) \neq \emptyset$ *for each* $\alpha_1, \alpha_2 > 0$.
(ii) *When* $N = 2$, *there exist* $0 < \underline{\alpha} \le \overline{\alpha} < \infty$ *such that* $\mathscr{M}_1(\varepsilon, \alpha_1, \alpha_2) \neq \emptyset$ *if* $\overline{\alpha} < \max\{\alpha_1, \alpha_2\}$, *and* $\mathscr{M}_1(\varepsilon, \alpha_1, \alpha_2) = \emptyset$ *if* $\max\{\alpha_1, \alpha_2\} < \underline{\alpha}$.
(iii) *When* $N = 3$, *there exist* $0 < \underline{\alpha} \le \overline{\alpha} < \infty$ *such that* $\mathscr{M}_1(\varepsilon, \alpha_1, \alpha_2) \neq \emptyset$ *if* $\overline{\alpha} < \min\{\alpha_1, \alpha_2\}$, *and* $\mathscr{M}_1(\varepsilon, \alpha_1, \alpha_2) = \emptyset$ *if* $\max\{\alpha_1, \alpha_2\} < \underline{\alpha}$.

Furthermore, for any $\alpha_1, \alpha_2 > 0$ (*with* $\overline{\alpha} < \max\{\alpha_1, \alpha_2\}$ *when* $N = 2$ *and* $\overline{\alpha} < \min\{\alpha_1, \alpha_2\}$ *when* $N = 3$), *there holds*

$$\begin{aligned} &\mathscr{M}_1(\varepsilon, \alpha_1, \alpha_2) \\ &\quad = \Big\{ \left(e^{i\theta_1} w_1(x-y), e^{i\theta_2} w_2(x-y), w_3(x-y)\right) : \theta_1, \theta_2 \in \mathbf{R},\ y \in \mathbf{R}^N, \\ &\qquad w_j\ (j = 1, 2, 3) \text{ is positive and radial solution of (2) with some } \lambda_1, \lambda_2 > 0 \Big\}. \end{aligned}$$

Remark 1 (i) We remark that in the existence result of Theorem 1, there is a slight difference between the cases $N=2$ and $N=3$. Indeed, we suppose that both of α_1 and α_2 are large when $N=3$. On the contrary, when $N=2$, we assume that one of α_1 and α_2 is large. This difference comes when we show that a minimizing sequence has a strongly convergence subsequence. See Proposition 4.

(ii) In [10], the authors consider another type of minimizing problem related to the nonlinear Schrödinger equations with nonzero conditions at infinity. They also study the existence and nonexistence result of minimizers and show that there is no minimizer if the minimum value is linear. For more precise statements, see [10].

(iii) In [28] the author studied a minimizing problem of nonlinear Schrödinger equations with general nonlinear terms and showed the orbital stability of the set of minimizers. He also studied the nonexistence of minimizers.

Next, we introduce another minimizing problem: For any $\alpha > 0$,

$$\begin{aligned} d_2(\varepsilon,\alpha) &:= \inf\bigl\{E(u) : u \in H,\ \|u_1\|_{L^2}^2 + \|u_2\|_{L^2}^2 = \alpha\bigr\}, \\ \mathcal{M}_2(\varepsilon,\alpha) &:= \bigl\{u \in H : \|u_1\|_{L^2}^2 + \|u_2\|_{L^2}^2 = \alpha,\ E(u) = d_2(\varepsilon,\alpha)\bigr\}. \end{aligned} \tag{4}$$

We remark that any minimizer for $d_2(\varepsilon,\alpha)$ is also a solution of (2) with some $\lambda_1 = \lambda_2 \in \mathbf{R}$. By definition, $d_2(\varepsilon,\alpha) \le d_1(\varepsilon,\alpha_1,\alpha_2)$ holds for every $\alpha,\alpha_1,\alpha_2 > 0$ with $\alpha = \alpha_1 + \alpha_2$.

First we state the solvability of (4).

Proposition 1 *Let $\varepsilon, \omega > 0$ and $\mu_1 \ge \mu_2 > 0$. Then:*

(i) $(N=1)$ *$\mathcal{M}_2(\varepsilon,\alpha) \ne \emptyset$ for all $\alpha > 0$.*
(ii) $(N=2,3)$ *There exist $0 < \underline{\alpha} \le \overline{\alpha} < \infty$ such that $\mathcal{M}_2(\varepsilon,\alpha) \ne \emptyset$ if $\overline{\alpha} < \alpha$ and $\mathcal{M}_2(\varepsilon,\alpha) = \emptyset$ if $\alpha < \underline{\alpha}$.*

Next we state the relationship between (3) and (4).

Theorem 2

(i) *Let $\mu_1 = \mu_2$. Then there exists an $\alpha_0 \ge 0$ such that if $\alpha > \alpha_0$ then $d_1(\varepsilon,\alpha_1,\alpha_2) = d_2(\varepsilon,\alpha)$ holds for any $\alpha_1,\alpha_2 > 0$ with $\alpha = \alpha_1 + \alpha_2$. In particular, $\mathcal{M}_1(\varepsilon,\alpha_1,\alpha_2) \subset \mathcal{M}_2(\varepsilon,\alpha)$ and*

$$\begin{aligned} &\mathcal{M}_2(\varepsilon,\alpha) \\ &\quad = \bigl\{\bigl(e^{i\theta_2} w_1(x-y)\cos\theta_1, e^{i\theta_3} w_1(x-y)\sin\theta_1, w_3(x-y)\bigr) : \theta_1,\theta_2,\theta_3 \in \mathbf{R}, \\ &\qquad y \in \mathbf{R}^N,\ w_1, w_3 \text{ are positive and radial solution of (5) below with some} \\ &\qquad \lambda > 0\bigr\}. \end{aligned}$$

$$\begin{cases} -\Delta u_1 + \lambda u_1 = \mu_1 u_1 u_3 & \text{in } \mathbf{R}^N, \\ -\varepsilon^2 \Delta u_3 + \omega u_3 = \mu_1 |u_1|^2 & \text{in } \mathbf{R}^N. \end{cases} \tag{5}$$

(ii) *Let* $\mu_1 > \mu_2$. *Then there exists an* $\alpha_0 \geq 0$ *such that for any* $\alpha > \alpha_0$,

$$\mathcal{M}_2(\varepsilon,\alpha) = \big\{ \big(e^{i\theta} w_1(x-y), 0, w_3(x-y)\big) : \theta \in \mathbf{R},\ y \in \mathbf{R}^N, \\ w_1, w_3 \text{ are positive and radial solution of (5) with some } \lambda > 0 \big\}.$$

In particular, if $\mu_1 > \mu_2$ *and* α_1, α_2 *satisfy* $\alpha = \alpha_1 + \alpha_2 > \alpha_0$, *then* $d_2(\varepsilon,\alpha) < d_1(\varepsilon,\alpha_1,\alpha_2)$ *holds.*

Remark 2 When $N = 1$, we can choose $\alpha_0 = 0$ in Theorem 2. On the other hand, when $N = 2, 3$, $d_1(\varepsilon,\alpha_1,\alpha_2) = 0 = d_2(\varepsilon,\alpha)$ hold for sufficiently small $\alpha_1,\alpha_2 > 0$ even though $\mu_1 > \mu_2$. See Corollary 1.

Lastly we consider the behavior of $d_1(\varepsilon,\alpha_1,\alpha_2)$ and $\mathcal{M}_1(\varepsilon,\alpha_1,\alpha_2)$ as $\varepsilon \to 0$. In this case, we assume $N = 1$ and the following equation plays an important role:

$$\begin{cases} i\partial_t \psi_1 + \partial_x^2 \psi_1 + \big(\tilde{\mu}_1 |\psi_1|^2 + \tilde{\beta} |\psi_2|^2\big)\psi_1 = 0 & \text{in } \mathbf{R}\times\mathbf{R}, \\ i\partial_t \psi_2 + \partial_x^2 \psi_2 + \big(\tilde{\beta} |\psi_1|^2 + \tilde{\mu}_2 |\psi_2|^2\big)\psi_2 = 0 & \text{in } \mathbf{R}\times\mathbf{R} \end{cases} \tag{6}$$

where $\tilde{\mu}_1, \tilde{\mu}_2, \tilde{\beta} > 0$ are constants. Equation (6) with $\tilde{\mu}_j = (\mu_j^2/\omega)$ and $\tilde{\beta} = (\mu_1\mu_2/\omega)$ appears formally when we set $\varepsilon = 0$ in (1) and substitute $\psi_3 = (\mu_1/\omega)|\psi_1|^2 + (\mu_2/\omega)|\psi_2|^2$ into the first and second equations in (1). Recently, many researchers have studied (6) regarding the existence of standing wave solutions and its stability. For instance, see [1–4, 6, 11, 13, 14, 18, 20–22, 24, 26, 29] and references therein.

We remark that (6) has also three conserved quantities: $\|\psi_1(t)\|_{L^2} = \|\psi_1(0)\|_{L^2}$, $\|\psi_2(t)\|_{L^2} = \|\psi_2(0)\|_{L^2}$ and $G(\psi(t)) = G(\psi(0))$ where $\psi(t,x) = (\psi_1(t,x), \psi_2(t,x))$ is a solution of (6) and for $u = (u_1,u_2)$ the functional $G(u)$ is defined by

$$G(u) := \frac{1}{2}\int_{\mathbf{R}} \big(|u_1'|^2 + |u_2'|^2\big)\,dx - \frac{1}{4}\int_{\mathbf{R}} \big(\tilde{\mu}_1|u_1|^4 + 2\tilde{\beta}|u_1u_2|^2 + \tilde{\mu}_2|u_2|^4\big)\,dx.$$

In a similar way to (3), we introduce the following minimizing problem: For any $\alpha_1, \alpha_2 > 0$,

$$d_3(\alpha_1,\alpha_2) := \inf\big\{ G(u) : u \in H^1 \times H^1,\ \|u_1\|_{L^2}^2 = \alpha_1,\ \|u_2\|_{L^2}^2 = \alpha_2 \big\},$$

and denote sets of minimizers for $d_3(\alpha_1,\alpha_2)$ by $\mathcal{M}_3(\alpha_1,\alpha_2)$.

The existence of a minimizer to $d_3(\alpha_1,\alpha_2)$ is studied in [9]. See also [22, 24, 26]. In [9], the authors showed $d_3(\alpha_1,\alpha_2) < 0$ and $\mathcal{M}_3(\alpha_1,\alpha_2) \neq \emptyset$ for all $\alpha_1,\alpha_2 > 0$.

Note that when $N = 1$, by Theorem 1, $\mathcal{M}_1(\varepsilon,\alpha_1,\alpha_2) \neq \emptyset$ holds for all $\varepsilon,\alpha_1,\alpha_2 > 0$. As asymptotic behaviors of $d_1(\varepsilon,\alpha_1,\alpha_2)$ and $\mathcal{M}_1(\varepsilon,\alpha_1,\alpha_2)$ as $\varepsilon \to 0$, we have the following.

Theorem 3 *Let* $N = 1$, $\tilde{\mu}_1 := \mu_1^2/\omega$, $\tilde{\mu}_2 := \mu_2^2/\omega$, $\tilde{\beta} = \mu_1\mu_2/\omega$ *and* $\alpha_1,\alpha_2 > 0$. *Then*

(i) *As* $\varepsilon \to 0$, $d_1(\varepsilon,\alpha_1,\alpha_2) \to d_3(\alpha_1,\alpha_2)$.

(ii) *For any* $\eta > 0$, *there exists an* $\varepsilon_\eta > 0$ *such that if* $\varepsilon \le \varepsilon_\eta$, *then*

$$\begin{aligned}&\mathrm{dist}_{H^1\times H^1}\big((u_1,u_2),\mathscr{M}_3(\alpha_1,\alpha_2)\big)\\&\quad+\big\|u_3-\big[(\mu_1/\omega)|u_1|^2+(\mu_2/\omega)|u_2|^2\big]\big\|_{H^1}\le\eta\end{aligned}$$

for all $(u_1,u_2,u_3)\in\mathscr{M}_1(\varepsilon,\alpha_1,\alpha_2)$.

In Sect. 2, we shall show basic properties of (3) and (4), and the nonexistence result. In Sect. 3, we prove Theorem 1 and Proposition 1. Section 4 is devoted to proving Theorem 2 and Theorem 3 is shown in Sect. 5.

2 Preliminaries

In this section, we prove basic properties related to the minimizing problems (3) and (4), and the nonexistence result.

In what follows, we use the following abbreviation: $L^p(\mathbf{R}^N)=L^p$ for $1\le p\le\infty$.

Next we introduce some notations. For any $f\in(H^1_{\mathbf{R}})^*$, the equation $-\varepsilon^2\Delta u+\omega u=f$ in $\mathbf{R}^N$ has a unique solution $u\in H^1_{\mathbf{R}}$. We denote it by $\Phi_{\varepsilon,\omega}[f]\in H^1_{\mathbf{R}}$. Since $1\le N\le 3$, for any $u_1,u_2\in H^1$, $\mu_1|u_1|^2+\mu_2|u_2|^2\in(H^1_{\mathbf{R}})^*$ holds. Moreover, $\Phi_{\varepsilon,\omega}$ has the following expression:

$$\Phi_{\varepsilon,\omega}\big[\mu_1|u_1|^2+\mu_2|u_2|^2\big](x)=\int_{\mathbf{R}^N}G_{N,\varepsilon,\omega}(x-y)\big(\mu_1\big|u_1(y)\big|^2+\mu_2\big|u_2(y)\big|^2\big)\,dy$$

where

$$G_{N,\varepsilon,\omega}(x):=\begin{cases}(2\varepsilon\sqrt{\omega})^{-1}\exp(-\sqrt{\omega}\,|x|/\varepsilon) & \text{if } N=1,\\(2\pi\varepsilon^2)^{-1}K_0(\sqrt{\omega}\,|x|/\varepsilon) & \text{if } N=2,\\(4\pi\varepsilon^2|x|)^{-1}\exp(-\sqrt{\omega}\,|x|/\varepsilon) & \text{if } N=3\end{cases}$$

and K_0 is the modified Bessel function defined by

$$K_0\big(|x|\big):=e^{-|x|}\big(2|x|\big)^{-1/2}\int_0^\infty e^{-t}t^{-1/2}\left(1+\frac{t}{2|x|}\right)^{-1/2}dt>0,$$

and $K_0(|x|)\sim-\log|x|$ as $|x|\to 0$. For instance, see [12]. Note that $G_{N,\varepsilon,\omega}\in L^p$ for $1\le p\le\infty$ $(N=1)$, $1\le p<\infty$ $(N=2)$ and $1\le p<3$ $(N=3)$.

We remark that $\Phi_{\varepsilon,\omega}[\mu_1|u_1|^2+\mu_2|u_2|^2](x)$ is characterized as the unique minimizer of

$$\begin{aligned}&\inf\big\{I_{u_1,u_2}(\varphi):\varphi\in H^1_{\mathbf{R}}\big\},\\&I_{u_1,u_2}(\varphi):=\frac{1}{2}\int_{\mathbf{R}^N}\big(\varepsilon^2|\nabla\varphi|^2+\omega\varphi^2\big)\,dx-\int_{\mathbf{R}^N}\big(\mu_1|u_1|^2+\mu_2|u_2|^2\big)\varphi\,dx.\end{aligned}$$

Hence we obtain

$$\begin{aligned} E(u_1,u_2,u_3) &= \frac{1}{2}\left(\|\nabla u_1\|_{L^2}^2 + \|\nabla u_2\|_{L^2}^2\right) + \frac{1}{2} I_{u_1,u_2}(u_3) \\ &\geq E\left(u_1,u_2,\Phi_{\varepsilon,\omega}\left[\mu_1|u_1|^2 + \mu_2|u_2|^2\right]\right) \end{aligned} \tag{7}$$

for every $(u_1,u_2,u_3) \in H$. Thus setting $F(u_1,u_2) := E(u_1,u_2,\Phi_{\varepsilon,\omega}[\mu_1|u_1|^2 + \mu_2|u_2|^2]) \in C^1(H^1 \times H^1, \mathbf{R})$, by (7) we have $E(u_1,u_2,u_3) \geq F(u_1,u_2)$ for each $(u_1,u_2,u_3) \in H$. Since $\Phi_{\varepsilon,\omega}[\mu_1|u_1|^2 + \mu_2|u_2|^2]$ is a solution of $-\varepsilon^2 \Delta u + \omega u = \mu_1|u_1|^2 + \mu_2|u_2|^2$ in $\mathbf{R}^N$, we also have

$$\begin{aligned} F(u_1,u_2) &= \frac{1}{2}\left(\|\nabla u_1\|_{L^2}^2 + \|\nabla u_2\|_{L^2}^2\right) \\ &\quad - \frac{1}{4}\int_{\mathbf{R}^N} \left(\mu_1|u_1|^2 + \mu_2|u_2|^2\right)\Phi_{\varepsilon,\omega}\left[\mu_1|u_1|^2 + \mu_2|u_2|^2\right] dx. \end{aligned} \tag{8}$$

Moreover, the following hold:

Lemma 1 *Let $(v_n)_{n=1}^\infty \subset H$ satisfy $\|v_{n,1}\|_{L^2}^2 = \alpha_1$, $\|v_{n,2}\|_{L^2}^2 = \alpha_2$ and $E(v_n) \to d_1(\varepsilon,\alpha_1,\alpha_2)$. Then $\|v_{n,3} - \Phi_{\varepsilon,\omega}[\mu_1|v_{n,1}|^2 + \mu_2|v_{n,2}|^2]\|_{H^1} \to 0$. Similarly, if (v_n) is a minimizing sequence for $d_2(\varepsilon,\alpha)$, then $\|v_{n,3} - \Phi_{\varepsilon,\omega}[\mu_1|v_{n,1}|^2 + \mu_2|v_{n,2}|^2]\|_{H^1} \to 0$.*

Proof We only consider for $d_1(\varepsilon,\alpha_1,\alpha_2)$. We argue by contradiction and suppose that there is an $\eta_0 > 0$ such that

$$\left\|v_{n,3} - \Phi_{\varepsilon,\omega}\left[\mu_1|v_{n,1}|^2 + \mu_2|v_{n,2}|^2\right]\right\|_{H^1} \geq \eta_0 \quad \text{for all } n \in \mathbf{N}. \tag{9}$$

Next, we consider a family of functionals defined by

$$\begin{aligned} I_n(\varphi) &:= \frac{1}{2}\int_{\mathbf{R}^N} \left(\varepsilon^2|\nabla\varphi|^2 + \omega\varphi^2\right) dx \\ &\quad - \int_{\mathbf{R}^N} \left(\mu_1|v_{n,1}|^2 + \mu_2|v_{n,2}|^2\right)\varphi \, dx \in C^1\left(H^1_{\mathbf{R}}, \mathbf{R}\right). \end{aligned}$$

Note that $I_n'(\Phi_{\varepsilon,\omega}[\mu_1|v_{n,1}|^2 + \mu_2|v_{n,2}|^2]) = 0$ and $I_n''(u)[\varphi,\varphi] = \int_{\mathbf{R}^N}(\varepsilon^2|\nabla\varphi|^2 + \omega\varphi^2)\,dx$ for all $u,\varphi \in H^1_{\mathbf{R}}$. Since $\Phi_{\varepsilon,\omega}[\mu_1|v_{n,1}|^2 + \mu_2|v_{n,2}|^2]$ is a unique minimizer of I_n in $H^1_{\mathbf{R}}$, it follows from (9) that there exists an $\eta_1 > 0$ such that $I_n(v_{n,3}) \geq I_n(\Phi_{\varepsilon,\omega}[\mu_1|v_{n,1}|^2 + \mu_2|v_{n,2}|^2]) + \eta_1$ for all $n \in \mathbf{N}$. Therefore, $E(v_n) \geq F(v_{n,1},v_{n,2}) + \eta_1/2$ holds for every $n \in \mathbf{N}$, which implies $d_1(\varepsilon,\alpha_1,\alpha_2) \geq d_1(\varepsilon,\alpha_1,\alpha_2) + \eta_1/2$. This is a contradiction and Lemma 1 holds. □

Next, we prove that any minimizing sequence of (3) (resp. (4)) is bounded in H.

Proposition 2 *For any $\eta > 0$, there exists an $M_{\eta,\alpha} > 0$ such that if $(u_1,u_2,u_3) \in H$ satisfies $E(u_1,u_2,u_3) \leq \eta$ and $\|u_1\|_{L^2}^2 + \|u_2\|_{L^2}^2 = \alpha$, then $\|u_1\|_{H^1} + \|u_2\|_{H^1} + \|u_3\|_{H^1} \leq M_{\eta,\alpha}$. Especially, $-\infty < d_1(\varepsilon,\alpha_1,\alpha_2)$ and $-\infty < d_2(\varepsilon,\alpha)$ hold for every $\alpha,\alpha_1,\alpha_2 > 0$.*

Proof Since E is bounded on any bounded set, we only show the boundedness of (u_1, u_2, u_3). Secondly, we remark that it is sufficient to show $\|u_1\|_{H^1} + \|u_2\|_{H^1} \le M_{\eta,\alpha}$ for all $u_1, u_2 \in H^1$ with $E(u_1, u_2, u_3) \le \eta$ and $\|u_1\|_{L^2}^2 + \|u_2\|_{L^2}^2 = \alpha$. In fact, from $E(u) \le \eta$, we infer that

$$\begin{aligned}\frac{1}{4}\int_{\mathbf{R}^N} &\left(\varepsilon^2|\nabla u_3|^2 + \omega u_3^2\right) dx \\ &\le \eta + \frac{1}{2}\left(\|\nabla u_1\|_{L^2}^2 + \|\nabla u_2\|_{L^2}^2\right) + \frac{1}{2}\int_{\mathbf{R}^N}\left(\mu_1|u_1|^2 + \mu_2|u_2|^2\right)u_3\,dx. \qquad (10)\end{aligned}$$

Thus, if we take an $M_{\eta,\alpha} > 0$ sufficiently large, then $\|u_3\|_{H^1} \le M_{\eta,\alpha}$ follows from (10).

Henceforth, we shall prove the boundedness for u_1 and u_2. First note that by Gagliardo-Nirenberg's inequality and Young's inequality, we have

$$\|\varphi\|_{L^{p_1}} \le C\|\nabla\varphi\|_{L^2}^a\|\varphi\|_{L^{q_1}}^{1-a}, \qquad (11)$$

$$\|(f * g)h\|_{L^1} \le \|f\|_{L^{p_2}}\|g\|_{L^{q_2}}\|h\|_{L^{r_2}} \qquad (12)$$

where $f * g$ stands for the convolution, $1/p_1 = a(1/2 - 1/N) + (1-a)/q_1$ and $1/p_2 + 1/q_2 + 1/r_2 = 2$.

Let $\|u_1\|_{L^2}^2 + \|u_2\|_{L^2}^2 = \alpha$ and $E(u_1, u_2, u_3) \le \eta$. Then by (8), (11) and (12) with $p_1 = 3$, $q_1 = 2$ and $p_2 = q_2 = r_2 = 3/2$, one obtains

$$\begin{aligned}\eta &\ge E(u_1, u_2, u_3) \ge F(u_1, u_2) \\ &\ge \frac{1}{2}\left(\|\nabla u_1\|_{L^2}^2 + \|\nabla u_2\|_{L^2}^2\right) - \frac{\|G_{N,\varepsilon,\omega}\|_{L^{3/2}}}{4}\left(\mu_1\|u_1\|_{L^3}^2 + \mu_2\|u_2\|_{L^3}^2\right)^2 \\ &\ge \frac{1}{2}\left(\|\nabla u_1\|_{L^2}^2 + \|\nabla u_2\|_{L^2}^2\right) \\ &\quad - C\left(\|\nabla u_1\|_{L^2}^{2N/3}\|u_1\|_{L^2}^{4-2N/3} + \|\nabla u_2\|_{L^2}^{2N/3}\|u_2\|_{L^2}^{4-2N/3}\right).\end{aligned}$$

When $N = 1, 2$, noting $2N/3 < 2$ and $\|u_j\|_{L^2}^2 \le \alpha$ for $j = 1, 2$, it holds that

$$\eta \ge \sum_{j=1}^{2}\left(\frac{\|\nabla u_j\|_{L^2}^2}{2} - \alpha^{2-N/3}C\|\nabla u_j\|_{L^2}^{2N/3}\right).$$

Thus there is an $\tilde{M}_{\eta,\alpha} > 0$ satisfying $\|\nabla u_1\|_{L^2} + \|\nabla u_2\|_{L^2} \le \tilde{M}_{\eta,\alpha}$ for every $u_1, u_2 \in H^1$ with $\|u_1\|_{L^2}^2 + \|u_2\|_{L^2}^2 = \alpha$ and $E(u_1, u_2, u_3) \le \eta$.

In the case where $N = 3$, we use the argument in [16]. For $R > 0$, we set $G_{R,1} := \chi_{B_R(0)}G_{N,\varepsilon,\omega}$ and $G_{R,2} := (1 - \chi_{B_R(0)})G_{N,\varepsilon,\omega}$ where $\chi_{B_R(0)}$ is the characteristic function of $B_R(0)$. Then we see $G_{R,1} \in L^p$ with $1 \le p < 3$ and $G_{R,2} \in L^\infty$ from the property of $G_{N,\varepsilon,\omega}$.

For each $u_1, u_2 \in H^1$, we also set

$$I_{R,j} := \int_{\mathbf{R}^6}|u_1(x)|^2 G_{R,j}(x-y)|u_2(y)|^2\,dx\,dy \quad \text{for } j = 1, 2.$$

For $I_{R,2}$, it follows that $I_{R,2} \le \|G_{R,2}\|_{L^\infty}\|u_1\|_{L^2}^2\|u_2\|_{L^2}^2$. On the other hand, by (12) with $p_2 = 1$, $q_2 = 3/2$ and $r_2 = 3$, one has $I_{R,1} \le \|u_1\|_{L^2}^2\|G_{R,1}\|_{L^{3/2}}\|u_2\|_{L^6}^2$. Thus it follows from Sobolev's inequality that $I_{R,1} \le C\|G_{R,1}\|_{L^{3/2}}\|u_1\|_{L^2}^2\|\nabla u_2\|_{L^2}^2$. Hence noting $\|u_j\|_{L^2}^2 \le \alpha$ for $j = 1, 2$ we have

$$\int_{\mathbf{R}^3} |u_1|^2 \Phi_{\varepsilon,\omega}\big[|u_2|^2\big]\, dx = I_{R,1} + I_{R,2} \le C\|G_{R,1}\|_{L^{3/2}}\alpha\|\nabla u_2\|_{L^2}^2 + \|G_{R,2}\|_{L^\infty}\alpha^2.$$

Thus we obtain

$$\begin{aligned} F(u_1, u_2) \ge{}& \frac{1}{2}\big(\|\nabla u_1\|_{L^2}^2 + \|\nabla u_2\|_{L^2}^2\big) \\ &- C\big\{\|G_{R,1}\|_{L^{3/2}}\alpha\big(\|\nabla u_1\|_{L^2}^2 + \|\nabla u_2\|_{L^2}^2\big) + \|G_{R,2}\|_{L^\infty}\alpha^2\big\}. \end{aligned}$$

Choosing an $R > 0$ so small that $C\|G_{R,1}\|_{L^{3/2}}\alpha \le 1/4$, one has

$$\eta \ge E(u_1, u_2, u_3) \ge F(u_1, u_2) \ge \frac{1}{4}\big(\|\nabla u_1\|_{L^2}^2 + \|\nabla u_2\|_{L^2}^2\big) - C\|G_{R,2}\|_{L^\infty}\alpha^2.$$

Hence, we have $\|u_1\|_{H^1} + \|u_2\|_{H^1} \le M_{\eta,\alpha}$ for every $u_1, u_2 \in H^1$ with $\|u_1\|_{L^2}^2 + \|u_2\|_{L^2}^2 = \alpha$ and $E(u_1, u_2, u_3) \le \eta$. This completes the proof. □

Next for any $\rho > 0$ and $\varphi \in H^1$, we set $\varphi_\rho(x) := \rho^{N/2}\varphi(\rho x)$. Remark that $\|\varphi_\rho\|_{L^2} = \|\varphi\|_{L^2}$ holds.

Lemma 2 *For all* $\varphi_1, \varphi_2 \in C_0^\infty(\mathbf{R}^N)$, *it holds that*

$$\begin{aligned} &\lim_{\rho\to 0} \frac{1}{\rho^N} \int_{\mathbf{R}^N} \Phi_{\varepsilon,\omega}\big[|\varphi_{1,\rho}|^2\big](x)\big|\varphi_{2,\rho}(x)\big|^2\, dx \\ &\quad = \|G_{N,\varepsilon,\omega}\|_{L^1} \int_{\mathbf{R}^N} \big|\varphi_1(x)\big|^2\big|\varphi_2(x)\big|^2\, dx. \end{aligned}$$

Proof Let $\varphi_1, \varphi_2 \in C_0^\infty(\mathbf{R}^N)$. Then, one has

$$\begin{aligned} &\frac{1}{\rho^N} \int_{\mathbf{R}^N} \Phi_{\varepsilon,\omega}\big[|\varphi_{1,\rho}|^2\big](x)\big|\varphi_{2,\rho}(x)\big|^2\, dx \\ &\quad = \frac{1}{\rho^N} \int_{\mathbf{R}^{2N}} \big|\varphi_1(y)\big|^2 G_{N,\varepsilon,\omega}\Big(\frac{x-y}{\rho}\Big)\big|\varphi_2(x)\big|^2\, dx\, dy. \end{aligned}$$

Thus it is sufficient to prove

$$\begin{aligned} &\lim_{\rho\to 0} \frac{1}{\rho^N} \int_{\mathbf{R}^N} G_{N,\varepsilon,\omega}\Big(\frac{x-y}{\rho}\Big)\big|\varphi_1(y)\big|^2\, dy \\ &\quad = \|G_{N,\varepsilon,\omega}\|_{L^1}\big|\varphi_1(x)\big|^2 \quad \text{uniformly w.r.t. } x \in \mathbf{R}^N. \end{aligned} \tag{13}$$

Since

$$\frac{1}{\rho^N} \int_{\mathbf{R}^N} G_{N,\varepsilon,\omega}\Big(\frac{x-y}{\rho}\Big)\big|\varphi_1(y)\big|^2\, dy = \int_{\mathbf{R}^N} G_{N,\varepsilon,\omega}(y)\big|\varphi_1(x-\rho y)\big|^2\, dy,$$

we have

$$\left|\frac{1}{\rho^N}\int_{\mathbf{R}^N} G_{N,\varepsilon,\omega}\left(\frac{x-y}{\rho}\right)|\varphi_1(y)|^2\,dy - \|G_{N,\varepsilon,\omega}\|_{L^1}|\varphi_1(x)|^2\right|$$
$$\le \int_{\mathbf{R}^N} G_{N,\varepsilon,\omega}(y)\big||\varphi_1(x-\rho y)|^2 - |\varphi_1(x)|^2\big|\,dy.$$

Noting φ_1 is uniformly continuous in $\mathbf{R}^N$, one sees that (13) holds. □

Next we define $e_j(\varepsilon,\alpha_j)$ by

$$\begin{aligned} e_j(\varepsilon,\alpha_j) &:= \inf\{E_j(u) : u \in H^1,\ \|u\|_{L^2}^2 = \alpha_j\},\\ E_j(u) &:= \frac{1}{2}\|\nabla u\|_{L^2}^2 - \frac{\mu_j^2}{4}\int_{\mathbf{R}^N} |u|^2 \Phi_{\varepsilon,\omega}[|u|^2]\,dx. \end{aligned} \tag{14}$$

From (8), we have $F(u_1,u_2) \le E_1(u_1) + E_2(u_2)$ for each $u_1, u_2 \in H^1$, which implies

$$d_1(\varepsilon,\alpha_1,\alpha_2) \le e_1(\varepsilon,\alpha_1) + e_2(\varepsilon,\alpha_2). \tag{15}$$

Next, we show some properties of $e_j(\varepsilon,\alpha_j)$.

Lemma 3

(i) *When* $N = 1$, $-\infty < e_j(\varepsilon,\alpha_j) < 0$ *for any* $\alpha_j > 0$.
(ii) *When* $N = 2, 3$, *there is an* $\alpha_{0,j} > 0$ *such that* $e_j(\varepsilon,\alpha_j) = 0$ *if* $0 < \alpha_j \le \alpha_{0,j}$ *and* $-\infty < e_j(\varepsilon,\alpha_j) < 0$ *if* $\alpha_{0,j} < \alpha_j$.

Proof First we remark that we can show $-\infty < e_j(\varepsilon,\alpha_j)$ in a similar way to Proposition 2.

Next we show $e_j(\varepsilon,\alpha_j) \le 0$ for all $\alpha_j > 0$. Let $\varphi \in C_0^\infty(\mathbf{R}^N)$ and $\|\varphi\|_{L^2}^2 = \alpha_j$. Then it follows from Lemma 2 that $\|\nabla\varphi_\rho\|_{L^2}^2 = \rho^2\|\nabla\varphi\|_{L^2}^2$ and $\int_{\mathbf{R}^N} |\varphi_\rho|^2 \times \Phi_{\varepsilon,\omega}[|\varphi_\rho|^2]\,dx = O(\rho^N)$. Hence $E_j(\varphi_\rho) \to 0$, which implies $e_j(\varepsilon,\alpha_j) \le 0$. Moreover, when $N = 1$, we can see $e_j(\varepsilon,\alpha_j) < 0$ for all $\alpha_j > 0$.

Next we consider the case where $N = 2, 3$. First we show $e_j(\varepsilon,\alpha_j) = 0$ for sufficiently small $\alpha_j > 0$. Choose $p_1 = 2N/(N-1)$, $q_1 = 2$ in (11) and $p_2 = r_2 = p_1/2$, $q_2 = N/2$ in (12). Then by (11) and (12), for all $u_j \in H^1$ with $\|u_j\|_{L^2}^2 = \alpha_j$, one has

$$\begin{aligned} E_j(u_j) &\ge \frac{\|\nabla u_j\|_{L^2}^2}{2} - \frac{\mu_j^2}{4}\|G_{N,\varepsilon,\omega}\|_{L^{q_2}}\|u_j\|_{L^{p_1}}^4\\ &\ge \frac{\|\nabla u_j\|_{L^2}^2}{2} - C\mu_j^2\|G_{N,\varepsilon,\omega}\|_{L^{q_2}}\|\nabla u_j\|_{L^2}^2\|u_j\|_{L^2}^2\\ &= \left(\frac{1}{2} - C\mu_j^2\|G_{N,\varepsilon,\omega}\|_{L^{q_2}}\alpha_j\right)\|\nabla u_j\|_{L^2}^2. \end{aligned}$$

Hence $e_j(\varepsilon,\alpha_j) = 0$ follows provided $\alpha_j > 0$ is sufficiently small.

Lastly we prove the existence of $\alpha_{0,j} > 0$ satisfying the desired property and use the argument in [25]. For any $u_j \in H^1$ and $t \geq 1$, we have

$$E_j(tu_j) = \frac{t^2}{2}\|\nabla u_j\|_{L^2}^2 - \frac{\mu_j^2 t^4}{4}\int_{\mathbf{R}^N} |u_j|^2 \Phi_{\varepsilon,\omega}\big[|u_j|^2\big]\,dx \leq t^2 E(u_j).$$

Thus noting $e_j(\varepsilon, \alpha_j) \leq 0$, $e_j(\varepsilon, t^2\alpha_j) \leq t^2 e_j(\varepsilon, \alpha_j) \leq e_j(\varepsilon, \alpha_j)$ follows for any $\alpha_j > 0$ and $t \geq 1$, which implies the map $\alpha_j \mapsto e_j(\varepsilon, \alpha_j)$ is monotone. Moreover, $e_j(\varepsilon, \alpha_j) < 0$ holds for all sufficiently large $\alpha_j > 0$. Thus set $\alpha_{0,j} := \inf\{\tilde{\alpha}_j : e_j(\varepsilon, \alpha_j) < 0 \text{ for all } \alpha_j \in (\tilde{\alpha}_j, \infty)\}$. The above arguments show $0 < \alpha_{0,j} < \infty$ and by the continuity of E_j, one can show $e_j(\varepsilon, \alpha_j) = 0$ if $0 < \alpha_j \leq \alpha_{0,j}$ and $e_j(\varepsilon, \alpha_j) < 0$ if $\alpha_{0,j} < \alpha_j$, which completes the proof. □

As a corollary to Lemma 3, we have the following.

Corollary 1

(i) *When* $N = 1$, $d_1(\varepsilon, \alpha_1, \alpha_2) < 0$ *and* $d_2(\varepsilon, \alpha) < 0$ *hold for every* $\alpha, \alpha_1, \alpha_2 > 0$.
(ii) *When* $N = 2, 3$, *there exist* $0 < \hat{\alpha}_1 \leq \hat{\alpha}_2 < \infty$ *and* $0 < \hat{\alpha}_3 \leq \hat{\alpha}_4 < \infty$ *such that*

$$\begin{aligned}
&d_1(\varepsilon, \alpha_1, \alpha_2) = 0, \quad && \mathscr{M}_1(\varepsilon, \alpha_1, \alpha_2) = \emptyset \quad && \textit{if } \max\{\alpha_1, \alpha_2\} < \hat{\alpha}_1,\\
&d_2(\varepsilon, \alpha) = 0, && \mathscr{M}_2(\varepsilon, \alpha) = \emptyset && \textit{if } \alpha < \hat{\alpha}_3,\\
&d_1(\varepsilon, \alpha_1, \alpha_2) < 0 && \big(\textit{resp. } d_2(\varepsilon, \alpha) < 0\big) && \textit{if } \hat{\alpha}_2 < \max\{\alpha_1, \alpha_2\}\ (\textit{resp. } \hat{\alpha}_4 < \alpha).
\end{aligned}$$

Proof The assertion (i) and the existence of $\hat{\alpha}_2, \hat{\alpha}_4$ follow from Lemma 3 and (15). So we prove the existence of $\hat{\alpha}_1$ and $\hat{\alpha}_3$. Repeating the arguments in Lemma 3, we have

$$F(u_1, u_2) \geq \sum_{j=1}^{2}\left(\frac{1}{2} - C\|G_{N,\varepsilon,\omega}\|_{L^{q_2}}\alpha_j\right)\|\nabla u_j\|_{L^2}^2 \tag{16}$$

where $q_2 = N/2$. Thus choose $\hat{\alpha}_1 > 0$ satisfying $0 \leq 1/2 - C\|G_{N,\varepsilon,\omega}\|_{L^{q_2}}\hat{\alpha}_1$. Then if $\max\{\alpha_1, \alpha_2\} < \hat{\alpha}_1$, by (16) we see $F(u_1, u_2) \geq 0$ for all $u_1, u_2 \in H^1$ with $\|u_j\|_{L^2}^2 = \alpha_j$ for $j = 1, 2$, which implies $d_1(\varepsilon, \alpha_1, \alpha_2) = 0$. Moreover, $\mathscr{M}_1(\varepsilon, \alpha_1, \alpha_2) = \emptyset$ holds from $d_1(\varepsilon, \alpha_1, \alpha_2) = 0$ and (16). Similarly we can find $\hat{\alpha}_3 > 0$ satisfying the desired property. □

3 Proof of Theorem 1 and Proposition 1

3.1 Solvability of (4)

In this subsection, we prove the existence of minimizers for (4). A proof is essentially same to the one in [25], however, for the reader's sake, we shall give the proof.

We begin with the following lemma which was essentially proven in [25] and we omit a proof.

Lemma 4 (Lemma 2.3 in [25]) *Assume* $d_2(\varepsilon,\alpha)<0$. *Then*

(i) *There exists a* $\delta_\alpha>0$ *such that*

$$d_2(\varepsilon,\alpha)=\inf\bigl\{F(u_1,u_2):\|u_1\|_{L^2}^2+\|u_2\|_{L^2}^2=\alpha,\\ \bigl\|\Phi_{\varepsilon,\omega}\bigl[\mu_1|u_1|^2+\mu_2|u_2|^2\bigr]\bigr\|_{H^1}\geq\delta_\alpha\bigr\}.$$

(ii) *For any* $t>1$, $d_2(\varepsilon,t\alpha)<td_2(\varepsilon,\alpha)$ *holds.*
(iii) *For each* $\hat{\alpha}\in(0,\alpha)$, $d_2(\varepsilon,\alpha)<d_2(\varepsilon,\hat{\alpha})+d_2(\varepsilon,\alpha-\hat{\alpha})$ *holds.*

Proposition 1 follows from Corollary 1 and

Proposition 3 *Suppose that* $d_2(\varepsilon,\alpha)<0$. *Then every minimizing sequence of* (4) *has a strongly convergent subsequence after suitable translations. In particular* $d_2(\varepsilon,\alpha)$ *is attained.*

Proof Let $(v_n)_{n=1}^\infty\subset H$ satisfy $\|v_{n,1}\|_{L^2}^2+\|v_{n,2}\|_{L^2}^2=\alpha$ and $E(v_n)\to d_2(\varepsilon,\alpha)$. Then by Lemma 1, $\|v_{n,3}-\Phi_{\varepsilon,\omega}[\mu_1|v_{n,1}|^2+\mu_2|v_{n,2}|^2]\|_{H^1}\to 0$. Therefore, it is sufficient to prove that $(v_{n,j})_{n=1}^\infty$ $(j=1,2)$ has a strongly convergent subsequence after suitable translations.

By the Concentration-Compactness Lemma (see [7, 19]), there is a subsequence $(v_{n_k,j})$ $(j=1,2)$ such that one of the following three cases occurs:

(a) (**compactness**) *There exists a sequence* $(x_k)_{k=1}^\infty\subset\mathbf{R}^N$ *satisfying that for any* $\varepsilon>0$ *there is an* $R>0$ *such that* $\int_{B_R(x_k)}|v_{n_k,1}(x)|^2+|v_{n_k,2}(x)|^2\,dx\geq\alpha-\varepsilon$ *for all* $k\geq 1$.
(b) (**vanishing**) *For all* $R>0$, $\lim_{k\to\infty}\sup_{y\in\mathbf{R}^N}\int_{B_R(y)}|v_{n_k,1}(x)|^2+|v_{n_k,2}(x)|^2\,dx=0$.
(c) (**dichotomy**) There exist $\hat{\alpha}\in(0,\alpha)$, $(x_k)_{k=1}^\infty\subset\mathbf{R}^N$, $(R_{k,1})_{k=1}^\infty$, $(R_{k,2})_{k=1}^\infty$ such that $3R_{k,1}\leq R_{k,2}$, $R_{k,1}\to\infty$ and

$$\int_{|x-x_k|\leq R_{k,1}}\bigl|v_{n_k,1}(x)\bigr|^2+\bigl|v_{n_k,2}(x)\bigr|^2\,dx\to\hat{\alpha},\\ \int_{|x-x_k|\geq R_{k,2}}\bigl|v_{n_k,1}(x)\bigr|^2+\bigl|v_{n_k,2}(x)\bigr|^2\,dx\to\alpha-\hat{\alpha}. \tag{17}$$

If (b) takes place, noting that $(v_{n_k,1})$ and $(v_{n_k,2})$ are bounded in H^1 from Proposition 2, we can show $v_{n_k,j}\to 0$ strongly in L^4 (see [7]) and $\mu_1|v_{n_k,1}|^2+\mu_2|v_{n_k,2}|^2\to 0$ strongly in $(H^1_{\mathbf{R}})^*$. Thus $\|\Phi_{\varepsilon,\omega}[\mu_1|v_{n_k,1}|^2+\mu_2|v_{n_k,2}|^2]\|_{H^1}\to 0$. Since $v_{n_k,3}\to 0$ strongly in H^1 from Lemma 1, we have $d_2(\varepsilon,\alpha)=\lim_{k\to\infty}E(v_{n_k})\geq 0$. This contradicts to $d_2(\varepsilon,\alpha)<0$.

Next suppose that (c) occurs. Then we choose $\varphi_{k,1}$ and $\varphi_{k,2}$ satisfying $0\leq\varphi_{k,\ell}(x)\leq 1$, $\|\nabla\varphi_{k,\ell}\|_{L^\infty}\to 0$ and

$$\varphi_{k,1}(x)=\begin{cases}1 & \text{if } |x-x_k|\leq R_{k,1},\\ 0 & \text{if } |x-x_k|\geq 2R_{k,1},\end{cases}\qquad \varphi_{k,2}(x)=\begin{cases}0 & \text{if } |x-x_k|\leq 2R_{k,1},\\ 1 & \text{if } |x-x_k|\geq R_{k,2}.\end{cases}$$

For $j,\ell=1,2$, set $v_{k,j}^{\ell}(x):=\varphi_{k,\ell}(x)v_{n_k,j}(x)$ and $v_{k,3}^{\ell}(x):=\Phi_{\varepsilon,\omega}[\mu_1|v_{k,1}^{\ell}|^2+\mu_2|v_{k,2}^{\ell}|^2](x)$. Then,

$$\|\nabla v_{n_k,j}\|_{L^2}^2\ge \|\nabla v_{k,j}^1\|_{L^2}^2+\|\nabla v_{k,j}^2\|_{L^2}^2+o(1),\qquad \|v_{n_k,j}-v_{k,j}^1-v_{k,j}^2\|_{L^p}\to 0 \tag{18}$$

for all $p\in[2,2^*)$ where $2^*=6$ if $N=3$ and $2^*=\infty$ if $N=1,2$. Hence we have $\|\Phi_{\varepsilon,\omega}[\mu_1|v_{n_k,1}|^2+\mu_2|v_{n_k,2}|^2]-v_{k,3}^1-v_{k,3}^2\|_{H^1}\to 0$ and

$$d_2(\varepsilon,\alpha)=\lim_{k\to\infty}E(v_{n_k})\ge\lim_{k\to\infty}F(v_{n_k,1},v_{n_k,2})\ge\liminf_{k\to\infty}\bigl(E\bigl(v_k^1\bigr)+E\bigl(v_k^2\bigr)\bigr). \tag{19}$$

On the other hand, from (17) and (18), we see $\|v_{k,1}^1\|_{L^2}^2+\|v_{k,2}^1\|_{L^2}^2\to\hat{\alpha}$ and $\|v_{k,1}^2\|_{L^2}^2+\|v_{k,2}^2\|_{L^2}^2\to\alpha-\hat{\alpha}$. Since $(v_{k,j}^{\ell})$ is bounded in H^1 and E is Lipschitz continuous in bounded sets, we have $E(v_k^1)\ge d_2(\varepsilon,\hat{\alpha})+o(1)$ and $E(v_k^2)\ge d_2(\varepsilon,\alpha-\hat{\alpha})+o(1)$. So from (19), $d_2(\varepsilon,\alpha)\ge d_2(\varepsilon,\hat{\alpha})+d_2(\varepsilon,\alpha-\hat{\alpha})$ holds and this contradicts to Lemma 4.

Thus (a) occurs. Set $w_{k,j}(x):=v_{n_k,j}(x+x_k)$. Since $(v_{n_k,j})$ is bounded in H^1, we may assume $w_{k,j}\rightharpoonup w_{0,j}$ weakly in H^1. From (a), we can show $w_{k,j}\to w_{0,j}$ strongly in L^2 and $\Phi_{\varepsilon,\omega}[\mu_1|w_{k,1}|^2+\mu_2|w_{k,2}|^2]\to\Phi_{\varepsilon,\omega}[\mu_1|w_{0,1}|^2+\mu_2|w_{0,2}|^2]$ strongly in H^1. Using the lower semicontinuity of norm, we have

$$d_2(\varepsilon,\alpha)\ge\lim_{k\to\infty}F(w_{k,1},w_{k,2})\ge F(w_{0,1},w_{0,2}).$$

Noting $\|w_{0,1}\|_{L^2}^2+\|w_{0,2}\|_{L^2}^2=\alpha$, we also obtain $d_2(\varepsilon,\alpha)=F(w_{0,1},w_{0,2})$ and $\|\nabla w_{k,j}\|_{L^2}^2\to\|\nabla w_{0,j}\|_{L^2}^2$. From the weak convergence in H^1, one has $w_{k,j}\to w_{0,j}$ strongly in H^1. □

3.2 Achievement of $d_1(\varepsilon,\alpha_1,\alpha_2)$ and Proof of Theorem 1

Next we shall prove that $d_1(\varepsilon,\alpha_1,\alpha_2)$ is attained under some suitable conditions and show Theorem 1. Here we follow the arguments in [9]. First we approximate the minimizing problem: for each $k\in\mathbf{N}$ we consider

$$\begin{aligned}d_{1,k}(\varepsilon,\alpha_1,\alpha_2):=\inf\bigl\{E(u):u_1,u_2\in H_0^1\bigl(B_k(0),\mathbf{C}\bigr),\ u_3\in H_0^1\bigl(B_k(0),\mathbf{R}\bigr),\\ \|u_1\|_{L^2}^2=\alpha_1,\ \|u_2\|_{L^2}^2=\alpha_2\bigr\}.\end{aligned}$$

For $d_{1,k}(\varepsilon,\alpha_1,\alpha_2)$, we have

Lemma 5

(i) *For every $\alpha_1,\alpha_2>0$, $d_{1,k}(\varepsilon,\alpha_1,\alpha_2)\to d_1(\varepsilon,\alpha_1,\alpha_2)$ as $k\to\infty$.*
(ii) *For any $\alpha_1,\alpha_2>0$ and $k\in\mathbf{N}$, there exists a minimizer $(u_{k,1},u_{k,2},u_{k,3})$ for $d_{1,k}(\varepsilon,\alpha_1,\alpha_2)$ which are nonnegative, radially symmetric and decreasing in $B_k(0)$.*

Proof By $d_1(\varepsilon,\alpha_1,\alpha_2) \le d_{1,k}(\varepsilon,\alpha_1,\alpha_2)$ for all $k \in \mathbf{N}$ and the density of $C_0^\infty(\mathbf{R}^N)$ in H^1, it is easy to see that the assertion (i) holds.

We consider the assertion (ii). Let $(u_m)_{m=1}^\infty$ be a minimizing sequence for $d_{1,k}(\varepsilon,\alpha_1,\alpha_2)$. Then by $H_0^1(B_k(0)) \subset H^1$ and Proposition 2, $(u_{m,j})_{m=1}^\infty$ is bounded in $H_0^1(B_k(0))$ for $j = 1,2,3$. So we may suppose $u_{m,j} \rightharpoonup u_{0,j}$ weakly in $H_0^1(B_k(0))$. By the compactness of the embedding $H_0^1(B_k(0)) \subset L^p(B_k(0))$ for $1 \le p < 2^*$, it holds $u_{k,j} \to u_{0,j}$ strongly in $L^p(B_k(0))$, which shows $\|u_{0,j}\|_{L^2}^2 = \alpha_j$ for $j = 1,2$. By the lower semicontinuity of norm, we get $E(u_0) \le d_{1,k}(\varepsilon,\alpha_1,\alpha_2)$. Hence u_0 is a minimizer for $d_{1,k}(\varepsilon,\alpha_1,\alpha_2)$.

Next let $v_j(x) := |u_{0,j}(x)|^*$ for $j = 1,2,3$ where φ^* stands for a Schwarz symmetrization of φ. Then v_j satisfies the following (see [17]): $v_j \in H_0^1(B_k(0))$, v_j is nonnegative, radially symmetric, decreasing, $\|\nabla v_j\|_{L^2} \le \|\nabla u_{0,j}\|_{L^2}$, $\|v_j\|_{L^2} = \|u_{0,j}\|_{L^2}$ and

$$\int_{B_k(0)} (\mu_1|u_1|^2 + \mu_2|u_2|^2)u_3\,dx \le \int_{B_k(0)} (\mu_1 v_1^2 + \mu_2 v_2^2)v_3\,dx.$$

Thus $E(v) \le E(u_0) = d_{1,k}(\varepsilon,\alpha_1,\alpha_2)$ and we conclude that v is also a minimizer for $d_{1,k}(\varepsilon,\alpha_1,\alpha_2)$ and satisfies the desired properties. □

Let $(u_k)_{k=1}^\infty$ be a sequence of minimizers for $d_{1,k}(\varepsilon,\alpha_1,\alpha_2)$ obtained in Lemma 5. Then there are the Lagrange multipliers $\lambda_{k,1}, \lambda_{k,2} \in \mathbf{R}$ such that

$$\begin{cases} -\Delta u_{k,1} + \lambda_{k,1}u_{k,1} = \mu_1 u_{k,1}u_{k,3} & \text{in } B_k(0), \\ -\Delta u_{k,2} + \lambda_{k,2}u_{k,2} = \mu_2 u_{k,2}u_{k,3} & \text{in } B_k(0), \\ -\varepsilon^2\Delta u_{k,3} + \omega u_{k,3} = \mu_1 u_{k,1}^2 + \mu_2 u_{k,2}^2 & \text{in } B_k(0). \end{cases} \tag{20}$$

From the constraint conditions, we see $u_{k,j} \not\equiv 0$ $(k \in \mathbf{N},\ j = 1,2,3)$. Thus the nonnegativity of $u_{k,j}$ and the maximum principle yield $u_{k,j}(x) > 0$ in $B_k(0)$ for $j = 1,2,3$.

Lemma 6

(i) *The sequences* $(\lambda_{k,1})_{k=1}^\infty$ *and* $(\lambda_{k,2})_{k=1}^\infty$ *are bounded.*
(ii) *Suppose* $d_1(\varepsilon,\alpha_1,\alpha_2) < 0$. *Then there exist* $C_0 > 0$ *and* $k_0 \in \mathbf{N}$ *such that* $\alpha_1\lambda_{k,1} + \alpha_2\lambda_{k,2} \ge C_0$ *for all* $k \ge k_0$.
(iii) *For* $j = 1,2$, $\liminf_{k\to\infty} \lambda_{k,j} \ge 0$ *holds.*

Proof (i) Noting $\|u_{k,j}\|_{L^2}^2 = \alpha_j$, from integration by parts and (20) it holds that

$$\begin{aligned} \lambda_{k,1}\alpha_1 &= \int_{B_k(0)} \mu_1 u_{k,1}^2 u_{k,3}\,dx - \|\nabla u_{k,1}\|_{L^2}^2, \\ \lambda_{k,2}\alpha_2 &= \int_{B_k(0)} \mu_2 u_{k,2}^2 u_{k,3}\,dx - \|\nabla u_{k,2}\|_{L^2}^2. \end{aligned} \tag{21}$$

By Proposition 2 and Lemma 5, $(u_{k,j})$ are bounded. Hence $(\lambda_{k,j})$ are also bounded by (21).

(ii) It follows from (21) that

$$\alpha_1\lambda_{k,1}+\alpha_2\lambda_{k,2}=-2d_{1,k}(\varepsilon,\alpha_1,\alpha_2)+\frac{1}{2}\int_{B_k(0)}\left(\mu_1u_{k,1}^2+\mu_2u_{k,2}^2\right)u_{k,3}\,dx.$$

Using $d_{1,k}(\varepsilon,\alpha_1,\alpha_2)\to d_1(\varepsilon,\alpha_1,\alpha_2)<0$, we infer that the assertion (ii) holds.

(iii) We argue by contradiction and suppose $\liminf_{k\to\infty}\lambda_{k,1}<0$. Taking a subsequence if necessary, we may assume $\lambda_{k_\ell,1}\to\lambda_{0,1}<0$. Choose an $\ell_0\in\mathbf{N}$ so that $\lambda_{k_{\ell_0},1}\le\lambda_{0,1}/2$.

Next, let (ν_ℓ,φ_ℓ) be the first eigenvalue and eigenfunction of $-\Delta$ on $B_{k_\ell}(0)$ under the Dirichlet boundary condition. We remark $\nu_\ell\to 0$. Enlarging ℓ_0, we may suppose $\nu_{\ell_0}<-\lambda_{0,1}/4$.

Noting $u_{k_{\ell_0},1}=\varphi_{\ell_0}=0$ on $\partial B_{k_{\ell_0}}(0)$ and

$$-\Delta u_{k_{\ell_0},1}+\lambda_{k_{\ell_0},1}u_{k_{\ell_0},1}=\mu_1u_{k_{\ell_0},1}u_{k_{\ell_0},3},\qquad -\Delta\varphi_{\ell_0}-\nu_{\ell_0}\varphi_{\ell_0}=0\quad\text{in } B_{k_{\ell_0}}(0),$$

from integration by parts, we obtain

$$(\lambda_{k_{\ell_0}}+\nu_{\ell_0})\int_{B_{k_{\ell_0}}(0)}\varphi_{\ell_0}u_{k_{\ell_0},1}\,dx=\mu_1\int_{B_{k_{\ell_0}}(0)}u_{k_{\ell_0},1}u_{k_{\ell_0},3}\varphi_{\ell_0}\,dx.$$

Since $\lambda_{k_{\ell_0},1}+\nu_{k_{\ell_0},1}<0$ and $u_{k_{\ell_0},1},u_{k_{\ell_0},3},\varphi_{\ell_0}>0$ in $B_{k_{\ell_0}}(0)$, we have a contradiction. Thus $\liminf_{k\to\infty}\lambda_{k,1}\ge 0$ and we can show for $\lambda_{k,2}$ similarly. □

By Lemma 6, we extract a subsequence $(\lambda_{k_\ell,j})$ $(j=1,2)$ satisfying $\lambda_{k_\ell,j}\to\lambda_{0,j}\ge 0$. Since $(u_{k,j})$ is bounded in H^1 from Proposition 2 and Lemma 5, we may also assume $u_{k_\ell,j}\rightharpoonup u_{0,j}$ weakly in H^1 for $j=1,2,3$.

Lemma 7 *If $\lambda_{0,j_0}>0$, then $u_{k_\ell,j_0}\to u_{0,j_0}$ strongly in L^2.*

Proof We only treat the case $j_0=1$ and assume $\lambda_{0,1}>0$. In order to prove our claim, it is enough to show that there are $C_1,C_2>0$ such that

$$u_{k_\ell,1}(x)\le C_1\exp\bigl(-C_2|x|\bigr)\quad\text{for all } x\in\mathbf{R}^N \text{ and } \ell\ge 1. \tag{22}$$

Since (u_{k_ℓ}) satisfies (20), by the elliptic regularity, $u_{k_\ell,j}\to u_{0,j}$ in $C^2_{\mathrm{loc}}(\mathbf{R}^N)$ for $j=1,2,3$. Since $u_{k_\ell,j}$ is radial and monotone, so is $u_{0,j}$. Further, one sees $u_{0,j}(x)\to 0$ as $|x|\to\infty$. Using these facts, we obtain $u_{k_\ell,j}\to u_{0,j}$ strongly in L^∞, which implies $u_{k_\ell,j}\to u_{0,j}$ strongly in L^p for $p\in(2,\infty]$. Thus by $\lambda_{0,1}>0$, there exist $R_0>0$ and ℓ_0 satisfying

$$u_{k_\ell,1}(R_0)\le 1,\qquad \lambda_{k_\ell,1}-\mu_1u_{k_\ell,3}(x)\ge\lambda_{0,1}/2\quad\text{for all } |x|\ge R_0 \text{ and } \ell\ge\ell_0.$$

On the other hand, it follows from (20) that in (R_0,k_ℓ) we have

$$\begin{aligned}0&=-u''_{k_\ell,1}(r)-\frac{N-1}{r}u'_{k_\ell,1}(r)+(\lambda_{k_\ell,1}-\mu_1u_{k_\ell,3})u_{k_\ell,1}\\&\ge -u''_{k_\ell,1}(r)-\frac{N-1}{r}u'_{k_\ell}(r)+\frac{\lambda_{0,1}}{2}u_{k_\ell,1}.\end{aligned}$$

Let $\delta>0$ satisfy $-\delta^2+\lambda_{0,1}/2>0$ and set $\varphi_0(x)=\exp(-\delta(|x|-R_0))$. A direct calculation shows $-\varphi_0''-(N-1)\varphi_0'/r+\lambda_{0,1}\varphi_0/2>0$ in (R_0,k_ℓ). By $u_{k_\ell,1}(k_\ell)=0$

and $u_{k_\ell,1}(R_0) \le 1 = \varphi_0(R_0)$, the comparison theorem yields $u_{k_\ell,1}(x) \le \varphi_0(x)$ for all $|x| \ge R_0$ and $\ell \ge \ell_0$. Thus (22) holds. □

Now we prove that (u_{k_ℓ}) has a strongly convergent subsequence, which means $d_1(\varepsilon, \alpha_1, \alpha_2)$ is attained.

Proposition 4 *Suppose $d_1(\varepsilon, \alpha_1, \alpha_2) < 0$ when $N = 1, 2$ and $e_j(\varepsilon, \alpha_j) < 0$ for $j = 1, 2$ when $N = 3$. Then $u_{k_\ell,j} \to u_{0,j}$ strongly in H^1 for $j = 1, 2, 3$.*

Proof By $d_1(\varepsilon, \alpha_1, \alpha_2) < 0$ and Lemma 6, we see $\max\{\lambda_{0,1}, \lambda_{0,2}\} > 0$. First we treat the case $\lambda_{0,1} > 0$.

Step 1. $u_{k_\ell,1} \to u_{0,1}$ and $u_{k_\ell,3} \to u_{0,3}$ strongly in H^1.

Since (u_{k_ℓ}) is a minimizing sequence for $d_1(\varepsilon, \alpha_1, \alpha_2)$, $\|\Phi_{\varepsilon,\omega}[\mu_1 u_{k_\ell,1}^2 + \mu_2 u_{k_\ell,2}^2] - u_{k_\ell,3}\|_{H^1} \to 0$ by Lemma 1. Furthermore, since $u_{k_\ell,j} \to u_{0,j}$ strongly in L^p for $p \in (2, \infty]$ and $j = 1, 2$ by the argument in the proof of Lemma 7, we have $\Phi_{\varepsilon,\omega}[\mu_1 u_{k_\ell,1}^2 + \mu_2 u_{k_\ell,2}^2] \to \Phi_{\varepsilon,\omega}[\mu_1 u_{0,1}^2 + \mu_2 u_{0,2}^2]$ strongly in H^1. Thus $\|u_{k_\ell,3} - \Phi_{\varepsilon,\omega}[\mu_1 u_{0,1}^2 + \mu_2 u_{0,2}^2]\|_{H^1} \to 0$. Noting $u_{k_\ell,3} \rightharpoonup u_{0,3}$ weakly in H^1, we get $u_{0,3} = \Phi_{\varepsilon,\omega}[\mu_1 u_{0,1}^2 + \mu_2 u_{0,2}^2]$ and $u_{k_\ell,3} \to u_{0,3}$ strongly in H^1.

Next, by (20), one has

$$\|\nabla u_{k_\ell,1}\|_{L^2}^2 = -\lambda_{k_\ell,1}\alpha_1 + \mu_1 \int_{\mathbf{R}^N} u_{k_\ell,1}^2 u_{k_\ell,3}\, dx \to -\lambda_{0,1}\alpha_1 + \mu_1 \int_{\mathbf{R}^N} u_{0,1}^2 u_{0,3}\, dx. \tag{23}$$

On the other hand, since $u_{k_\ell,1} \to u_{0,1}$ strongly in L^2 by Lemma 7, $\|u_{0,1}\|_{L^2}^2 = \alpha_1$ holds. Noting also $-\Delta u_{0,1} + \lambda_{0,1} u_{0,1} = \mu_1 u_{0,1} u_{0,3}$ in $\mathbf{R}^N$ and (23), we obtain $\|\nabla u_{k_\ell,1}\|_{L^2}^2 \to \|\nabla u_{0,1}\|_{L^2}^2$. Therefore $u_{k_\ell,1} \to u_{0,1}$ strongly in H^1 from the weak convergence of $(u_{k_\ell,1})$.

Step 2. $\lambda_{0,2} > 0$.

We prove indirectly and assume $\lambda_{0,2} = 0$. Then $u_{0,2}$ satisfies $-\Delta u_{0,2} = \mu_2 u_{0,2} u_{0,3}$ in $\mathbf{R}^N$, which implies

$$\|\nabla u_{0,2}\|_{L^2}^2 = \mu_2 \int_{\mathbf{R}^N} u_{0,2}^2 u_{0,3}\, dx. \tag{24}$$

In the case where $N = 3$, since $\|u_{0,1}\|_{L^2}^2 = \alpha_1$ and $u_{0,3} = \Phi_{\varepsilon,\omega}[\mu_1 u_{0,1}^2 + \mu_2 u_{0,2}^2]$, it follows from (24) that

$$\begin{aligned} E(u_0) = F(u_{0,1}, u_{0,2}) &= \frac{1}{2}\|\nabla u_{0,1}\|_{L^2}^2 - \frac{\mu_1}{4}\int_{\mathbf{R}^N} u_{0,1}^2 u_{0,3}\, dx + \frac{\mu_2}{4}\int_{\mathbf{R}^N} u_{0,2}^2 u_{0,3}\, dx \\ &= E_1(u_{0,1}) + \frac{\mu_2^2}{4}\int_{\mathbf{R}^N} u_{0,2}^2 \Phi_{\varepsilon,\omega}\big[u_{0,2}^2\big]\, dx \ge E_1(u_{0,1}) \ge e_1(\varepsilon, \alpha_1). \end{aligned}$$

On the other hand, by the lower semicontinuity of norm and $u_{k_\ell,3} \to u_{0,3}$ strongly in H^1, one has $E(u_0) \le \liminf_{\ell\to\infty} E(u_{k_\ell}) = d_1(\varepsilon,\alpha_1,\alpha_2)$. Combining (15), we see $e_1(\varepsilon,\alpha_1) \le e_1(\varepsilon,\alpha_1)+e_2(\varepsilon,\alpha_2)$, which contradicts to the condition $e_2(\varepsilon,\alpha_2) < 0$. Hence $\lambda_{0,2} > 0$.

In the case where $N = 1, 2$, we set $v_\ell(x) := u_{k_\ell,2}(x)/u_{k_\ell,2}(0)$. Since $u_{k_\ell,2}$ is radially symmetric and decreasing, $0 \le v_\ell(x) \le 1 = v_\ell(0)$ for $0 \le |x| \le k_\ell$. Furthermore, v_ℓ satisfies $-\Delta v_\ell + \lambda_{k_\ell,2} v_\ell = \mu_2 v_\ell u_{k_\ell,3}$ in $B_{k_\ell}(0)$. By the elliptic regularity, taking a subsequence if necessary, we may assume $v_\ell \to v_0$ in $C^2_{\mathrm{loc}}(\mathbf{R}^N)$ where v_0 satisfies $v_0(0) = 1$ and $-\Delta v_0 = \mu_2 v_0 u_{0,3}$ in $\mathbf{R}^N$. We remark that by Liouville's theorem, the inequality $-\Delta u \ge 0$ in $\mathbf{R}^N$ has no bounded positive solution except for constant functions when $N = 1, 2$. In fact, when $N = 1$, it is easy to show. When $N = 2$, by Liouville's theorem, we can prove it. See [27].

By $v_0(0) = 1$, we have $v_0(x) \equiv 1$. However, this contradicts to $-\Delta v_0 = \mu_2 v_0 u_{0,3}$ in $\mathbf{R}^N$ since $u_{0,3}(x) > 0$ in $\mathbf{R}^N$. Therefore $\lambda_{0,2} > 0$.

From Step 2, $\lambda_{0,2} > 0$ holds. Then using Lemma 7, we can show $u_{k_\ell,2} \to u_{0,2}$ strongly in H^1 as in Step 1. Since other case can be shown in a similar way, we complete the proof. □

Next we shall prove the characterization of $\mathcal{M}_1(\varepsilon,\alpha_1,\alpha_2)$.

Proposition 5 *Suppose $d_1(\varepsilon,\alpha_1,\alpha_2) < 0$ when $N = 1, 2$ and $e_j(\varepsilon,\alpha_j) < 0$ $(j = 1, 2)$ when $N = 3$. Then*

$$\mathcal{M}_1(\varepsilon,\alpha_1,\alpha_2) = \big\{\big(e^{i\theta_1} w_1(x-y), e^{i\theta_2} w_2(x-y), w_3(x-y)\big)\colon \theta_1,\theta_2 \in \mathbf{R},\ y \in \mathbf{R}^N,$$
$$w_j \textit{ is positive and radially symmetric solution of } (2) \textit{ with some } \lambda_1,\lambda_2 > 0\big\}.$$

Proof First we remark that $\mathcal{M}_1(\varepsilon,\alpha_1,\alpha_2) \ne \emptyset$ holds from Lemma 5 and Proposition 4. Let $u \in \mathcal{M}_1(\varepsilon,\alpha_1,\alpha_2)$ and set $v_j(x) := |u_j(x)|$ for $j = 1, 2, 3$. Then we observe $v \in \mathcal{M}_1(\varepsilon,\alpha_1,\alpha_2)$ and v satisfies (2) for some $\lambda_1,\lambda_2 \in \mathbf{R}$. By the elliptic regularity, $v_j \in C^2(\mathbf{R}^N)$ and $v_j(x) > 0$ in $\mathbf{R}^N$ for $j = 1, 2, 3$ from the nonnegativity and the maximum principle.

Next we shall prove $\lambda_1,\lambda_2 > 0$. If $\lambda_1 < 0$, then let (ν_R,φ_R) be the first eigenvalue and eigenfunction of $-\Delta$ on $B_R(0)$ under the Dirichlet boundary condition. Choose $R > 0$ so large that $\nu_R < -\lambda_1$ holds. From (2), $\varphi_R(R) = 0$ and $-\Delta\varphi_R = \nu_R\varphi_R$ in $B_R(0)$, we obtain

$$\int_{B_R(0)} (\nabla v_1 \cdot \nabla\varphi_R + \lambda_1 v_1 \varphi_R)\,dx = \mu_1 \int_{B_R(0)} v_1 v_3 \varphi_R\,dx,$$
$$\int_{B_R(0)} (\nabla v_1 \cdot \nabla\varphi_R - \nu_R v_1 \varphi_R)\,dx = \int_{\partial B_R(0)} \nabla\varphi_R \cdot \frac{x}{R} v_1\,dS.$$

Since φ is radial, positive and $\nabla\varphi_R(x)\cdot x/R = \varphi_R'(R) < 0$ on $\partial B_R(0)$, it holds that

$$0 > (\lambda_1 + \nu_R)\int_{B_R(0)} v_1\varphi_R\,dx = \mu_1 \int_{B_R(0)} v_1 v_3 \varphi_R\,dx - \int_{\partial B_R(0)} \varphi'(R) v_1\,dS > 0.$$

This is a contradiction and $\lambda_1 \ge 0$. Similarly, $\lambda_2 \ge 0$ holds.

Next, we show $\lambda_1, \lambda_2 > 0$. If $\lambda_1 = 0$, then $-\Delta v_1 = \mu_1 v_1 v_3$ in $\mathbf{R}^N$. Therefore, we can show that the contradiction occurs as in Step 2 of Proposition 4. Thus $\lambda_1, \lambda_2 > 0$ hold.

Since $\lambda_1, \lambda_2 > 0$, we may apply the result of [5] and see that there are radially symmetric decreasing functions (w_1, w_2, w_3) and $y \in \mathbf{R}$ such that $v_j(x) = w_j(x-y)$ hold for $j = 1, 2, 3$. In [5], the authors only consider the case where $N \geq 2$, however, we can apply their result for (2) with $\lambda_1, \lambda_2 > 0$ in the case $N = 1$. See also [14] when $N = 1$.

Next we prove that if $u \in \mathscr{M}_1(\varepsilon, \alpha_1, \alpha_2)$, then there are $\theta_1, \theta_2 \in \mathbf{R}$ such that $u_j(x) = e^{i\theta_j}|u_j(x)|$ for all $x \in \mathbf{R}^N$. If we can prove this claim, combining the above argument, Proposition 5 holds. To prove our claim, we follow the argument of [23] (cf. [7] and [8]). Set $u_j(x) = u_{j,1}(x) + iu_{j,2}(x)$ and $v_j(x) := |u_j(x)|$ for $j = 1, 2$ where $u_{j,\ell}$ is a real-valued function. Since u_j satisfies (2), $u_{j,\ell}$ are smooth. Moreover, by the above argument, $v_j(x) > 0$ in $\mathbf{R}^N$. Thus for all $x \in \mathbf{R}^N$,

$$\nabla v_j(x) = \frac{1}{v_j(x)}\big(u_{j,1}(x)\nabla u_{j,1}(x) + u_{j,2}(x)\nabla u_{j,2}(x)\big), \quad \big|\nabla v_j(x)\big| \leq \big|\nabla u_j(x)\big|. \tag{25}$$

On the other hand, since $(u_1, u_2, v_3), (v_1, v_2, v_3) \in \mathscr{M}_1(\varepsilon, \alpha_1, \alpha_2)$, it holds that $\|\nabla u_j\|_{L^2}^2 = \|\nabla v_j\|_{L^2}^2$ for $j = 1, 2$. Therefore, by (25), we get

$$0 = \big|\nabla u_j(x)\big|^2 - \big|\nabla v_j(x)\big|^2 = \frac{1}{v_j^2(x)}\big|u_{j,2}(x)\nabla u_{j,1}(x) - u_{j,1}(x)\nabla u_{j,2}(x)\big|^2,$$

which implies $u_{j,2}(x)\nabla u_{j,1}(x) = u_{j,1}(x)\nabla u_{j,2}(x)$ for all $x \in \mathbf{R}^N$.

Next we compute $\partial_{x_k}(u_j(x)^2/v_j(x)^2)$. It is easily seen that

$$\mathrm{Re}\left(\frac{\partial}{\partial x_k}\frac{u_j^2}{v_j^2}\right) = \frac{4}{v_j^4}\left(-u_{j,1}^2 u_{j,2}\frac{\partial u_{j,2}}{\partial x_k} + u_{j,1}u_{j,2}^2\frac{\partial u_{j,1}}{\partial x_k}\right) = 0, \quad \mathrm{Im}\left(\frac{\partial}{\partial x_k}\frac{u_j^2}{v_j^2}\right) = 0.$$

Hence $u_j(x)/v_j(x)$ is a constant function and there are $\theta_1, \theta_2 \in \mathbf{R}$ such that $u_1(x) = e^{i\theta_1}v_1(x)$ and $u_2(x) = e^{i\theta_2}v_2(x)$ for all $x \in \mathbf{R}^N$. Thus we complete the proof. □

Proposition 6 *Suppose that $d_2(\varepsilon, \alpha) < 0$. Then the following hold*:

$$\begin{aligned}\mathscr{M}_2(\varepsilon, \alpha) = \big\{&\big(e^{i\theta_1}w_1(x-y), e^{i\theta_2}w_2(x-y), w_3(x-y)\big) : \theta_1, \theta_2 \in \mathbf{R},\ y \in \mathbf{R}^N,\\ &w_j \text{ is nonnegative and radially symmetric solution (5) with some}\\ &\lambda_1 = \lambda_2 > 0\big\}.\end{aligned}$$

Since we can show in a similar way to Proposition 5, we omit a proof.

Now we give a proof for Theorem 1.

Proof of Theorem 1 By Lemma 3 and Corollary 1, there exist $0 < \underline{\alpha} \leq \bar{\alpha} < \infty$ such that

(i) $d_1(\varepsilon, \alpha_1, \alpha_2) < 0$ holds for any $\alpha_1, \alpha_2 > 0$ when $N = 1$,

(ii) $\mathcal{M}_1(\varepsilon,\alpha_1,\alpha_2)=\emptyset$ if $\max\{\alpha_1,\alpha_2\}<\underline{\alpha}$ when $N=2,3$,
(iii) $d_1(\varepsilon,\alpha_1,\alpha_2)<0$ if $\bar{\alpha}<\max\{\alpha_1,\alpha_2\}$ when $N=2$,
(iv) $e_j(\varepsilon,\alpha_j)<0$ for $j=1,2$ if $\bar{\alpha}<\min\{\alpha_1,\alpha_2\}$ when $N=3$.

Thus from Propositions 4 and 5, Theorem 1 holds. □

4 Proof of Theorem 2

In this section, we shall prove Theorem 2.

Proof of Theorem 2 Choose an $\alpha_0>0$ so that $e_j(\varepsilon,\alpha)<0$ for any $\alpha>\alpha_0$ and $j=1,2$. We remark that from Lemma 3 we can choose $\alpha_0=0$ when $N=1$. Since $d_2(\varepsilon,\alpha)\le\min\{e_1(\varepsilon,\alpha),e_2(\varepsilon,\alpha)\}<0$, one sees $\mathcal{M}_2(\varepsilon,\alpha)\neq\emptyset$ by Proposition 3. First we show the assertion (ii) and suppose $\mu_1>\mu_2$ and $(u_1,u_2,u_3)\in\mathcal{M}_2(\varepsilon,\alpha)$. By Proposition 6, we may also suppose $u_j(x)=e^{i\theta_j}w_j(x-y)$ for some $\theta_j\in\mathbf{R}$ and $y\in\mathbf{R}^N$ where w_j is a nonnegative solution of (2) with $\lambda_1=\lambda_2>0$. Furthermore, by the maximum principle, $w_j>0$ holds if $w_j\not\equiv 0$. Multiply the first equation (resp. the second equation) by w_2 (resp. w_1) and integrate over $\mathbf{R}^N$, we obtain

$$0=(\mu_1-\mu_2)\int_{\mathbf{R}^N}w_1w_2w_3\,dx.$$

Since $\mu_1>\mu_2$ and $w_3>0$, we have either $w_1\equiv 0$ or $w_2\equiv 0$. We remark that since $e_j(\varepsilon,\alpha)<0$, we can prove that $e_j(\varepsilon,\alpha)$ is attained as in Proposition 3. Thus from $\mu_1>\mu_2$, we have $e_1(\varepsilon,\alpha)<e_2(\varepsilon,\alpha)$ and conclude that $w_2\equiv 0$. Thus the assertion (ii) holds.

Next we show the assertion (i) and suppose $\mu_1=\mu_2=\mu$. Let $(u_1,u_2,u_3)\in\mathcal{M}_2(\varepsilon,\alpha)$ and u_j be nonnegative, radially symmetric and decreasing for $j=1,2,3$. This is possible by Proposition 6.

Set $v(x):=\sqrt{u_1^2(x)+u_2^2(x)}$. Then we can show $|\nabla v(x)|^2\le|\nabla u_1(x)|^2+|\nabla u_2(x)|^2$ and $\|v\|_{L^2}^2=\alpha$. Thus, noting $\Phi_{\varepsilon,\omega}[\mu v^2]=u_3$, we have $d_2(\varepsilon,\alpha)\le E(v,0,u_3)\le E(u_1,u_2,u_3)=d_2(\varepsilon,\alpha)$, which implies $(v,0,u_3)$ is also a minimizer for $d_2(\varepsilon,\alpha)$.

For each $\alpha_1,\alpha_2>0$ with $\alpha=\alpha_1+\alpha_2$, we define v_1,v_2 by $v_1(x):=\sqrt{\alpha_1/\alpha}\,v(x)$ and $v_2(x):=\sqrt{\alpha_2/\alpha}\,v(x)$. Then it follows that $\|v_j\|_{L^2}^2=\alpha_j$ for $j=1,2$ and $E(v_1,v_2,u_3)=E(v,0,u_3)=d_2(\varepsilon,\alpha)$. Hence we have $d_1(\varepsilon,\alpha_1,\alpha_2)\le d_2(\varepsilon,\alpha)$, so $d_2(\varepsilon,\alpha)=d_1(\varepsilon,\alpha_1,\alpha_2)$ holds.

Moreover, using the same arguments in the proof of Proposition 5, we see

$$\nabla\left(\frac{u_j^2(x)}{v^2(x)}\right)=0\quad\text{for } j=1,2 \text{ and } x\in\mathbf{R}^N.$$

Hence, there exist $c_1,c_2\ge 0$ such that $u_1(x)=c_1v(x)$ and $u_2(x)=c_2v(x)$ for all $x\in\mathbf{R}$. Noting $v^2(x)=u_1^2(x)+u_2^2(x)$, we have $c_1^2+c_2^2=1$, which implies $c_1=\cos\theta$ and $c_2=\sin\theta$ for some $\theta\in\mathbf{R}$. Thus, from Proposition 6, we conclude that

$$\mathscr{M}_2(\varepsilon,\alpha)=\big\{\big(e^{i\theta_2}w_1(x-y)\cos\theta_1, e^{i\theta_3}w_1(x-y)\sin\theta_1, w_3(x-y)\big):$$
$$\theta_1,\theta_2,\theta_3\in\mathbf{R},\ y\in\mathbf{R}^N\big\}.$$

□

5 Proof of Theorem 3

In this section, we will prove Theorem 3. Throughout this section, we always assume $N=1$, $\tilde{\mu}_1=\mu_1^2/\omega$, $\tilde{\mu}_2=\mu_2^2/\omega$ and $\tilde{\beta}=\mu_1\mu_2/\omega$.

Lemma 8

(i) *There exists a* $C_1>0$ *such that* $-C_1\le d_1(\varepsilon,\alpha_1,\alpha_2)$ *holds for every* $\varepsilon>0$.
(ii) $\limsup_{\varepsilon\to 0} d_1(\varepsilon,\alpha_1,\alpha_2)\le d_3(\alpha_1,\alpha_2)<0$.

Proof First we prove the assertion (ii). Choose $\varphi_j\in H^1$ so that $\|\varphi_j\|_{L^2}=\alpha_j$. Then it is easy to see $G(\varphi_\rho)<0$ for sufficiently small $\rho>0$ where $\varphi_{\rho,j}(x)=\rho^{1/2}\varphi(\rho x)$. Thus $d_3(\alpha_1,\alpha_2)<0$.

Next, let $\eta>0$ and $\varphi_1,\varphi_2\in C_0^\infty(\mathbf{R})$ satisfy $\|\varphi_j\|_{L^2}^2=\alpha_j$ and $G(\varphi_1,\varphi_2)\le d_3(\alpha_1,\alpha_2)+\eta$. Then $d_1(\varepsilon,\alpha_1,\alpha_2)\le F(\varphi_1,\varphi_2)$. Thus to show the assertion (ii), it is sufficient to prove

$$\lim_{\varepsilon\to 0}\Phi_{\varepsilon,\omega}\big[\mu_j|\varphi_j|^2\big]\to\mu_j|\varphi_j|^2/\omega\quad\text{strongly in } L^\infty(\mathbf{R})\text{ as }\varepsilon\to 0. \tag{26}$$

In fact, it holds from (26) that $F(\varphi_1,\varphi_2)\to G(\varphi_1,\varphi_2)$. Hence one obtains $\limsup_{\varepsilon\to 0} d_1(\varepsilon,\alpha_1,\alpha_2)\le d_3(\alpha_1,\alpha_2)+\eta$. Since $\eta>0$ is arbitrary, we get (ii).

We prove (26). Since $\int_{\mathbf{R}} e^{-|y|}dy=2$ and

$$\begin{aligned}\Phi_{\varepsilon,\omega}\big[\mu_j|\varphi_j|^2\big](x)&=\frac{\mu_j}{2\varepsilon\sqrt{\omega}}\int_{\mathbf{R}} e^{-\sqrt{\omega}|y|/\varepsilon}\big|\varphi_j(x-y)\big|^2dy\\&=\frac{\mu_j}{2\omega}\int_{\mathbf{R}} e^{-|y|}\big|\varphi_j(x-\varepsilon y/\sqrt{\omega})\big|^2dy,\end{aligned}$$

it follows from the uniform continuity of φ_j that

$$\begin{aligned}&\left|\Phi_{\varepsilon,\omega}\big[\mu_j|\varphi_j|^2\big](x)-\frac{\mu_j}{\omega}\big|\varphi_j(x)\big|^2\right|\\&\quad\le\frac{\mu_j}{2\omega}\int_{\mathbf{R}} e^{-|y|}\Big|\big|\varphi_j(x-\varepsilon y/\sqrt{\omega})\big|^2-\big|\varphi_j(x)\big|^2\Big|dy\to 0\end{aligned}$$

uniformly with respect to $x\in\mathbf{R}$ as $\varepsilon\to 0$. Thus (26) holds.

Next we show the assertion (i). For all $u_1,u_2\in H^1$ with $\|u_j\|_{L^2}^2=\alpha_j$, by Young's inequality, one has

$$\begin{aligned}&\int_{\mathbf{R}}\Phi_{\varepsilon,\omega}\big[\mu_1|u_1|^2+\mu_2|u_2|^2\big]\big(\mu_1|u_1|^2+\mu_2|u_2|^2\big)\,dx\\&\quad\le\|G_{\varepsilon,\omega}\|_{L^1}\big(\mu_1\|u_1\|_{L^4}^2+\mu_2\|u_2\|_{L^4}^2\big)^2.\end{aligned}$$

On the other hand, by Gagliardo-Nirenberg's inequality, $\|u_j\|_{L^4} \le C\|u_j'\|_{L^2}^{1/4}\|u_j\|_{L^2}^{3/4}$. Noting $\|G_{\varepsilon,\omega}\|_{L^1} = 1/\omega$ and $\|u_j\|_{L^2}^2 = \alpha_j$, we obtain

$$F(u_1,u_2) \ge \left(\|u_1'\|_{L^2}^2 + \|u_2'\|_{L^2}^2\right)/2 - C\left(\alpha_1^{3/2}\|u_1'\|_{L^2} + \alpha_2^{3/2}\|u_2'\|_{L^2}\right) \ge -C_1 \quad (27)$$

for some $C_1 > 0$ independent of ε and u_1, u_2. Hence, $-C_1 \le d_1(\varepsilon,\alpha_1,\alpha_2)$ holds for any $\varepsilon > 0$. □

For each $\alpha_1, \alpha_2 > 0$, let (w_ε) be a minimizer for $d_1(\varepsilon,\alpha_1,\alpha_2)$ and a positive, even and monotone decreasing solution of

$$\begin{cases} -w_{\varepsilon,1}'' + \lambda_{\varepsilon,1} w_{\varepsilon,1} = \mu_1 w_{\varepsilon,1} w_{\varepsilon,3}, \\ -w_{\varepsilon,2}'' + \lambda_{\varepsilon,2} w_{\varepsilon,2} = \mu_2 w_{\varepsilon,2} w_{\varepsilon,3}, \\ -\varepsilon^2 w_{\varepsilon,3}'' + \omega w_{\varepsilon,3} = \mu_1 w_{\varepsilon,1}^2 + \mu_2 w_{\varepsilon,2}^2 \end{cases} \quad (28)$$

in $\mathbf{R}$ for some $\lambda_{\varepsilon,1}, \lambda_{\varepsilon,2} > 0$. Such a pair exists due to Proposition 5. Then we shall show the following:

Proposition 7 *There exist subsequences $(w_{\varepsilon_k})_{k=1}^\infty$, $(\lambda_{\varepsilon_k,j})_{k=1}^\infty$ and $(w_{0,1}, w_{0,2}) \in \mathscr{M}_3(\alpha_1,\alpha_2)$, $\lambda_{0,1}, \lambda_{0,2} > 0$ such that $w_{\varepsilon_k,1} \to w_{0,1}$, $w_{\varepsilon_k,2} \to w_{0,2}$, $w_{\varepsilon_k,3} \to \mu_1 w_{0,1}^2/\omega + \mu_2 w_{0,2}^2/\omega$ strongly in H^1 and $\lambda_{\varepsilon_k,1} \to \lambda_{0,1}$, $\lambda_{\varepsilon_k,2} \to \lambda_{0,2} > 0$.*

Proof First, by Lemma 8 and (27), $(w_{\varepsilon,1})$ and $(w_{\varepsilon,2})$ are bounded in H^1. Secondly, by $w_{\varepsilon,3}(x) = \Phi_{\varepsilon,\omega}[\mu_1 w_{\varepsilon,1}^2 + \mu_2 w_{\varepsilon,2}^2]$, Young's inequality yields $\|w_{\varepsilon,3}\|_{L^p} \le \|G_{\varepsilon,\omega}\|_{L^1}(\mu_1\|w_{\varepsilon,1}\|_{L^{2p}}^2 + \mu_2\|w_{\varepsilon,2}\|_{L^{2p}}^2)$ for every $p \in [1,\infty]$. Combining $\|G_{\varepsilon,\omega}\|_{L^1} = 1/\omega$, $(w_{\varepsilon,3})$ is bounded in L^p for all $p \in [1,\infty]$. Furthermore, it follows from (28) and the elliptic estimate that $(\lambda_{\varepsilon,1})$, $(\lambda_{\varepsilon,2})$ are bounded and $(w_{\varepsilon,j})$ is bounded in H^2 for $j = 1, 2$. Using Young's inequality again, we can show $(w_{\varepsilon,3})$ is bounded in $W^{1,p}$ for any $p \in [1,\infty]$. Using the same procedure, we see that $(w_{\varepsilon,j})$ $(j = 1,2)$ is bounded in H^3 and $(w_{\varepsilon,3})$ in $W^{2,p}$ for any $p \in [1,\infty]$.

Now from the boundedness of (w_ε) and (28), there exist a subsequence (ε_k), $w_{0,1}, w_{0,2} \in H^2$ and $\lambda_{0,1}, \lambda_{0,2} \ge 0$ such that $\lambda_{\varepsilon_k,j} \to \lambda_{0,j}$, $w_{\varepsilon_k,j} \to w_{0,j}$ strongly in $C^2_{\mathrm{loc}}(\mathbf{R})$ for $j = 1, 2$. Furthermore, noting the monotonicity of $w_{\varepsilon,j}$, one sees $w_{\varepsilon_k,j} \to w_{0,j}$ strongly in L^∞ as in Lemma 7. Thus from (28), we see $w_{\varepsilon_k,3} \to \mu_1 w_{0,1}^2/\omega + \mu_2 w_{0,2}^2/\omega$ strongly in L^p for every $p \in (1,\infty]$.

By $d_3(\alpha_1,\alpha_2) < 0$, Lemma 8 and the similar arguments in Lemmas 6 and 7, we see that one of $\lambda_{0,1}$ and $\lambda_{0,2}$ is strictly positive and $w_{\varepsilon_k,j} \to w_{0,j}$ strongly in L^2 provided $\lambda_{0,j} > 0$. Now we assume $\lambda_{0,1} > 0$. From the convergence of $(w_{\varepsilon_k,3})$ and the maximum principle, $w_{0,1}$ is a positive solution to $-w_{0,1}'' + \lambda_{0,1} w_{0,1} = \tilde{\mu}_1 w_{0,1}^3 + \tilde{\beta} w_{0,1} w_{0,2}^2$ in $\mathbf{R}$. Furthermore, by the arguments of Step 2 in the proof of Proposition 4, $\lambda_{0,2} > 0$ holds. Thus $w_{\varepsilon_k,2} \to w_{0,2}$ strongly in L^2. Therefore, using (28) and the elliptic estimate again, one has $w_{\varepsilon_k,1} \to w_{0,1}$ and $w_{\varepsilon_k,2} \to w_{0,2}$ strongly in H^1. Since $w_{\varepsilon_k,3}(x) = \Phi_{\varepsilon,\omega}[\mu_1 w_{\varepsilon_k,1}^2 + \mu_2 w_{\varepsilon_k,2}^2]$ holds, $w_{\varepsilon_k,3} \to \mu_1 w_{0,1}^2/\omega + \mu_2 w_{0,2}^2/\omega$ strongly in H^1. If we assume $\lambda_{0,2} > 0$ first, then we can show the same conclusion similarly.

Lastly we show that $(w_{0,1}, w_{0,2})$ is a minimizer for $d_3(\alpha_1, \alpha_2)$. Since $\|w_{0,j}\|_{L^2}^2 = \alpha_j$, we have $d_3(\alpha_1, \alpha_2) \le G(w_{0,1}, w_{0,2})$. On the other hand, $d_1(\varepsilon_k, \alpha_1, \alpha_2) = F(w_{\varepsilon_k,1}, w_{\varepsilon_k,2}) \to G(w_{0,1}, w_{0,2})$ holds. Thus by Lemma 8, $G(w_{0,1}, w_{0,2}) \le d_3(\alpha_1, \alpha_2)$, which implies that $G(w_0) = d_3(\alpha_1, \alpha_2)$ and $(w_{0,1}, w_{0,2}) \in \mathscr{M}_3(\alpha_1, \alpha_2)$. □

Now we prove Theorem 3.

Proof of Theorem 3 Since the assertion (i) follows from the assertion (ii), we only prove the assertion (ii). We argue by contradiction and suppose that there exists a sequence (u_k) such that $\varepsilon_k \to 0$, $u_k \in \mathscr{M}_1(\varepsilon_k, \alpha_1, \alpha_2)$ and

$$\begin{aligned}&\mathrm{dist}_{H^1\times H^1}\big((u_{k,1}, u_{k,2}), \mathscr{M}_3(\alpha_1, \alpha_2)\big)\\&\quad+ \big\|u_{k,3} - \big[(\mu_1/\omega)|u_{k,1}|^2 + (\mu_2/\omega)|u_{k,2}|^2\big]\big\|_{H^1} \ge \eta_0\end{aligned}$$

for some $\eta_0 > 0$. Then by Proposition 5, there exist (w_k), $y_k \in \mathbf{R}$ and $\theta_{k,j} \in [0, 2\pi]$ such that $u_{k,j}(x) = e^{i\theta_{k,j}} w_{k,j}(x - y_k)$ for $j = 1, 2$ and $u_{k,3}(x) = w_{k,3}(x - y_k)$. Applying Proposition 7, we may assume that there exists $(w_{0,1}, w_{0,2}) \in \mathscr{M}_3(\alpha_1, \alpha_2)$ such that $w_{k,j} \to w_{0,j}$ strongly in H^1. Here we remark that $(e^{i\theta_{k,1}} w_{0,1}(x - y_k), e^{i\theta_{k,2}} w_{0,2}(x - y_k)) \in \mathscr{M}_3(\alpha_1, \alpha_2)$ for each $k \in \mathbf{N}$. Therefore,

$$\begin{aligned}&\mathrm{dist}_{H^1\times H^1}\big((u_{k,1}, u_{k,2}), \mathscr{M}_3(\alpha_1, \alpha_2)\big)\\&\quad+ \big\|u_{k,3} - \big[(\mu_1/\omega)|u_{k,1}|^2 + (\mu_2/\omega)|u_{k,2}|^2\big]\big\|_{H^1} \to 0,\end{aligned}$$

which is a contradiction. Thus (ii) holds. □

Acknowledgements The author would like to thank Professor Maris and Professor Shibata for letting him know Refs. [10] and [28], and fruitful discussions. The author was supported by Grant-in-Aid for JSPS Fellows 22-1561 and "Japanese-German Graduate Externship, International Research Training Group" associated with Waseda University and Technische Universität Darmstadt.

References

1. Ambrosetti, A., Colorado, E.: Standing waves of some coupled nonlinear Schrödinger equations. J. Lond. Math. Soc. **75**(1), 67–82 (2007)
2. Bartsch, T., Wang, Z.-Q.: Note on ground states of nonlinear Schrödinger systems. J. Partial Differ. Equ. **19**(3), 200–207 (2006)
3. Bartsch, T., Wang, Z.-Q., Wei, J.: Bound states for a coupled Schrödinger system. Fixed Point Theory Appl. **2**(2), 353–367 (2007)
4. Bartsch, T., Dancer, N., Wang, Z.-Q.: A Liouville theorem, a-priori bounds, and bifurcating branches of positive solutions for a nonlinear elliptic system. Calc. Var. Partial Differ. Equ. **37**(3–4), 345–361 (2010)
5. Busca, J., Sirakov, B.: Symmetry results for semilinear elliptic systems in the whole space. J. Differ. Equ. **163**(1), 41–56 (2000)
6. Chang, J., Liu, Z.: Ground states of nonlinear Schrödinger systems. Proc. Am. Math. Soc. **138**(2), 687–693 (2010)

7. Cazenave, T.: Semilinear Schrödinger Equations. Courant Lecture Notes in Mathematics, vol. 10. New York University, Courant Institute of Mathematical Sciences/American Mathematical Society, New York/Providence (2003), xiv+323 pp.
8. Cazenave, T., Lions, P.-L.: Orbital stability of standing waves for some nonlinear Schrödinger equations. Commun. Math. Phys. **85**(4), 549–561 (1982)
9. Cao, D., Chern, I.-L., Wei, J.-C.: On ground state of spinor Bose-Einstein condensates. NoDEA Nonlinear Differ. Equ. Appl. **18**(4), 427–445 (2011)
10. Chiron, D., Maris, M.: Traveling waves for nonlinear Schrödinger equations with nonzero conditions at infinity, II. arXiv:1203.1912v1 [math.AP]
11. de Figueiredo, D.G., Lopes, O.: Solitary waves for some nonlinear Schrödinger systems. Ann. Inst. Henri Poincaré, Anal. Non Linéaire **25**(1), 149–161 (2008)
12. Gidas, B., Ni, W.M., Nirenberg, L.: Symmetry of positive solutions of nonlinear elliptic equations in $\mathbf{R}^N$. In Mathematical Analysis and Applications, Part A. Adv. in Math. Suppl. Stud., vol. 7a, pp. 369–402 (1981)
13. Hirano, N., Shioji, N.: Multiple existence of solutions for coupled nonlinear Schrödinger equations. Nonlinear Anal. **68**(12), 3845–3859 (2008)
14. Ikoma, N.: Uniqueness of positive solutions for a nonlinear elliptic system. NoDEA Nonlinear Differ. Equ. Appl. **16**(5), 555–567 (2009)
15. Kartashov, Y.V., Torner, L., Vysloukh, V.A., Mihalache, D.: Multipole vector solitons in nonlocal nonlinear media. Opt. Lett. **31**(10), 1483–1485 (2006)
16. Lieb, E.H.: Existence and uniqueness of the minimizing solution of Choquard's nonlinear equation. Stud. Appl. Math. **57**(2), 93–105 (1977)
17. Lieb, E.H., Loss, M.: Analysis, 2nd edn. Graduate Studies in Mathematics, vol. 14. American Mathematical Society, Providence (2001)
18. Lin, T.-C., Wei, J.: Ground state of N coupled nonlinear Schrödinger equations in $\mathbf{R}^n$, $n \leq 3$. Commun. Math. Phys. **255**(3), 629–653 (2005), and Commun. Math. Phys. **277**(2), 573–576 (2008)
19. Lions, P.-L.: The concentration-compactness principle in the calculus of variations. The locally compact case. I. Ann. Inst. Henri Poincaré, Anal. Non Linéaire **1**(2), 109–145 (1984)
20. Maia, L.A., Montefusco, E., Pellacci, B.: Positive solutions for a weakly coupled nonlinear Schrödinger system. J. Differ. Equ. **229**(2), 743–767 (2006)
21. Maia, L.A., Montefusco, E., Pellacci, B.: Infinitely many nodal solutions for a weakly coupled nonlinear Schrödinger system. Commun. Contemp. Math. **10**(5), 651–669 (2008)
22. Maia, L.A., Montefusco, E., Pellacci, B.: Orbital stability property for coupled nonlinear Schrödinger equations. Adv. Nonlinear Stud. **10**(3), 681–705 (2010)
23. Montefusco, E., Pellacci, B., Squassina, M.: Soliton dynamics for CNLS systems with potentials. Asymptot. Anal. **66**(2), 61–86 (2010)
24. Nguyen, N.V., Wang, Z.-Q.: Orbital stability of solitary waves for a nonlinear Schrödinger system. Adv. Differ. Equ. **16**(9–10), 977–1000 (2011)
25. Ohta, M.: Stability of stationary states for the coupled Klein-Gordon-Schrödinger equations. Nonlinear Anal. **27**(4), 455–461 (1996)
26. Ohta, M.: Stability of solitary waves for coupled nonlinear Schrödinger equations. Nonlinear Anal. **26**(5), 933–939 (1996)
27. Protter, M.H., Weinberger, H.F.: Maximum Principles in Differential Equations. Springer, New York (1984). Corrected reprint of the 1967 original
28. Shibata, M.: Rearrangement inequality and stability of standing waves of nonlinear Schrödinger equations. Preprint
29. Sirakov, B.: Least energy solitary waves for a system of nonlinear Schrödinger equations in $\mathbf{R}^n$. Commun. Math. Phys. **271**(1), 199–221 (2007)

Decay Rate of L^q Norms of Critical Schrödinger Heat Semigroups

Kazuhiro Ishige and Yoshitsugu Kabeya

Abstract Let $H := -\Delta + V$ be a critical Schrödinger operator on $L^2(\mathbf{R}^N)$, where $N \geq 3$ and V is a radially symmetric function decaying quadratically at the space infinity. We study the optimal decay rate of the operator norm of the Schrödinger heat semigroup e^{-tH} from $L^2(\mathbf{R}^N)$ to $L^q(\mathbf{R}^N)$ $(2 \leq q \leq \infty)$.

Keywords Schrödinger heat semigroup · Critical Schrödinger operator · $L^2 - L^q$ estimate

1 Introduction

Let $H := -\Delta + V$ be a Schrödinger operator on $L^2(\mathbf{R}^N)$, where $N \geq 3$ and $V \in L^p_{\mathrm{loc}}(\mathbf{R}^N)$ with $p > N/2$. The operator H is said to be nonnegative if

$$\int_{\mathbf{R}^N} \left\{ |\nabla\phi|^2 + V\phi^2 \right\} dx \geq 0 \quad \text{for all } \phi \in C_0^\infty(\mathbf{R}^N).$$

In addition, the operator H is said to be

- subcritical if H is nonnegative and for any $W \in C_0^\infty(\mathbf{R}^N)$, $H - \varepsilon W$ is nonnegative for small enough $\varepsilon > 0$;
- critical if H is nonnegative but not subcritical.

Furthermore the operator H is subcritical if and only if there exists, for any $y \in \mathbf{R}^N$, a positive solution $G(x, y)$ satisfying

$$\big(-\Delta + V(x)\big) G(x, y) = \delta(x - y) \quad \text{in } \mathbf{R}^N,$$

K. Ishige (✉)
Mathematical Institute, Tohoku University, Aoba, Sendai 980-8578, Japan
e-mail: ishige@math.tohoku.ac.jp

Y. Kabeya
Department of Mathematical Sciences, Osaka Prefecture University, Sakai 599-8531, Japan
e-mail: kabeya@ms.osakafu-u.ac.jp

R. Magnanini et al. (eds.), *Geometric Properties for Parabolic and Elliptic PDE's*, Springer INdAM Series 2, DOI 10.1007/978-88-470-2841-8_11,

where δ is the Dirac function (see for example [10, Sect. 2]). In this paper we focus on nonnegative Schrödinger operators $H = -\Delta + V$ with a radially symmetric smooth potential V satisfying

$$V(x) = \omega|x|^{-2}\big(1+o(1)\big) \quad \text{as } |x| \to \infty \text{ with } -\left(\frac{N-2}{2}\right)^2 < \omega \le 0, \tag{1}$$

and study the decay rate of

$$\big\|e^{-tH}\big\|_{q,2} := \sup_{\phi\in L^2(\mathbf{R}^N)\setminus\{0\}} \frac{\|e^{-tH}\phi\|_{L^q(\mathbf{R}^N)}}{\|\phi\|_{L^2(\mathbf{R}^N)}} \quad (2 \le q \le \infty),$$

as $t \to \infty$. Here $\|e^{-tH}\|_{q,2}$ means the operator norm of the Scrödinger heat semigroup e^{-tH} from $L^2(\mathbf{R}^N)$ to $L^q(\mathbf{R}^N)$.

Nonnegative Schrödinger operators have been studied extensively by many mathematicians since the pioneering work due to Simon [14] and [15] (see [1, 3, 5–13, 16–18], and references therein). Among others, in [10], Murata studied the structure and the behavior of the positive harmonic functions for nonnegative Schrödinger operators. In particular, in the case of (1), under some additional assumptions, he proved that the positive harmonic function U for the operator $H = -\Delta + V$ satisfies

$$C^{-1}\big(1+|x|\big)^{A(\omega)} \le U(x) \le C\big(1+|x|\big)^{A(\omega)}, \quad x \in \mathbf{R}^N,$$

for some positive constant C, where

$$A(\omega) = \begin{cases} \alpha(\omega) & \text{if } H \text{ is subcritical,} \\ \beta(\omega) & \text{if } H \text{ is critical,} \end{cases} \tag{2}$$

with

$$\alpha(\omega) := \frac{-(N-2)+\sqrt{(N-2)^2+4\omega}}{2},$$
$$\beta(\omega) := \frac{-(N-2)-\sqrt{(N-2)^2+4\omega}}{2}.$$

(See also Sect. 2.1.) Subsequently, in [1], Davies and Simon studied the large time behavior of $\|e^{-tH}\|_{q,p}$ for subcritical operators $H = -\Delta + V$ by using positive harmonic functions for the operator H. Their results imply that, if V satisfies (1) and $H = -\Delta + V$ is subcritical, then, for any $\varepsilon > 0$, there exists a positive constant C such that

$$C^{-1}(1+t)^{-\frac{N}{4}-\frac{\alpha(\omega)}{2}-\varepsilon} \le \big\|e^{-tH}\big\|_{\infty,2} \le C(1+t)^{-\frac{N}{4}-\frac{\alpha(\omega)}{2}+\varepsilon} \tag{3}$$

for all sufficiently large t. Recently, in [3], the authors of this paper considered nonnegative Schrödinger operators $H = -\Delta + V$ under the condition

$$(V)\quad \begin{cases} \text{(i)} & V = V(r) \in C^1\big([0,\infty)\big); \\ \text{(ii)} & \text{there exist constants } \omega \in (-\omega_*, 0] \text{ and } \theta > 0 \text{ such that} \\ & V(r) = \omega r^{-2} + O\big(r^{-2-\theta}\big) \text{ as } r \to \infty, \text{ where } \omega_* = (N-2)^2/4; \\ \text{(iii)} & \sup_{r\ge 1} r^3\big|V'(r)\big| < \infty, \end{cases}$$

and gave the following theorem, which exhibits the large time decay estimates of $\|e^{-tH}\|_{q,2}$ $(2 \le q \le \infty)$ not only for the subcritical case but also for the critical case and which improves the estimate (3). Put

$$\eta_q(t) := \begin{cases} (1+t)^{-\frac{N}{2}(\frac{1}{2}-\frac{1}{q})} & \text{if } q < \infty \text{ and } qA(\omega)+N>0, \\ (1+t)^{-\frac{N}{4}-\frac{A(\omega)}{2}}[\log(2+t)]^{\frac{1}{q}} & \text{if } q < \infty \text{ and } qA(\omega)+N=0, \\ (1+t)^{-\frac{N}{4}-\frac{A(\omega)}{2}} & \text{if } q=\infty \text{ or } qA(\omega)+N<0, \end{cases}$$

for any $t>0$, where $q \in [2,\infty]$. See Theorem 1.2 in [3].

Theorem 1 *Let $N \ge 3$. Assume condition (V) and $V \le 0$. Let $H := -\Delta + V$ be a nonnegative Schrödinger operator on $L^2(\mathbf{R}^N)$. Then, for any $q \in [2,\infty]$, there holds the following*:

(i) *If H is subcritical, then there exists a positive constant C_1 such that*

$$C_1^{-1}\eta_q(t) \le \left\|e^{-tH}\right\|_{q,2} \le C_1\eta_q(t) \tag{4}$$

for all sufficiently large t.

(ii) *If H is critical and $\beta(\omega) > -N/2$, then there exists a positive constant C_2 such that*

$$\left\|e^{-tH}\right\|_{q,2} \le C_2\eta_q(t) \tag{5}$$

for all sufficiently large t.

(iii) *If H is critical and $\beta(\omega) \le -N/2$, then there exists a positive constant C_3 such that*

$$\left\|e^{-tH}\right\|_{q,2} \le C_3 t^{-\frac{N+2\beta(\omega)}{2-N-2\beta(\omega)}(\frac{1}{2}-\frac{1}{q})}$$

for all sufficiently large t.

Under condition (V), Theorem 1 gives the exact decay rate of $\|e^{-tH}\|_{q,2}$ for the subcritical case, however gives only upper decay estimates of $\|e^{-tH}\|_{q,2}$ for the critical case. As far as we know, there are no results giving the exact decay rate of $\|e^{-tH}\|_{q,2}$ for critical Schrödinger operators.

In this paper, as a continuation of [3], under condition (V), we consider again nonnegative Schrödinger operators $H=-\Delta+V$, in particular, critical Schrödinger operators for the case $\beta(\omega) > -N/2$, and prove the following theorem. Theorem 2 can give the exact decay rate of $\|e^{-tH}\|_{q,2}$ $(2 \le q \le \infty)$ as $t \to \infty$ for critical Schrödinger operators and implies that the decay estimate (5) is optimal.

Theorem 2 *Let $N \ge 3$. Assume condition (V), and let $H := -\Delta + V$ be a nonnegative Schrödinger operator on $L^2(\mathbf{R}^N)$ such that $A(\omega) > -N/2$. Then, for any $q \in [2,\infty]$, there holds*

$$C_1^{-1}\eta_q(t) \le \left\|e^{-tH}\right\|_{q,2} \le C_1\eta_q(t) \tag{6}$$

for all sufficiently large t.

We remark that, if H is subcritical, then automatically $A(\omega) = \alpha(\omega) > -N/2$. Furthermore we remark that Theorem 2 does not require the nonpositivity of the potential V (compare with Theorem 1).

The main difficulty of proving Theorem 2 is to prove the left hand side of the inequality (6) for the critical case. In [3] the authors of this paper obtained the precise large time behavior of $e^{-tH}\phi$ with $\phi \in L^2(\mathbf{R}^N, e^{|x|^2/4}\,dx)$, and proved the left hand of the inequality (4). However the argument is based on the one in [2] and is not applicable to the critical case since $\beta(\omega) < -(N-2)/2$ (see the proof of Proposition 3.1 in [2] and Proposition 4.1 in [3]). Let $u(t) = e^{-tH}\phi$. In order to prove Theorem 2, we first use the parabolic Harnack inequality and the property

$$\int_{\mathbf{R}^N} u(x,t)U(x)\,dx = \int_{\mathbf{R}^N} \phi(x)U(x)\,dx, \quad t > 0$$

(see (22)), and obtain some upper bounds of $e^{-tH}\phi$. Then we can find positive constants L and ε satisfying

$$\int_{\varepsilon(1+t)^{1/2}\le|x|\le L(1+t)^{1/2}} u(x,t)U(x)\,dx \ge \frac{1}{2}\int_{\mathbf{R}^N} \phi(x)U(x)\,dx > 0$$

for all sufficiently large t (see Lemma 3). Next, by using the parabolic Harnack inequality again, we obtain some lower estimates of u (see (44)). Then, applying an argument used in [3] with the aid of the comparison principle, we obtain lower estimates of $\|u(t)\|_{L^q(\mathbf{R}^N)}$, and complete the proof of Theorem 2.

The rest of this paper is organized as follows. In Sect. 2 we recall some preliminary results on the positive harmonic functions and the Schrödinger heat semigroups e^{-tH}. In Sect. 3 we prove the inequality (6), and complete the proof of Theorem 2.

2 Preliminaries

In this section we give some notation used in this paper. Furthermore we consider nonnegative Schrödinger operators $H = -\Delta + V$ under condition (V), and recall some preliminary results on the positive harmonic functions and the properties of e^{-tH}.

We introduce some notation. Let

$$\|\cdot\|_p := \|\cdot\|_{L^p(\mathbf{R}^N)} \quad \text{and} \quad \|\cdot\| := \|\cdot\|_{L^2(\mathbf{R}^N,\rho\,dx)},$$

where $p \in [1,\infty]$ and $\rho(x) = e^{|x|^2/4}$. On the other hand, for any sets Λ and Σ, let $f = f(\lambda,\sigma)$ and $h = h(\lambda,\sigma)$ be maps from $\Lambda \times \Sigma$ to $(0,\infty)$. Then we say

$$f(\lambda,\sigma) \preceq h(\lambda,\sigma)$$

for all $\lambda \in \Lambda$ if, for any $\sigma \in \Sigma$, there exists a positive constant C such that $f(\lambda,\sigma) \le Ch(\lambda,\sigma)$ for all $\lambda \in \Lambda$. In addition, we say $f(\lambda,\sigma) \asymp h(\lambda,\sigma)$ for all $\lambda \in \Lambda$ if $f(\lambda,\sigma) \preceq h(\lambda,\sigma)$ and $f(\lambda,\sigma) \succeq h(\lambda,\sigma)$ for all $\lambda \in \Lambda$.

2.1 Positive Harmonic Functions

We first consider the ordinary differential equation

$$z'' + \frac{N-1}{r}z' - V(r)z = 0 \quad \text{in } (0, \infty). \tag{7}$$

It is known that there exists a positive solution U of (7) satisfying

$$0 < \lim_{r \to 0} U(r) < \infty, \qquad U(r) = r^{A(\omega)}\big(1 + o(1)\big) \quad \text{as } r \to \infty, \tag{8}$$

where $A(\omega)$ is the constant given in (2). See Theorem 5.7 in [10]. (See also Theorem 1.1 in [3].) Then the function $U(|x|)$ is a positive harmonic function for the operator $H := -\Delta + V$, that is, there hold

$$U\big(|x|\big) > 0 \quad \text{and} \quad -\Delta U + V\big(|x|\big)U = 0 \tag{9}$$

for all $x \in \mathbf{R}^N$.

Next we consider the ordinary differential equation

$$U'' + \frac{N-1}{r}U' - V(r)U = f(r) \quad \text{in } (0, \infty), \tag{10}$$

where $f \in C([0, \infty))$. Then, by [3, Lemma 2.2], we see that, for any solution v of (10) satisfying $\limsup_{r \to 0} |v(r)| < \infty$, there exists a constant c such that

$$v(r) = cU(r) + F[f](r)$$

for all $r \geq 0$, where

$$F[f](r) := U(r) \int_0^r s^{1-N}\big[U(s)\big]^{-2} \left(\int_0^s \tau^{N-1} U(\tau) f(\tau)\, d\tau \right) ds.$$

2.2 Preliminary Results

Let $\phi \in C_0(\mathbf{R}^N)$ such that $\phi \geq (\not\equiv) 0$ in $\mathbf{R}^N$, and put $u(t) = e^{-tH}\phi$. We first give some estimates of $u(t)$ by using the explicit representation of $e^{t\Delta}\phi$ and the parabolic Harnack inequality. The function $u = u(x,t)$ satisfies

$$\begin{cases} \partial_t u = \Delta u - V\big(|x|\big)u & \text{in } \mathbf{R}^N \times (0, \infty), \\ u(x, 0) = \phi(x) & \text{in } \mathbf{R}^N, \end{cases} \tag{11}$$

and by the comparison principle we have

$$0 \leq u(x,t) \leq e^{t\|V\|_\infty} \big(e^{t\Delta}\phi\big)(x)$$

for all $(x, t) \in \mathbf{R}^N \times (0, \infty)$. Then, since

$$\big(e^{t\Delta}\phi\big)(x) = (4\pi t)^{-\frac{N}{2}} \int_{\mathbf{R}^N} e^{-\frac{|x-y|^2}{4t}} \phi(y)\, dy, \tag{12}$$

we have

$$0 \le u(x,t) \preceq t^{-\frac{N}{4}} e^{t\|V\|_\infty} \|\phi\|_2 \tag{13}$$

for all $(x,t) \in \mathbf{R}^N \times (0,\infty)$. Put

$$w(y,s) := (1+t)^{\frac{N}{2}} u(x,t), \quad y := (1+t)^{-\frac{1}{2}} x, \ s := \log(1+t). \tag{14}$$

Then, by (11) we see that the function w satisfies

$$\begin{cases} \partial_s w = \Delta w + \dfrac{y}{2} \cdot \nabla_y w - \tilde{V}(y,s)w + \dfrac{N}{2} w \\ \quad = \dfrac{1}{\rho} \mathrm{div}(\rho \nabla_y w) - \tilde{V}(y,s)w + \dfrac{N}{2} w & \text{in } \mathbf{R}^N \times (0,\infty), \\ w(y,0) = \phi(y) & \text{in } \mathbf{R}^N, \end{cases} \tag{15}$$

where $\tilde{V}(y,s) = e^s V(e^{\frac{s}{2}} y)$. Furthermore, by (12) and (13) we have

$$\sup_{0<s<S} \|w(s)\| < \infty \tag{16}$$

for any $S > 0$.

On the other hand, by condition (V) we have

$$|\tilde{V}(y,s)| \preceq |y|^{-2}, \tag{17}$$

$$\left| \tilde{V}(y,s) - \frac{\omega}{|y|^2} \right| \preceq \frac{e^s}{(e^{s/2}|y|)^{2+\theta}} \le |y|^{-(2+\theta)} e^{-\frac{\theta}{2}s} \tag{18}$$

for all $(y,s) \in \mathbf{R}^N \times (0,\infty)$, where θ is the constant given in condition (V) (ii). Then, by (17), applying the parabolic Harnack inequality to (15), we see that, for any $0 < \varepsilon < R$, there exists a constant C such that

$$\max_{\varepsilon \le |y| \le R} w(y,s) \le C \min_{\varepsilon \le |y| \le R} w(y,s+1) \tag{19}$$

for all $s \ge 2$. Remark that, due to (17) and the term $(y/2) \cdot \nabla_y w$ in (15), the constant C possibly depends on ε and R. The inequality (19) together with (14) implies that

$$\max_{\varepsilon(1+t)^{1/2} \le |x| \le R(1+t)^{1/2}} u(x,t) \le C \min_{\varepsilon(1+t)^{1/2} \le |x| \le R(1+t)^{1/2}} u\big(x, e(t+1)-1\big) \tag{20}$$

for all $t \ge e^2 - 1$.

Next we recall the following two lemmas given in [3] (see Lemmas 3.1 and 3.2 in [3]). For any $\varepsilon > 0$ and $T \ge 0$, let

$$\begin{aligned} D_\varepsilon(T) &:= \big\{(x,t) \in \mathbf{R}^N \times (T,\infty) : |x| < \varepsilon(1+t)^{1/2}\big\}, \\ \Gamma_\varepsilon(T) &:= \big\{(x,t) \in \mathbf{R}^N \times (T,\infty) : |x| = \varepsilon(1+t)^{1/2}\big\} \\ &\quad \cup \big\{(x,T) \in \mathbf{R}^N \times \{T\} : |x| < \varepsilon(1+t)^{1/2}\big\}. \end{aligned}$$

Lemma 1 *Assume the same conditions as in Theorem* 2. *For any* $\gamma \ge 0$, *put*

$$\zeta(t) := (1+t)^{-\gamma - \frac{A(\omega)}{2}}.$$

Then, for any $T > 0$ and any sufficiently small $\varepsilon > 0$, there exist a constant C and a function $W(x,t)$ such that

$$\begin{aligned} &\partial_t W \geq \Delta W - V(|x|)W && \text{in } \mathbf{R}^N \times (0,\infty), \\ &W(x,t) \leq C\zeta(t)U(|x|) && \text{in } D_\varepsilon(T), \\ &W(x,t) \geq (1+t)^{-\gamma} && \text{in } \Gamma_\varepsilon(T). \end{aligned} \tag{21}$$

Lemma 2 *Assume the same conditions as in Theorem 2. Fix $T > 0$ and ε being a sufficiently small positive constant. Let $u = e^{-tH}\phi$ be a solution of (1) satisfying*

$$\|u(t)\|_2 \leq C_1(1+t)^{-d}\|\phi\|_2, \quad t > 0,$$

for some constants $C_1 > 0$ and $d \geq 0$. Then there exists a constant C_2 such that

$$|u(x,t)| \leq C_2(1+t)^{-d-\frac{N}{4}}\|\phi\|_2$$

for all $x \in \mathbf{R}^N$ and $t > T$ with $|x| \geq \varepsilon(1+t)^{1/2}$. Furthermore there exists a constant C_3 such that

$$|u(x,t)| \leq C_3(1+t)^{-d-\frac{N}{4}-\frac{A(\omega)}{2}}\|\phi\|_2 U(|x|)$$

for all $(x,t) \in D_\varepsilon(T)$.

3 Proof of Theorem 2

In this section we prove the inequality (6), and complete the proof of Theorem 2. The right hand side of the inequality (6) is proved by the same argument as in the proof of Theorem 1.2 in [3] with the aid of Lemma 2.2. So it suffices to prove the left hand side of the inequality (6).

Let ϕ be a radially symmetric, nonnegative function such that $\phi \in C_0(\mathbf{R}^N)$ and $\|\phi\| = 1$. Put $u(t) = e^{-tH}\phi$ and

$$M := \int_{\mathbf{R}^N} \phi(x)U(|x|)\,dx > 0.$$

Then, by (9) we have

$$\frac{d}{dt}\int_{\mathbf{R}^N} u(x,t)U(|x|)\,dx = 0,$$

and obtain

$$\int_{\mathbf{R}^N} u(x,t)U(|x|)\,dx = M \tag{22}$$

for all $t > 0$. Then, due to (22) and the parabolic Harnack inequality, we have:

Lemma 3 *Assume the same conditions as in Theorem 2. Then there exist positive constants C, T, and ε_0 such that*

$$0 < u(x,t) < C(1+t)^{-\frac{N}{2}-A(\omega)}U(|x|) \tag{23}$$

for all $(x,t) \in D_{\varepsilon_0}(T)$.

Proof By (8) and the positivity of U we have

$$U(r) \asymp (1+r)^{A(\omega)} \tag{24}$$

for all $r \geq 0$. Then, by (14), (22), and (24) we have

$$\begin{aligned} M &= \int_{\mathbf{R}^N} u(x,t)U\big(|x|\big)\,dx \geq \int_{(1+t)^{1/2}<|x|<2(1+t)^{1/2}} u(x,t)U\big(|x|\big)\,dx \\ &\succeq (1+t)^{\frac{A(\omega)}{2}} \int_{(1+t)^{1/2}<|x|<2(1+t)^{1/2}} u(x,t)\,dx \\ &\succeq (1+t)^{\frac{N}{2}+\frac{A(\omega)}{2}} \min_{(1+t)^{1/2}\leq|x|\leq 2(1+t)^{1/2}} u(x,t) = e^{\frac{A(\omega)}{2}s} \min_{1\leq|y|\leq 2} w(y,s) \end{aligned} \tag{25}$$

for all $t \geq 1$ and $s \geq \log 2$ with $s = \log(1+t)$. Let ε_0 be a sufficiently small constant. Then, by (19) we have

$$\max_{\varepsilon_0\leq|y|\leq 2} w(y,s) \leq C_1 \min_{\varepsilon_0\leq|y|\leq 2} w(y,s+1)$$

for all $s \geq 2$, where C_1 is a constant. This together with (14) and (25) implies that

$$\begin{aligned} (1+t)^{\frac{N}{2}}\big|u(x,t)\big| &\leq \max_{\varepsilon_0\leq|y|\leq 2} w(y,s) \preceq \min_{\varepsilon_0\leq|y|\leq 2} w(y,s+1) \\ &\preceq e^{-\frac{A(\omega)}{2}(s+1)} \preceq (1+t)^{-\frac{A(\omega)}{2}} \end{aligned} \tag{26}$$

for all $(x,t) \in \mathbf{R}^N \times (T,\infty)$ with $|x| = \varepsilon_0(1+t)^{1/2}$, where $T = e^2 - 1$. On the other hand, since $\|\phi\|_2 \leq \|\phi\| = 1$, by (13) we have

$$\big\|u(T)\big\|_\infty \leq C_2\|\phi\|_2 \leq C_2 \tag{27}$$

for some constant C_2.

Let W be a function defined in Lemma 1 such that $\varepsilon = \varepsilon_0$, $\gamma = N/2 + A(\omega) > 0$. By (21), (26), and (27) we take a sufficiently large constant C_3 so that

$$u(x,t) \leq C_3 W(x,t)$$

for all $(x,t) \in \Gamma_{\varepsilon_0}(T)$. Then, by Lemma 1 we apply the comparison principle to obtain

$$0 < u(x,t) \leq C_4(1+t)^{-\frac{N}{2}-A(\omega)}U\big(|x|\big), \quad (x,t) \in D_{\varepsilon_0}(T)$$

for some constant C_4. Thus we have (23), and Lemma 3 follows. □

Next we prove the following lemma.

Lemma 4 *Assume the same conditions as in Theorem* 2. *Then there exists a constant C such that*

$$\big\|w(s)\big\| \leq Ce^{-\frac{A(\omega)}{2}s} \tag{28}$$

for all sufficiently large s.

Proof For the subcritical case, the inequality (28) is given in Proposition 4.1 in [3]. So it suffices to prove the inequality (28) for the critical case.

Let H be a critical Schrödinger operator. Then

$$A(\omega) = \beta(\omega) < \alpha(\omega). \tag{29}$$

Let ε_0 and T be constants given in Lemma 3 and $R > 0$. Lemma 3 implies

$$\begin{aligned} \big|w(y,s)\big| &\preceq e^{-\frac{\beta(\omega)}{2}s}|y|^{\beta(\omega)} && \text{for all } y \in B(0,\varepsilon_0),\ s \geq S, \\ \big|w(y,s)\big| &\preceq e^{-\beta(\omega)s} && \text{for all } y \in B\big(0, Re^{-s/2}\big),\ s \geq S, \end{aligned} \tag{30}$$

where $S = \log(1+T)$. By (15), (18), and (30) we apply the regularity theorems for parabolic equations (see [4]), and obtain

$$\big|\big(\nabla_y^l w\big)(y,s)\big| + \big|(\partial_s w)(y,s)\big| \preceq e^{-\frac{\beta(\omega)}{2}s}, \quad l = 0, 1, 2, \tag{31}$$

for all (y,s) with $\varepsilon_0/4 \leq |y| \leq \varepsilon_0/2$ and $s \geq S+1$.

Let ε_0 be the constant given in Lemma 3. Put $\varepsilon = \varepsilon_0/8$ and

$$I(s) := \int_{|y|\geq 2\varepsilon} |w|^2 \rho\, dy.$$

Then, by (15), (18), and (31) we have

$$\begin{aligned} I'(s) &= 2\int_{|y|\geq 2\varepsilon} w\left[\operatorname{div}(\rho\nabla w) - e^s V\big(e^{s/2}y\big)w\rho + \frac{N}{2}w\rho\right]dy \\ &= 2\int_{|y|=2\varepsilon} w\partial_\nu w\rho\, d\sigma - 2\int_{|y|\geq 2\varepsilon}\left[|\nabla w|^2 + \frac{\omega}{|y|^2}w^2\right]\rho\, dy \\ &\quad + NI(s) - 2\int_{|y|\geq 2\varepsilon}\left[e^s V\big(e^{s/2}y\big) - \frac{\omega}{|y|^2}\right]w^2\rho\, dy \\ &\leq -2\int_{|y|\geq 2\varepsilon}\left[|\nabla w|^2 + \frac{\omega}{|y|^2}w^2\right]\rho\, dy \\ &\quad + O\big(e^{-\beta(\omega)s}\big) + NI(s) + O\big(e^{-\frac{\theta}{4}s}\big)I(s) \end{aligned} \tag{32}$$

for all $s \geq S+1$. Let ξ be a smooth function on $[0,\infty)$ such that

$$0 \leq \xi(r) \leq 1 \quad \text{on } [0,\infty), \qquad \xi(r) = 1 \quad \text{on } [0, 3\varepsilon), \qquad \xi(r) = 0 \quad \text{on } [4\varepsilon, \infty).$$

Since $\omega > -(N-2)^2/4$, we can take a sufficiently small $\delta > 0$ so that

$$\omega_\delta := \frac{1+\delta}{1-\delta}\omega > -\frac{(N-2)^2}{4}. \tag{33}$$

By (31) we have

$$\begin{aligned} \int_{|y|\geq 2\varepsilon} |\nabla w|^2\rho\, dy &= \int_{|y|\geq 2\varepsilon} \big|\nabla\big[w(1-\xi) + w\xi\big]\big|^2 \rho\, dy \\ &\geq (1-\delta)\int_{|y|\geq 2\varepsilon}\big|\nabla\big[w(1-\xi)\big]\big|^2\rho\, dy - C_1\int_{|y|\geq 2\varepsilon}\big|\nabla[w\xi]\big|^2\rho\, dy \\ &\geq (1-\delta)\int_{\mathbf{R}^N}\big|\nabla\big[w(1-\xi)\big]\big|^2\rho\, dy - C_2 e^{-\beta(\omega)s} \end{aligned} \tag{34}$$

for all $s \geq S+1$, where C_1 and C_2 are constants. Similarly, since $\omega < 0$, by (31) we have

$$\begin{aligned}
&\int_{|y|\geq 2\varepsilon} \frac{\omega}{|y|^2} w^2 \rho\, dy \\
&\quad \geq (1+\delta) \int_{|y|\geq 2\varepsilon} \frac{\omega}{|y|^2} \big[w(1-\xi)\big]^2 \rho\, dy - C_3 \int_{|y|\geq 2\varepsilon} \frac{1}{|y|^2} [w\xi]^2 \rho\, dy \\
&\quad \geq (1+\delta) \int_{\mathbf{R}^N} \frac{\omega}{|y|^2} \big[w(1-\xi)\big]^2 \rho\, dy - C_4 e^{-\beta(\omega)s} \qquad (35)
\end{aligned}$$

for all $s \geq S+1$, where C_3 and C_4 are constants. Therefore, by (32), (34), and (35) we have

$$\begin{aligned}
I'(s) &\leq -2(1-\delta)\bigg[\int_{\mathbf{R}^N} \big|\nabla\big[w(1-\xi)\big]\big|^2 \rho\, dy + \int_{\mathbf{R}^N} \frac{\omega_\delta}{|y|^2} \big[w(1-\xi)\big]^2 \rho\, dy\bigg] \\
&\quad + NI(s) + O\big(e^{-\frac{\theta}{4}s}\big) I(s) + O\big(e^{-\beta(\omega)s}\big) \qquad (36)
\end{aligned}$$

for all $s \geq S+1$. On the other hand, by Lemma 4.2 in [3] and (33) we have

$$\begin{aligned}
&\int_{\mathbf{R}^N} \big|\nabla\big[w(1-\xi)\big]\big|^2 \rho\, dy + \int_{\mathbf{R}^N} \frac{\omega_\delta}{|y|^2} \big[w(1-\xi)\big]^2 \rho\, dy \\
&\quad \geq \left(\frac{\alpha(\omega_\delta)}{2} + \frac{N}{2}\right) \int_{\mathbf{R}^N} \big|w(1-\xi)\big|^2 \rho\, dy.
\end{aligned}$$

This together with (31) and (36) implies that

$$\begin{aligned}
I'(s) &\leq -(1-\delta)\big[\alpha(\omega_\delta) + N\big] \int_{\mathbf{R}^N} \big|w(1-\xi)\big|^2 \rho\, dy \\
&\quad + NI(s) + O\big(e^{-\frac{\theta}{4}s}\big) I(s) + O\big(e^{-\beta(\omega)s}\big) \\
&\leq -(1-\delta)\big[\alpha(\omega_\delta) + N\big] I(s) + O\big(e^{-\frac{\theta}{4}s}\big) I(s) + NI(s) + O\big(e^{-\beta(\omega)s}\big) \\
&= \big[-(1-\delta)\alpha(\omega_\delta) + \delta N\big] I(s) + O\big(e^{-\frac{\theta}{4}s}\big) I(s) + O\big(e^{-\beta(\omega)s}\big) \qquad (37)
\end{aligned}$$

for all $s \geq S+1$. Taking a sufficiently small δ if necessary, by (29) we have

$$-(1-\delta)\alpha(\omega_\delta) + \delta N < -\beta(\omega) - \delta,$$

and by (37) we obtain

$$I'(s) \leq \big[-\beta(\omega) - \delta + C_5 e^{-\frac{\theta}{4}s}\big] I(s) + C_5 e^{-\beta(\omega)s}, \quad s \geq S+1,$$

for some constant $C_5 > 0$. This together with (16) yields

$$I(s) \leq C_6 e^{-\beta(\omega)s - \delta s} I(S+1) + C_6 e^{-\beta(\omega)s} \leq C_7 e^{-\beta(\omega)s} \qquad (38)$$

for all $s \geq S+1$, where C_6 and C_7 are constants. On the other hand, since $\beta(\omega) > -N/2$, by (30) we have

$$\int_{|y|\leq 2\varepsilon} w(y,s)^2\, dy \preceq e^{-\beta(\omega)} \int_{|y|\leq 2\varepsilon} |y|^{2\beta(\omega)}\, dy \preceq e^{-\beta(\omega)} \qquad (39)$$

for all $s \geq S+1$. Therefore, by (38) and (39) we have

$$\left\| w(s) \right\|^2 = O\big(e^{-\beta(\omega)s}\big)$$

for all sufficiently large s, and obtain (28). Thus Lemma 4 follows. □

By Lemma 4 with (14) we have

$$\left\| u(t) \right\|_2 = O\big(t^{-\frac{N}{4}-\frac{A(\omega)}{2}}\big) \tag{40}$$

for all sufficiently large t. This implies that

$$\left\| \partial_t u(t) \right\|_2 = O\big(t^{-\frac{N}{4}-\frac{A(\omega)}{2}-1}\big)$$

for all sufficiently large t (see Lemma 4.1 in [3]). Then, since

$$(\partial_t u)(t) = e^{-(t-1)H}(\partial_t u)(1),$$

due to Lemma 4, there exist positive constants T and ε such that

$$\big|(\partial_t u)(x,t)\big| \preceq (1+t)^{-\frac{N}{2}-\beta(\omega)-1} U\big(|x|\big)$$

for all $(x,t) \in D_\varepsilon(T)$. Furthermore, by Lemma 3.2 we have:

Lemma 5 *Assume the same conditions as in Theorem 2. Then there exist positive constants L and ε such that*

$$\int_{\varepsilon(1+t)^{1/2} \leq |x| \leq L(1+t)^{1/2}} u(x,t) U\big(|x|\big)\, dx \geq \frac{M}{2} > 0 \tag{41}$$

for all sufficiently large t.

Proof For any $L > 0$, by (14) and (28) we have

$$\begin{aligned}
&\int_{|x| \geq L(1+t)^{1/2}} u(x,t) U\big(|x|\big)\, dx \\
&\quad = \int_{|y| \geq L} w(y,s) U\big(e^{s/2}|y|\big)\, dy \leq \left\| w(s) \right\| \left(\int_{|y| \geq L} U\big(e^{s/2}|y|\big)^2 \rho^{-1}(y)\, dy \right)^{1/2} \\
&\quad \preceq e^{\frac{A(\omega)}{2}s} \left\| w(s) \right\| \left(\int_{|y| \geq L} |y|^{2A(\omega)} \rho^{-1}(y)\, dy \right)^{1/2} \\
&\quad \preceq \left(\int_{|y| \geq L} |y|^{2A(\omega)} \rho^{-1}(y)\, dy \right)^{1/2}
\end{aligned}$$

for all sufficiently large t. Then we can take a sufficiently large L so that

$$\int_{|x| \geq L(1+t)^{1/2}} u(x,t) U\big(|x|\big)\, dx \leq \frac{M}{4} \tag{42}$$

for all sufficiently large t. On the other hand, since $A(\omega) > -N/2$, by Lemma 3 we take a sufficiently small $\varepsilon > 0$ so that

$$\begin{aligned}
&\int_{|x|\le\varepsilon(1+t)^{1/2}} u(x,t)U\big(|x|\big)\,dx \\
&\quad\le C_1(1+t)^{-\frac{N}{2}-A(\omega)}\int_{|x|\le\varepsilon(1+t)^{1/2}} U\big(|x|\big)^2\,dx \\
&\quad\le C_2(1+t)^{-\frac{N}{2}-A(\omega)}\int_{|x|\le\varepsilon(1+t)^{1/2}} |x|^{2A(\omega)}\,dx \\
&\quad\le C_3\varepsilon^{2A(\omega)+N}\le\frac{M}{4}
\end{aligned}\tag{43}$$

for all sufficiently large t. Then, by (22), (42), and (43) we have (41), and Lemma 5 follows. □

Now we are ready to complete the proof of Theorem 2.

Proof Due to Lemma 5 and (24), there exist positive constants ε and L such that

$$\begin{aligned}
\frac{M}{2}&\le\int_{\varepsilon(1+t)^{1/2}\le|x|\le L(1+t)^{1/2}} u(x,t)U\big(|x|\big)\,dx \\
&\preceq(1+t)^{\frac{N}{2}+\frac{A(\omega)}{2}}\max_{\varepsilon(1+t)^{1/2}\le|x|\le L(1+t)^{1/2}} u(x,t)
\end{aligned}$$

for all sufficiently large t. Then, by (20) we can find positive constants C_1 and T satisfying

$$\min_{\varepsilon(1+t)^{1/2}\le|x|\le L(1+t)^{1/2}} u(x,t)\ge C_1(1+t)^{-\frac{N}{2}-\frac{A(\omega)}{2}}\tag{44}$$

for all $t\ge T$. Put

$$H(x,t):=(1+t)^{-\frac{N}{2}-A(\omega)}U\big(|x|\big).$$

By (24) we have

$$H(x,t)\preceq(1+t)^{-\frac{N}{2}-\frac{A(\omega)}{2}}$$

for all $x\in\mathbf{R}^N$ and $t>T$ with $|x|=\varepsilon(1+t)^{1/2}$. This together with (44) and the positivity of the solution u implies that

$$C_2H(x,t)\le u(x,t),\quad (x,t)\in\Gamma_\varepsilon(T),\tag{45}$$

for some positive constant C_2. Furthermore, by (9) and $A(\omega)>-N/2$ we have

$$\partial_t H-\Delta H+V\big(|x|\big)H=\left(-\frac{N}{2}-A(\omega)\right)(1+t)^{-\frac{N}{2}-A(\omega)-1}U\big(|x|\big)<0\tag{46}$$

in $\mathbf{R}^N\times(0,\infty)$. Therefore, by (45) and (46) we apply the comparison principle to obtain

$$u(x,t)\ge C_2H(x,t)=C_2(1+t)^{-\frac{N}{2}-A(\omega)}U\big(|x|\big)\tag{47}$$

for all $(x,t)\in D_\varepsilon(T)$. This together with (24) implies that

$$
\begin{aligned}
\|u(t)\|_q &\geq \left(\int_{|x|\leq\varepsilon(1+t)^{1/2}} |u(x,t)|^q\,dx\right)^{1/q}\\
&\succeq (1+t)^{-\frac{N}{2}-A(\omega)}\left(\int_{|x|\leq\varepsilon(1+t)^{1/2}} U(|x|)^q\,dx\right)^{1/q}\\
&\succeq (1+t)^{-\frac{N}{2}-A(\omega)}\left(\int_{|x|\leq\varepsilon(1+t)^{1/2}} (1+|x|)^{qA(\omega)}\,dx\right)^{1/q}\\
&\succeq \begin{cases} (1+t)^{-\frac{N}{2}(1-\frac{1}{q})-\frac{A(\omega)}{2}} & \text{if } q<\infty \text{ and } qA(\omega)+N>0,\\ (1+t)^{-\frac{N}{2}-A(\omega)}(\log t)^{\frac{1}{q}} & \text{if } q<\infty \text{ and } qA(\omega)+N=0,\\ (1+t)^{-\frac{N}{2}-A(\omega)} & \text{otherwise,}\end{cases}
\end{aligned} \tag{48}
$$

for all sufficiently large t, where $q \in [2,\infty)$. Then, by (40) and (48) we have

$$
\left\|e^{-tH}\right\|_{q,2} \geq \frac{\|e^{-2tH}\phi\|_q}{\|e^{-tH}\phi\|_2} = \frac{\|u(2t)\|_q}{\|u(t)\|_2} \succeq \eta_q(t) \tag{49}
$$

for all sufficiently large t, where $q \in [2,\infty)$. Similarly, by (40) and (47) we have

$$
\left\|e^{-tH}\right\|_{\infty,2} \geq \frac{\|e^{-2tH}\phi\|_\infty}{\|e^{-tH}\phi\|_2} \geq \frac{u(0,2t)}{\|u(t)\|_2} \succeq t^{-\frac{N}{4}-\frac{A(\omega)}{2}} \asymp \eta_\infty(t) \tag{50}
$$

for all sufficiently large t. Therefore, by (49) and (50), for any $q \in [2,\infty]$, we have the left hand side of the inequality (6), and the proof of Theorem 2 is complete. □

References

1. Davies, E.B., Simon, B.: L^p norms of noncritical Schrödinger semigroups. J. Funct. Anal. **102**, 95–115 (1991)
2. Ishige, K., Kabeya, Y.: Large time behaviors of hot spots for the heat equation with a potential. J. Differ. Equ. **244**, 2934–2962 (2008). Corrigendum in J. Differ. Equ. **245**, 2352–2354 (2008)
3. Ishige, K., Kabeya, Y.: L^p norms of nonnegative Schrödinger heat semigroup and the large time behavior of hot spots. J. Funct. Anal. **262**, 2695–2733 (2012)
4. Ladyzenskaja, O.A., Solonnikov, V.A., Ural'ceva, N.N.: Linear and Quasilinear Equations of Parabolic Type. Am. Math. Soc., Providence (1968)
5. Milman, P.D., Semenov, Y.A.: Global heat kernel bounds via desingularizing weights. J. Funct. Anal. **212**, 373–398 (2004)
6. Moschini, L., Tesei, A.: Harnack inequality and heat kernel estimates for the Schrödinger operator with hardy potential. Rend. Mat. Acc. Lincei **16**, 171–180 (2005)
7. Moschini, L., Tesei, A.: Parabolic Harnack inequality for the heat equation with inverse-square potential. Forum Math. **19**, 407–427 (2007)
8. Murata, M.: Positive solutions and large time behaviors of Schrödinger semigroups, Simon's problem. J. Funct. Anal. **56**, 300–310 (1984)
9. Murata, M.: Large time asymptotics for fundamental solutions of diffusion equations. Tohoku Math. J. **37**, 151–195 (1985)
10. Murata, M.: Structure of positive solutions to $(-\Delta + V)u = 0$ in $\mathbf{R}^n$. Duke Math. J. **53**, 869–943 (1986)
11. Pinchover, Y.: On positive solutions of second-order elliptic equations, stability results, and classification. Duke Math. J. **57**, 955–980 (1988)

12. Pinchover, Y.: Criticality and ground states for second-order elliptic equations. J. Differ. Equ. **80**, 237–250 (1989)
13. Pinchover, Y.: On criticality and ground states of second order elliptic equations, II. J. Differ. Equ. **87**, 353–364 (1990)
14. Simon, B.: Large time behavior of the L^p norm of Schrödinger semigroups. J. Funct. Anal. **40**, 66–83 (1981)
15. Simon, B.: Schrödinger semigroups. Bull. Am. Math. Soc. **7**, 447–526 (1982)
16. Zhang, Q.S.: Global bounds of Schrödinger heat kernels with negative potentials. J. Funct. Anal. **182**, 344–370 (2001)
17. Zhao, Z.: Subcriticality, positivity and gaugeability of the Schrödinger operator. Bull. Am. Math. Soc. **23**, 513–517 (1990)
18. Zhao, Z.: Subcriticality and gaugeability of the Schrödinger operator. Trans. Am. Math. Soc. **334**, 75–96 (1992)

Hadamard Variation for Electromagnetic Frequencies

Shuichi Jimbo

Abstract A regular variation of a bounded domain in the Euclidean space is considered. The perturbation formula for the eigenvalue of an operator arising in the Maxwell equation under this type of domain variation is given.

Keywords Hadamard variation · Eigenvalues · Perturbation formula · Maxwell equation

1 Introduction

In this paper we deal with the harmonic oscillation in the Maxwell equation in a bounded domain (under a certain boundary condition) and consider the smooth dependency of the eigenfrequency under the domain perturbation. The set of the eigenfrequencies of the oscillation depends on the geometric feature of the domain and it is one of the important quantities of the domain. The purpose of this paper is to consider the regular deformation of the domain and to give a perturbation formula for each eigenfrequency.

Such kind of studies are called a problem of Hadamard variation or a domain variation problem. The most famous work is the Hadamard's variational formula for eigenvalues of Laplacian (and Green function). After Hadamard's pioneering work (cf. [6]), there have been a lot of studies for the case of the Laplace operator and other elliptic operator (cf. [4, 16, 18, 20] for related works). As well as regular perturbation of domains, there are a lot of studies on singularly perturbed domains concerning the eigenvalues of elliptic operators in many different situations (see the references or those of [8, 10, 11]). The Courant-Hilbert book [1] gave a study on global properties of the set of the eigenvalues with their continuous dependence on the domains.

We study the perturbation of the eigenvalues of the elliptic operator which arises in the Maxwell equation for regular variation of the domain. Hirakawa, a Japanese

S. Jimbo (✉)
Department of Mathematics, Hokkaido University, Sapporo 060-0810, Japan
e-mail: jimbo@math.sci.hokudai.ac.jp

R. Magnanini et al. (eds.), *Geometric Properties for Parabolic and Elliptic PDE's*, Springer INdAM Series 2, DOI 10.1007/978-88-470-2841-8_12,

physicist raised this problem in [7] and carried out a heuristic calculation and deduced some variational formula (it is of different type from our result). It seems that there have been only a few results for this subject from PDE point of view. This is one of the motivations of our study.

The electric-magnetic phenomena is modeled by the Maxwell equation (with an appropriate boundary condition) in the classical theory of Electromagnetism. The Maxwell equation is a coupled system of the electric field E and the magnetic field H. Let Ω be a bounded domain in $\mathbb{R}^3$ with C^3 boundary and consider the Maxwell equation

$$\varepsilon_0 \frac{\partial E}{\partial t} - \operatorname{rot} H = \mathbf{0}, \qquad \mu_0 \frac{\partial H}{\partial t} + \operatorname{rot} E = \mathbf{0}, \qquad \operatorname{div} E = 0, \qquad \operatorname{div} H = 0 \quad (1)$$

with some boundary condition (cf. (2)). Here $\varepsilon_0 > 0$ is the dielectric constant and $\mu_0 > 0$ is the magnetic permeability of the space where the electric-magnetic wave occurs (cf. [7]). We impose the boundary condition so that the space is surrounded by a perfect conductor. It gives the following condition

$$E \times \nu = \mathbf{0}, \qquad H \cdot \nu = 0 \quad \text{on } \partial\Omega. \quad (2)$$

Here ν is the outward unit normal vector on the boundary and the dot $\cdot$ is the standard inner product.

Time harmonic solutions are very fundamental in the sense that general solutions can be expressed by superposition (infinite linear combination) of the time harmonic solutions which are written in the following form

$$E(t,x) = \exp(i\omega t)\widetilde{E}(x), \qquad H(t,x) = \exp(i\omega t)\widetilde{H}(x) \quad (3)$$

where $\omega > 0$ is a parameter. Substitute these functions into (1) and get

$$i\varepsilon_0\omega\widetilde{E} - \operatorname{rot}\widetilde{H} = \mathbf{0}, \qquad i\mu_0\omega\widetilde{H} + \operatorname{rot}\widetilde{E} = \mathbf{0}, \qquad \operatorname{div}\widetilde{E} = 0, \qquad \operatorname{div}\widetilde{H} = 0. \quad (4)$$

Applying the operator rot on the second equation and use the first equation, we get

$$-\mu_0\varepsilon_0\omega^2\widetilde{E} + \operatorname{rot}\operatorname{rot}\widetilde{E} = \mathbf{0} \quad \text{in } \Omega, \qquad \widetilde{E} \times \nu = \mathbf{0} \quad \text{on } \partial\Omega. \quad (5)$$

The eigenfrequency is the value ω, for which (5) allows a nontrivial solution $\widetilde{E}$. By a scale transform of the space variable, we can assume $\mu_0\varepsilon_0 = 1$ without loss of a mathematical generality. Replace the symbol of the variable of the vector field $\widetilde{E}$ by $\boldsymbol{\Phi}$ and put $\lambda = \omega^2$. Thus we get the following eigenvalue problem,

$$\operatorname{rot}\operatorname{rot}\boldsymbol{\Phi} - \lambda\boldsymbol{\Phi} = \mathbf{0}, \qquad \operatorname{div}\boldsymbol{\Phi} = 0 \quad \text{in } \Omega, \qquad \boldsymbol{\Phi} \times \nu = \mathbf{0} \quad \text{on } \partial\Omega. \quad (6)$$

Here the unknown function $\boldsymbol{\Phi}$ is $\mathbb{R}^3$-valued in Ω.

It is easy to see that any eigenvalue is a nonnegative real number. Actually, take the inner product of (6) and $\boldsymbol{\Phi}$, and integrate in Ω, we have

$$\begin{aligned} 0 &= \int_\Omega \langle \operatorname{rot}\operatorname{rot}\boldsymbol{\Phi}, \boldsymbol{\Phi}\rangle\, dx - \lambda \int_\Omega \langle \boldsymbol{\Phi}, \boldsymbol{\Phi}\rangle\, dx \\ &= \int_{\partial\Omega} \langle \nu \times \operatorname{rot}\boldsymbol{\Phi}, \boldsymbol{\Phi}\rangle\, dS + \int_\Omega \langle \operatorname{rot}\boldsymbol{\Phi}, \operatorname{rot}\boldsymbol{\Phi}\rangle\, dx - \lambda \int_\Omega \langle \boldsymbol{\Phi}, \boldsymbol{\Phi}\rangle\, dx \\ &= \int_{\partial\Omega} \langle \operatorname{rot}\boldsymbol{\Phi}, \boldsymbol{\Phi} \times \nu\rangle\, dS + \int_\Omega |\operatorname{rot}\boldsymbol{\Phi}|^2\, dx - \lambda \int_\Omega \langle \boldsymbol{\Phi}, \boldsymbol{\Phi}\rangle\, dx. \end{aligned}$$

Using the boundary condition $\boldsymbol{\Phi} \times \nu = \mathbf{0}$ on $\partial\Omega$, we get

$$\int_\Omega \left|\operatorname{rot} \boldsymbol{\Phi}(x)\right|^2 dx = \lambda \int_\Omega \left|\boldsymbol{\Phi}(x)\right|^2 dx. \tag{7}$$

This implies that λ is nonnegative if $\boldsymbol{\Phi} \not\equiv \mathbf{0}$ in Ω. Hence we see that any eigenvalue is non-negative.

Zero Eigenspace We consider Eq. (6) for $\lambda = 0$. From (7), if $\boldsymbol{\Phi} \not\equiv \mathbf{0}$ in Ω, we have $\operatorname{rot} \boldsymbol{\Phi} = 0$ in Ω. This implies that $\boldsymbol{\Phi}$ has an expression $\boldsymbol{\Phi}(x) = \nabla\eta$. Here we should note that the function η could be multi-valued function (in the case Ω is not simply connected). From the boundary condition $\boldsymbol{\Phi} \times \nu = \nabla\eta \times \nu = \mathbf{0}$, $\nabla\eta$ is parallel to the normal vector ν at any boundary point. This implies that η is constant in any connected component of $\partial\Omega$. So η is necessarily single-valued function. On the other hand, take any function $\eta \in H^1(\Omega)$ which is constant on any connected component of $\partial\Omega$ and put $\Phi = \nabla\eta$ and then it becomes an eigenfunction for $\lambda = 0$.

Summing up the above arguments, we know that the zero eigenspace is the following:

$$X_0 = \left\{\nabla\eta \mid \eta \in C^2(\overline{\Omega}),\ \Delta\eta = 0 \text{ in } \Omega,\ \eta \text{ is constant in each component of } \partial\Omega\right\}. \tag{8}$$

Existence of Eigenvalues We define a basic function space for the argument to prove the existence of the eigenvalues. Put

$$X = \left\{\boldsymbol{\Phi} \in H^1\left(\Omega; \mathbb{R}^3\right) \mid \operatorname{div} \boldsymbol{\Phi} = 0 \text{ in } \Omega,\ \boldsymbol{\Phi} \times \nu = \mathbf{0} \text{ on } \partial\Omega\right\}. \tag{9}$$

It is easy to see that $\dim X_0 = \sharp(\text{components of } \partial\Omega) - 1$. It is known that X is a closed subspace of $H^1(\Omega; \mathbb{R}^3)$ and X is also closed in the sense of weak convergence in $H^1(\Omega; \mathbb{R}^3)$ (because X is linear).

As we know about the zero eigenvalue and the corresponding eigenfunctions, we deal with the positive eigenvalues from now.

Proposition 1 *The eigenvalue problem* (6) *has the positive eigenvalues* $\{\Lambda_k\}_{k=1}^\infty$ *with* $\lim_{k\to\infty} \Lambda_k = \infty$.

Proof To prove the existence of the eigenvalues, we can carry out a completely similar argument as the case of the Laplacian and the Schrödinger operator (cf. [2, 3]). So we only give a sketch of the argument. Hereafter the symbol $\perp$ means the orthogonality in $L^2(\Omega; \mathbb{R}^3)$. Put

$$\Lambda_1 = \inf\left\{\mathscr{R}(\boldsymbol{\phi}) \mid \boldsymbol{\phi} \in X,\ \boldsymbol{\phi} \perp X_0\right\}, \quad \text{where } \mathscr{R}(\boldsymbol{\phi}) = \int_\Omega \left|\operatorname{rot} \boldsymbol{\phi}\right|^2 dx \Big/ \int_\Omega |\boldsymbol{\phi}|^2 dx.$$

$\mathscr{R}$ attains the minimum Λ_1 with a minimizer $\boldsymbol{\Phi}^{(1)} \in X$ which is an eigenfunction corresponding to the eigenvalue Λ_1. This is proved as follows. Take a minimizing sequence $\{\boldsymbol{\phi}_\ell\}_{\ell=1}^\infty$ with $\|\boldsymbol{\phi}_\ell\|_{L^2(\Omega;\mathbb{R}^3)} = 1$. It is bounded also in $H^1(\Omega; \mathbb{R}^3)$ due to Lemma 2 and Lemma 3 below. This sequence contains a weakly convergent subsequence in $H^1(\Omega; \mathbb{R}^3)$ which is also strongly convergent in $L^2(\Omega; \mathbb{R}^3)$. Since X

is closed, the limit $\boldsymbol{\Phi}^{(1)}$ of the subsequence belongs to X and satisfies $\boldsymbol{\Phi}^{(1)} \perp X_0$. From the lower semicontinuity of $\mathscr{R}$ in X, $\boldsymbol{\Phi}^{(1)}$ becomes a minimizer in $X_0^{\perp} \cap X$. Taking the variation of $\mathscr{R}$ at $\boldsymbol{\Phi}^{(1)}$ (minimizer), we get

$$\mathrm{rot}\,\mathrm{rot}\,\boldsymbol{\Phi}^{(1)} - \Lambda_1 \boldsymbol{\Phi}^{(1)} = \mathbf{0} \quad \text{in } \Omega.$$

Carry out this argument in the space $X \cap (X_0 \oplus L.H.[\boldsymbol{\Phi}^{(1)}])^{\perp}$, we get the second positive eigenvalue Λ_2 as the minimum of $\mathscr{R}$ with the eigenfunction (minimizer) $\boldsymbol{\Phi}^{(2)} \in X$ with $\boldsymbol{\Phi}^{(2)} \in (X_0 \oplus L.H.[\boldsymbol{\Phi}^{(1)}])^{\perp}$. We can repeat this argument and get the sequence $0 < \Lambda_1 \leqq \Lambda_2 \leqq \Lambda_3 \leqq \cdots$. We also note that this sequence is unbounded. The eigenfunctions obtained above are sufficiently regular if $\partial\Omega$ is regular. This can be proved by the arguments in Chap. 7 in [13], where the harmonic forms in the smooth manifold with a boundary, are studied. The regularity of $\Phi^{(k)}$ inside Ω is proved by the argument in [5] for each component. For the regularity near the boundary, the technique in [13] is applied. The higher regularity estimates of the eigenfunction are also obtain in this process. □

Lemma 1 (Trace inequality) *For any $\eta > 0$, there exists $c(\eta) > 0$ such that*

$$\int_{\partial\Omega} \phi(x)^2 \, dS \leqq \eta \int_{\Omega} |\nabla \phi(x)|^2 \, dx + c(\eta) \int_{\Omega} \phi(x)^2 \, dx \quad (\phi \in H^1(\Omega)).$$

See Mizohata [12, Chap. 3] for the proof.

Lemma 2 *If $\boldsymbol{\Psi} \in H^1(\Omega; \mathbb{R}^3)$ and $\boldsymbol{\Psi} \times \nu = \mathbf{0}$ on $\partial\Omega$, then*

$$\int_{\Omega} |\,\mathrm{rot}\,\boldsymbol{\Psi}|^2 \, dx + \int_{\Omega} |\,\mathrm{div}\,\boldsymbol{\Psi}|^2 \, dx = \int_{\Omega} |\nabla \boldsymbol{\Psi}|^2 \, dx + \int_{\partial\Omega} H(x) \big|\boldsymbol{\Psi}(x)\big|^2 \, dS.$$

Here $H(x)$ is the mean curvature at $x \in \partial\Omega$ with respect to the unit outward normal vector ν.

Proof The proof is carried out through the straightforward calculation. □

Proposition 2 (Max-min principle) *The k-th positive eigenvalue Λ_k is characterized by the following formula*:

$$\Lambda_k = \sup_{E \subset X_0^{\perp},\, \dim E \leqq k-1} \inf\big\{\mathscr{R}(\boldsymbol{\Phi}) \,\big|\, \Phi \in X,\ \boldsymbol{\Phi} \perp X_0,\ \boldsymbol{\Phi} \perp E\big\}. \tag{10}$$

Here E is a subspace of $L^2(\Omega; \mathbb{R}^3)$.

For Max-Min principle for more general frame work of selfadjoint elliptic operators in Hilbert spaces (cf. [1–3, 17]).

Formulation of the Domain Variation and Eigenvalue Problem Assume that $\Omega \subset \mathbb{R}^3$ is a bounded domain with C^3 boundary $\partial\Omega$. Let $\rho = \rho(\xi)$ be a C^1 function in $\partial\Omega$. Put the set

$$\partial\Omega(\varepsilon) = \big\{\xi + \varepsilon\rho(\xi)\nu(\xi) \in \mathbb{R}^3 \,\big|\, \xi \in \partial\Omega\big\} \tag{11}$$

when $|\varepsilon|$ is small, there exists a unique bounded domain such that $\Omega(\varepsilon)$ is homeomorphic to Ω and its boundary agrees to $\partial\Omega(\varepsilon)$.

For this domain $\Omega(\varepsilon)$, we consider the following eigenvalue problem,

$$\begin{cases} \operatorname{rot}\operatorname{rot}\boldsymbol{\Phi} - \lambda\boldsymbol{\Phi} = \mathbf{0}, & \\ \operatorname{div}\boldsymbol{\Phi} = 0 & \text{in } \Omega(\varepsilon), \\ \boldsymbol{\Phi}\times\nu = \mathbf{0} & \text{on } \partial\Omega(\varepsilon). \end{cases} \tag{12}$$

From the formula $\operatorname{rot}\operatorname{rot} = \nabla\operatorname{div} - \Delta$, the eigenvalue problem (12) is also written as

$$\begin{cases} \Delta\boldsymbol{\Phi} + \lambda\boldsymbol{\Phi} = \mathbf{0}, & \\ \operatorname{div}\boldsymbol{\Phi} = 0 & \text{in } \Omega(\varepsilon), \\ \boldsymbol{\Phi}\times\nu = \mathbf{0} & \text{on } \partial\Omega(\varepsilon). \end{cases} \tag{13}$$

The set of the eigenvalues is a discrete unbounded sequence of real (nonnegative) values.

Definition 1 Let $\{\lambda_k(\varepsilon)\}_{k=1}^{\infty}$ be the set of positive eigenvalues (of (12)) which are arranged in increasing order with counting multiplicity.

Definition 2 Let $\{\boldsymbol{\Phi}_\varepsilon^{(k)}\}_{k=1}^{\infty}$ be the corresponding system of the eigenfunctions, which is orthonormal as

$$\big(\boldsymbol{\Phi}_\varepsilon^{(p)}, \boldsymbol{\Phi}_\varepsilon^{(q)}\big)_{L^2(\Omega(\varepsilon);\mathbb{R}^3)} = \delta(p,q) \quad (p,q \geqq 1).$$

We note that $\Lambda_k = \lambda_k(0)$.

Theorem 1 *Assume that the k-th eigenvalue Λ_k in* (6) *is simple. Then $\lambda_k(\varepsilon)$ is differentiable at $\varepsilon = 0$ and its derivative is given by the following formula*:

$$\begin{aligned} \frac{d\lambda_k(\varepsilon)}{d\varepsilon}\bigg|_{\varepsilon=0} &= \int_{\partial\Omega}\left(|\nabla\boldsymbol{\Phi}_0^{(k)}|^2 - 2\left|\frac{\partial\boldsymbol{\Phi}_0^{(k)}}{\partial\nu}\right|^2 + \big(2K(x)-\Lambda_k\big)|\boldsymbol{\Phi}_0^{(k)}|^2\right)\rho\, dS \\ &\quad + 2\int_{\partial\Omega}\langle\boldsymbol{\Phi}_0^{(k)},\nu\rangle\langle\operatorname{rot}\boldsymbol{\Phi}_0^{(k)}\times\nabla\rho,\nu\rangle\, dS. \end{aligned} \tag{14}$$

Here $K(x)$ is the Gaussian curvature of $\partial\Omega$ at x. $\nabla\rho$ is the gradient vector in the tangent space of $\partial\Omega$.

2 Transformation of the Problem

The method of the proof is to make a transformation (diffeomorphism) $\gamma_\varepsilon : \Omega \longrightarrow \Omega(\varepsilon)$ and to transform the problem to one on the fixed domain (through the change of the variable $x = \gamma_\varepsilon(y)$). So the problem on the ε-dependent variable domain reduces to the equations which includes ε in the coefficients.

So we prepare the transformation map and calculate the equation in a fixed domain Ω.

Lemma 3 *There exists* $\delta_0 > 0$ *and a smooth diffeomorphism map*

$$\gamma_\varepsilon : \overline{\Omega} \longrightarrow \overline{\Omega(\varepsilon)}$$

such that γ_ε *depends smoothly on* $\varepsilon \in (-\delta_0, \delta_0)$ *and*

$$\gamma_\varepsilon\big(\xi + t\nu(\xi)\big) = \xi + (t+\varepsilon)\rho(\xi)\nu(\xi) \quad \textit{for}\ \xi \in \partial\Omega,\ |t| < \delta_0,\ |\varepsilon| < \delta_0. \tag{15}$$

Proof Prepare a coordinate near the boundary $\partial\Omega$ and consider the map which moves a point of the δ_0-neighborhood of $\partial\Omega$ as in (15). We can construct a smooth map γ_ε with this property using a smooth cut-off and extension up to the whole Ω. □

We prepare some notation. The variation of the map γ_ε under perturbation by ε is given by a vector field g as follows,

$$g(y) = \big(g_1(y), g_2(y), g_3(y)\big)^t = \frac{\partial \gamma_\varepsilon(y)}{\partial \varepsilon}\bigg|_{\varepsilon=0} \qquad (y \in \Omega).$$

From the condition (15), we have

$$\frac{d\gamma_\varepsilon}{d\varepsilon}\big(\xi + t\nu(\xi)\big)_{\varepsilon=0} = \rho(\xi)\nu(\xi) \quad \text{for } \xi \in \partial\Omega,\ |t| < \delta_0. \tag{16}$$

This formula is also written as

$$g\big(\xi + t\nu(\xi)\big) = \rho(\xi)\nu(\xi) \quad (\xi \in \partial\Omega,\ |t| < \delta_0).$$

Take the derivative of the both side of this expression with respect to t and put $t = 0$, we have the following property of g.

Lemma 4 $(\partial g/\partial y)\nu = 0$ *on* $\partial\Omega$.

We start the calculation of the variational equation. We denote the unknown variable by $\boldsymbol{\Phi}$ and the transformed unknown variable by $\widetilde{\boldsymbol{\Phi}}$. Their relation is

$$\widetilde{\boldsymbol{\Phi}}(y) = (\boldsymbol{\Phi} \circ \gamma_\varepsilon)(y) \quad (y \in \Omega).$$

We express the unknown variable $\boldsymbol{\Phi}$ by its components as follows:

$$\boldsymbol{\Phi}(x) = \big(\Phi_1(x), \Phi_2(x), \Phi_3(x)\big)^t, \qquad \widetilde{\boldsymbol{\Phi}}(y) = \big(\widetilde{\Phi}_1(y), \widetilde{\Phi}_2(y), \widetilde{\Phi}_3(y)\big)^t.$$

Accordingly we have

$$\widetilde{\Phi}_i(y) = (\Phi_i \circ \gamma_\varepsilon)(y) \quad (y \in \Omega,\ i = 1, 2, 3).$$

We calculate the system of equations for $\widetilde{\boldsymbol{\Phi}}$ with the boundary condition. $\nu = (\nu_1, \nu_2, \nu_3)$ is the unit outward normal vector on $\partial\Omega$. We extend this field ν up to some neighborhood of $\partial\Omega$ for later convenience such that $\nu(x) = \nu(\xi)$ for $x = \xi + t\nu(\xi)$ with $\xi \in \partial\Omega$, $|t| < \delta_0$. A direct calculation gives

$$\nabla_y \widetilde{\Phi}_i(y) = \nabla_x \Phi_i(x)\left(\frac{\partial \gamma_\varepsilon}{\partial y}(y)\right), \qquad \nabla_x \Phi_i = \left(\frac{\partial \Phi_i}{\partial x_1}, \frac{\partial \Phi_i}{\partial x_2}, \frac{\partial \Phi_i}{\partial x_3}\right),$$

$$\nabla_y \widetilde{\Phi}_i = \left(\frac{\partial \widetilde{\Phi}_i}{\partial y_1}, \frac{\partial \widetilde{\Phi}_i}{\partial y_2}, \frac{\partial \widetilde{\Phi}_i}{\partial y_3}\right).$$

$$\gamma_\varepsilon(y) = \left(\gamma_{1,\varepsilon}(y), \gamma_{2,\varepsilon}(y), \gamma_{3,\varepsilon}(y)\right)^t,$$

$$\frac{\partial \gamma_\varepsilon}{\partial y}(y) = \begin{pmatrix} \partial\gamma_{1,\varepsilon}/\partial y_1 & \partial\gamma_{1,\varepsilon}/\partial y_2 & \partial\gamma_{1,\varepsilon}/\partial y_3 \\ \partial\gamma_{2,\varepsilon}/\partial y_1 & \partial\gamma_{2,\varepsilon}/\partial y_2 & \partial\gamma_{2,\varepsilon}/\partial y_3 \\ \partial\gamma_{3,\varepsilon}/\partial y_1 & \partial\gamma_{3,\varepsilon}/\partial y_2 & \partial\gamma_{3,\varepsilon}/\partial y_3 \end{pmatrix}.$$

We get the transformed equation. For $i = 1, 2, 3$,

$$\mathrm{div}_y\left(\det\left(\frac{\partial\gamma_\varepsilon}{\partial y}\right)\nabla_y\widetilde{\Phi}_i\left[\frac{\partial\gamma_\varepsilon}{\partial y}\right]^{-1}\left(\left[\frac{\partial\gamma_\varepsilon}{\partial y}\right]^{-1}\right)^t\right) + \lambda\det\left(\frac{\partial\gamma_\varepsilon}{\partial y}\right)\widetilde{\Phi}_i = 0 \quad \text{in } \Omega. \tag{17}$$

The "div-free" condition is written as

$$\sum_{k=1}^{3}\sum_{\ell=1}^{3}\frac{\partial\widetilde{\Phi}_k}{\partial y_\ell}\left(\left[\frac{\partial\gamma_\varepsilon}{\partial y}(y)\right]^{-1}\right)_{\ell k} = 0 \quad \text{in } \Omega. \tag{18}$$

Here, for the matrix M, $M_{\ell k}$ denotes the (ℓ, k) component of M and M^t is the transpose of M.

We calculate the boundary condition for $\widetilde{\boldsymbol{\Phi}}$ on $\partial\Omega$. The unit outward normal vector ν_ε at $x = \gamma_\varepsilon(y)$ on $\partial\Omega(\varepsilon)$ is given by

$$\nu_\varepsilon(\gamma_\varepsilon(y)) = \left[(\partial\gamma_\varepsilon/\partial y)^t(y)\right]^{-1}\nu(y)/\left|\left[(\partial\gamma_\varepsilon/\partial y)^t(y)\right]^{-1}\nu(y)\right| \quad \text{for } y \in \partial\Omega.$$

Since $\Phi(\gamma_\varepsilon(y)) = \widetilde{\Phi}(y)$ at $y \in \partial\Omega$ is parallel to $\nu_\varepsilon(\gamma_\varepsilon)$, $[(\partial\gamma_\varepsilon/\partial y)^t(y)]\widetilde{\Phi}(y)$ is parallel to $\nu(y)$ due to the above expression of $\nu_\varepsilon(\gamma_\varepsilon(y))$. So we get the following boundary condition for $\widetilde{\Phi}$.

$$\left[(\partial\gamma_\varepsilon/\partial y)^t(y)\right]\widetilde{\boldsymbol{\Phi}}(y) \times \nu(y) = \mathbf{0} \quad \text{on } \partial\Omega.$$

We write each component as follows.

$$\widetilde{\Phi}_1\left(-\nu_2(y)\frac{\partial\gamma_{1,\varepsilon}}{\partial y_1} + \nu_1(y)\frac{\partial\gamma_{1,\varepsilon}}{\partial y_2}\right) + \widetilde{\Phi}_2\left(-\nu_2(y)\frac{\partial\gamma_{2,\varepsilon}}{\partial y_1} + \nu_1(y)\frac{\partial\gamma_{2,\varepsilon}}{\partial y_2}\right)$$
$$+ \widetilde{\Phi}_3\left(-\nu_2(y)\frac{\partial\gamma_{3,\varepsilon}}{\partial y_1} + \nu_1(y)\frac{\partial\gamma_{3,\varepsilon}}{\partial y_2}\right) = 0 \quad \text{on } \partial\Omega, \tag{19}$$

$$\widetilde{\Phi}_1\left(\nu_3(y)\frac{\partial\gamma_{1,\varepsilon}}{\partial y_1} - \nu_1(y)\frac{\partial\gamma_{1,\varepsilon}}{\partial y_3}\right) + \widetilde{\Phi}_2\left(\nu_3(y)\frac{\partial\gamma_{2,\varepsilon}}{\partial y_1} - \nu_1(y)\frac{\partial\gamma_{2,\varepsilon}}{\partial y_3}\right)$$
$$+ \widetilde{\Phi}_3\left(\nu_3(y)\frac{\partial\gamma_{3,\varepsilon}}{\partial y_1} - \nu_1(y)\frac{\partial\gamma_{3,\varepsilon}}{\partial y_3}\right) = 0 \quad \text{on } \partial\Omega, \tag{20}$$

$$\widetilde{\Phi}_1\left(-\nu_3(y)\frac{\partial\gamma_{1,\varepsilon}}{\partial y_2} + \nu_2(y)\frac{\partial\gamma_{1,\varepsilon}}{\partial y_3}\right) + \widetilde{\Phi}_2\left(-\nu_3(y)\frac{\partial\gamma_{2,\varepsilon}}{\partial y_2} + \nu_2(y)\frac{\partial\gamma_{2,\varepsilon}}{\partial y_3}\right)$$
$$+ \widetilde{\Phi}_3\left(-\nu_3(y)\frac{\partial\gamma_{3,\varepsilon}}{\partial y_2} + \nu_2(y)\frac{\partial\gamma_{3,\varepsilon}}{\partial y_3}\right) = 0 \quad \text{on } \partial\Omega. \tag{21}$$

The above system (17)–(21) is the equation in Ω.

3 The Perturbation in the Eigenvalue Problem as a Toy Model

The main purpose of this paper is study the ε-dependence of the system (17)–(21). We explain the line of the proof for the perturbation formula of the eigenvalue $\lambda_k(\varepsilon)$ by using a simple toy model, which is a similar finite dimensional problem. Perturbation theory for linear operators with analysis on the spectra is extensively studied in [9]. The continuous dependence of the eigenvalue on the operator is studied in a rather general framework.

Let W be a finite dimensional space with an inner product $(\cdot,\cdot)$. Let $\mathscr{A}(\varepsilon)$ be a linear transformation in W which depends on the parameter $\varepsilon \in \mathbb{R}$ smoothly. We assume that $\mathscr{A}(\varepsilon)$ is self-adjoint for each ε. Assume also that $\mathscr{A}(0)$ has a simple eigenvalue $\mu(0)$ with an eigenvector $\boldsymbol{v}_0$. From [9], there exists exactly one eigenvalue $\mu(\varepsilon)$ of $\mathscr{A}(\varepsilon)$ when $|\varepsilon|$ is small and it approaches $\mu(0)$ for $\varepsilon \to 0$. We study the differentiability of $\mu(\varepsilon)$ at $\varepsilon = 0$.

We consider the eigenvalue problem

$$\mathscr{A}(\varepsilon)\boldsymbol{v} + \mu\boldsymbol{v} = 0. \tag{22}$$

Formal Calculus First we carry out a heuristic argument. Denote $\boldsymbol{v}_\varepsilon \in X$ be the eigenvector corresponding to an eigenvalue $\mu(\varepsilon)$ and assume that $\mu(\varepsilon)$ and $\boldsymbol{v}_\varepsilon$ are differentiable in ε.

$$\mathscr{A}(\varepsilon)\boldsymbol{v}_\varepsilon + \mu(\varepsilon)\boldsymbol{v}_\varepsilon = 0. \tag{23}$$

Differentiate the both sides and we get

$$\mathscr{A}(\varepsilon)\dot{\boldsymbol{v}}_\varepsilon + \dot{A}(\varepsilon)\boldsymbol{v}_\varepsilon + \mu(\varepsilon)\dot{\boldsymbol{v}}_\varepsilon + \dot{\mu}(\varepsilon)\boldsymbol{v}_\varepsilon = 0.$$

Here the dot means the derivative with respect to ε. Put $\varepsilon = 0$ and we have

$$\big(\mathscr{A}(0) + \mu(0)\big)\dot{\boldsymbol{v}}_0 = -\dot{\mathscr{A}}(0)\boldsymbol{v}_0 - \dot{\mu}(0)\boldsymbol{v}_0. \tag{24}$$

Regard this equation for the unknown variable $\dot{\boldsymbol{v}}_0$ and then we get the condition for the right hand side term. We see that the kernel of $\mathscr{A}(0) + \mu(0)I$ is orthogonal to the right hand side from the self-adjointness of the operator. This is saying

$$\big(\dot{\mathscr{A}}(0)\boldsymbol{v}_0 + \dot{\mu}(0)\boldsymbol{v}_0\big) \perp \boldsymbol{v}_0 \quad \Rightarrow \quad \dot{\mu}(0)\|\boldsymbol{v}_0\|^2 = -\big(\dot{A}(0)\boldsymbol{v}_0, \boldsymbol{v}_0\big).$$

This consideration gives the candidate for $\dot{\mu}(0)$ which is the necessary condition of the existence of $\dot{\boldsymbol{v}}_0$. The heuristic argument is finished.

Justification First we put

$$d = -\big(\dot{\mathscr{A}}(0)\boldsymbol{v}_0, \boldsymbol{v}_0\big)/\|\boldsymbol{v}_0\|^2 \tag{25}$$

and consider the following equation (with the unknown variable $\boldsymbol{w}$)

$$\big(\mathscr{A}(0) + \mu(0)\big)\boldsymbol{w} = -\dot{\mathscr{A}}(0)\boldsymbol{v}_0 + d\boldsymbol{v}_0.$$

We recall that $\mathscr{A}(\varepsilon)$ is self-adjoint. From the definition of d (orthogonality condition of the right hand side), it has a solution and it is unique if the condition $\boldsymbol{w} \perp \boldsymbol{v}_0$

is prescribed. We consider the following vector as an approximate eigenvector for "ε-problem"

$$\widetilde{\boldsymbol{v}}_\varepsilon = \boldsymbol{v}_0 + \varepsilon \boldsymbol{w}. \tag{26}$$

Take the inner product of $\widetilde{\boldsymbol{v}}_\varepsilon$ and the eigenvalue equation (23) and get

$$\big(\mathscr{A}(\varepsilon)\boldsymbol{v}_\varepsilon + \mu(\varepsilon)\boldsymbol{v}_\varepsilon, \boldsymbol{v}_0 + \varepsilon \boldsymbol{w}\big) = 0,$$
$$\big(\mathscr{A}(\varepsilon)\boldsymbol{v}_\varepsilon, \boldsymbol{v}_0\big) + \mu(\varepsilon)(\boldsymbol{v}_\varepsilon, \boldsymbol{v}_0) + \varepsilon\big(\mathscr{A}(\varepsilon)\boldsymbol{v}_\varepsilon + \mu(\varepsilon)\boldsymbol{v}_\varepsilon, \boldsymbol{w}\big) = 0.$$

Here we assume that the eigenvector is normalized $\|\boldsymbol{v}_\varepsilon\| = 1$. Consider the Taylor expansion of $\mathscr{A}(\varepsilon)$ in ε at 0

$$\mathscr{A}(\varepsilon) = \mathscr{A}(0) + \varepsilon \dot{A}(0) + o(\varepsilon) \quad \text{(Taylor expansion)},$$

and we calculate the quantity

$$\begin{aligned}
\big(\mathscr{A}(\varepsilon)\boldsymbol{v}_\varepsilon, \boldsymbol{v}_0\big) &= \big(\boldsymbol{v}_\varepsilon, \mathscr{A}(\varepsilon)\boldsymbol{v}_0\big) = \big(\boldsymbol{v}_\varepsilon, \mathscr{A}(0)\boldsymbol{v}_0\big) + \varepsilon\big(\boldsymbol{v}_\varepsilon, \dot{\mathscr{A}}(0)\boldsymbol{v}_0\big) + \big(\boldsymbol{v}_\varepsilon, o(\varepsilon)\boldsymbol{v}_0\big) \\
&= -\mu(0)(\boldsymbol{v}_\varepsilon, \boldsymbol{v}_0) + \varepsilon\big(\boldsymbol{v}_\varepsilon, \dot{\mathscr{A}}(0)\boldsymbol{v}_0\big) + \big(\boldsymbol{v}_\varepsilon, o(\varepsilon)\boldsymbol{v}_0\big), \\
\big(\mathscr{A}(\varepsilon)\boldsymbol{v}_\varepsilon, \boldsymbol{w}\big) &= \big(\boldsymbol{v}_\varepsilon, \mathscr{A}(\varepsilon)\boldsymbol{w}\big) = \big(\boldsymbol{v}_\varepsilon, \mathscr{A}(0)\boldsymbol{w}\big) + \varepsilon\big(\boldsymbol{v}_\varepsilon, \dot{\mathscr{A}}(0)\boldsymbol{w}\big) + \big(\boldsymbol{v}_\varepsilon, o(\varepsilon)\boldsymbol{w}\big) \\
&= -\big(\boldsymbol{v}_\varepsilon, \big(\dot{\mathscr{A}}(0) + d\big)\boldsymbol{v}_0\big) - \mu(0)(\boldsymbol{v}_\varepsilon, \boldsymbol{w}) + \varepsilon\big(\boldsymbol{v}_\varepsilon, \dot{\mathscr{A}}(0)\boldsymbol{w}\big) \\
&\quad + \big(\boldsymbol{v}_\varepsilon, o(\varepsilon)\boldsymbol{w}\big).
\end{aligned}$$

Substitute it to the equation and get

$$\begin{aligned}
\frac{\mu(\varepsilon) - \mu(0)}{\varepsilon}(\boldsymbol{v}_\varepsilon, \boldsymbol{v}_0) = &-\big(\boldsymbol{v}_\varepsilon, \dot{\mathscr{A}}(0)\boldsymbol{v}_0\big) - \big(\boldsymbol{v}_\varepsilon, o(1)\boldsymbol{v}_0\big) \\
&- \big(\mu(\varepsilon) - \mu(0)\big)(\boldsymbol{v}_\varepsilon, \boldsymbol{w}) + \big(\boldsymbol{v}_\varepsilon, \big(\dot{\mathscr{A}}(0) + d\big)\boldsymbol{v}_0\big) \\
&- \varepsilon\big(\boldsymbol{v}_\varepsilon, \dot{\mathscr{A}}(0)\boldsymbol{w}\big) - \big(\boldsymbol{v}_\varepsilon, o(\varepsilon)\boldsymbol{w}\big).
\end{aligned}$$

Take any sequence $\{\varepsilon(m)\}_{m=1}^\infty$ which approaches 0 for $m \to \infty$. Then there exists a subsequence $\{\varepsilon(m(p))\}_{p=1}^\infty$ and a unit vector $\boldsymbol{v}' \in X$ such that

$$\lim_{p\to\infty} \boldsymbol{v}_{\varepsilon(m(p))} = \boldsymbol{v}', \quad \mathscr{A}(0)\boldsymbol{v}' + \mu(0)\boldsymbol{v}' = 0.$$

As we assumed that the eigenvalue $\mu(0)$ is simple, we have $\boldsymbol{v}' = \boldsymbol{v}_0$ or $\boldsymbol{v}' = -\boldsymbol{v}_0$.

$$\lim_{p\to\infty} \frac{\mu(\varepsilon(m(p))) - \mu(0)}{\varepsilon(m(p))}(\boldsymbol{v}', \boldsymbol{v}_0) = d\big(\boldsymbol{v}', \boldsymbol{v}_0\big).$$

The sequence $\{\varepsilon(m)\}_{m=1}^\infty$ was arbitrary and d is defined independently of the choice of this sequence and so we conclude the formula

$$\lim_{\varepsilon\to 0} \frac{\mu(\varepsilon) - \mu(0)}{\varepsilon} = d = -\frac{(\dot{\mathscr{A}}(0)\boldsymbol{v}_0, \boldsymbol{v}_0)}{\|\boldsymbol{v}_0\|^2}.$$

This implies that $\mu(\varepsilon)$ is differentiable in ε at 0 and the derivative agrees exactly to the value d in (25).

4 Analysis of the Perturbation of the Electromagnetic Eigenvalue

We consider the positive eigenvalues $\{\lambda_k(\varepsilon)\}_{k=1}^{\infty}$ and the corresponding eigenfunctions $\{\boldsymbol{\Phi}_\varepsilon^{(k)}\}_{k=1}^{\infty}$ of (12), which are orthonormal in $L^2(\Omega(\varepsilon);\mathbb{R}^3)$ and satisfy

$$\begin{aligned}&\left(\boldsymbol{\Phi}_\varepsilon^{(p)},\boldsymbol{\Phi}_\varepsilon^{(q)}\right)_{L^2(\Omega(\varepsilon);\mathbb{R}^3)}=\delta(p,q),\\&\left(\mathrm{rot}\,\boldsymbol{\Phi}_\varepsilon^{(p)},\mathrm{rot}\,\boldsymbol{\Phi}_\varepsilon^{(q)}\right)_{L^2(\Omega(\varepsilon);\mathbb{R}^3)}=\delta(p,q)\lambda_p(\varepsilon).\end{aligned}\tag{27}$$

By the aid of the max-min principle (Proposition 2) for $\lambda_k(\varepsilon)$ in (12) with the (almost) test functions $\boldsymbol{\Phi}_0^{(k)}(\gamma_\varepsilon^{-1}(x))$ $(k \geqq 1)$, we can derive an **upper estimate**

$$\lambda_k(\varepsilon) \leqq \Lambda_k + O(\varepsilon). \tag{28}$$

To obtain the lower estimate, we first note that there exists a constant $\delta_k > 0$ and $c_k > 0$ (from (27), (28)) such that

$$\left\|\boldsymbol{\Phi}_\varepsilon^{(k)}\right\|_{H^1(\Omega(\varepsilon);\mathbb{R}^3)} \leqq c_k \quad \text{for } |\varepsilon| \leqq \delta_k. \tag{29}$$

Recall $\widetilde{\boldsymbol{\Phi}}_\varepsilon^{(k)}(y) = (\boldsymbol{\Phi}_\varepsilon^{(k)} \circ \gamma_\varepsilon)(y)$. As the transformation $x = \gamma_\varepsilon(y)$ smoothly approach the identity map, we have the following estimates (with the aid of Lemmas 1 and 2).

Lemma 5 *For each* $k \in \mathbb{N}$, *there exists a constant* $c(k) > 0$ *such that*

$$\left\|\widetilde{\boldsymbol{\Phi}}_\varepsilon^{(k)}\right\|_{H^1(\Omega;\mathbb{R}^3)} \leqq c(k) \quad \textit{for small } \varepsilon > 0.$$

As $\Omega(\varepsilon)$ depends smoothly on ε, we can apply the regularity argument for $\boldsymbol{\Phi}_\varepsilon^{(k)}$ in the boundary value problem (13) which is developed in the famous Morrey's book [13]. We can have the following regularity.

Lemma 6 *For each* $k \in \mathbb{N}$, *there exists a constant* $c'(k) > 0$ *such that*

$$\left\|\widetilde{\boldsymbol{\Phi}}_\varepsilon^{(k)}\right\|_{C^2(\overline{\Omega};\mathbb{R}^3)} \leqq c'(k) \quad \textit{for small } \varepsilon > 0.$$

Take an arbitrary sequence $\{\varepsilon(p)\}_{p\geqq 1}$ such that $\lim_{p\to\infty}\varepsilon(p) = 0$. Then, there exists a subsequence $\{\varepsilon(p(m))\}_{m=1}^{\infty}$ and an orthonormal system $\{\Theta^{(k)}\}_{k=1}^{\infty}$ in $L^2(\Omega;\mathbb{R}^3)$ such that

$$\widetilde{\boldsymbol{\Phi}}_{\varepsilon(p(m))}^{(k)} \longrightarrow \Theta^{(k)} \quad (m \to \infty) \tag{30}$$

strongly in $L^2(\Omega;\mathbb{R}^3)$ and weakly in $H^1(\Omega;\mathbb{R}^3)$ and $\mathrm{div}\,\Theta^{(k)} = 0$ in Ω, $\Theta^{(k)} \times \nu = \mathbf{0}$ on $\partial\Omega$. From (28), (30), we have

$$\begin{aligned}\Lambda_k &\geqq \liminf_{m\to\infty}\lambda_k\big(\varepsilon\big(p(m)\big)\big) = \liminf_{m\to\infty}\int_{\Omega(\varepsilon(p(m)))}\left|\mathrm{rot}\,\boldsymbol{\Phi}_{\varepsilon(p(m))}^{(k)}(x)\right|^2 dx\\&= \liminf_{m\to\infty}\int_{\Omega}\left|\mathrm{rot}\,\widetilde{\boldsymbol{\Phi}}_{\varepsilon(p(m))}^{(k)}(y)\right|^2 dy \geqq \int_{\Omega}\left|\mathrm{rot}\,\Theta^{(k)}(y)\right|^2 dy.\end{aligned}\tag{31}$$

From the orghogonality of $\{\Theta^{(k)}\}_{k=1}^{\infty}$ in $L^2(\Omega;\mathbb{R}^3)\cap X_0^{\perp}$ with (31), we have $\int_\Omega |\operatorname{rot}\Theta^{(k)}|^2\,dy = \Lambda_k$ for $k \geqq 1$. This implies $\Theta^{(k)}$ is necessarily a k-th eigenfunction. Eventually we get the convergence $\lim_{m\to\infty}\lambda_k(\varepsilon(p(m))) = \Lambda_k$. Since $\{\varepsilon(p)\}$ was arbitrary, we have the following result.

Proposition 3 $\lim_{\varepsilon\to 0}\lambda_k(\varepsilon) = \Lambda_k$ $(k \geqq 1)$.

We study the detailed asymptotics of $\lambda_k(\varepsilon)$ for $\varepsilon \to 0$. We follow the line of argument given in the previous section. By a formal perturbation argument, we first find a candidate of $(d\lambda_k(\varepsilon)/d\varepsilon)|_{\varepsilon=0}$ by a formal calculus.

To calculate the derivative of the equation of (17)–(18) and the boundary condition (19)–(21), we prepare some formulas.

Lemma 7 *Let $A(\varepsilon)$ be an invertible square matrix which is differentiable in ε. Then we have*

$$\frac{d}{d\varepsilon}A(\varepsilon)^{-1} = -A(\varepsilon)^{-1}\frac{d}{d\varepsilon}A(\varepsilon)A(\varepsilon)^{-1}. \tag{32}$$

Moreover, if $A(0) = I$ (Identity matrix), then

$$\frac{d}{d\varepsilon}\det A(\varepsilon)\bigg|_{\varepsilon=0} = \operatorname{Tr}\left(\frac{dA(\varepsilon)}{d\varepsilon}\bigg|_{\varepsilon=0}\right). \tag{33}$$

Proof This is proved by a direct calculation. □

Since $\gamma_0(y) = y$ (Identity map), it follows $(\partial\gamma_0/\partial y) = I$. Hence we can apply the above formulas (32), (33) to the Jacobian matrix $\partial\gamma_\varepsilon/\partial y$, we have

$$\frac{d}{d\varepsilon}\left(\frac{\partial\gamma_\varepsilon}{\partial y}\right)^{-1}\bigg|_{\varepsilon=0} = -\frac{\partial g(y)}{\partial y}. \tag{34}$$

$$\frac{d}{d\varepsilon}\det\left(\frac{\partial\gamma_\varepsilon}{\partial y}\right)\bigg|_{\varepsilon=0} = \operatorname{div}_y g(y) = \sum_{j=1}^{3}\frac{\partial g_j(y)}{\partial y_j}. \tag{35}$$

Variational Equation Fix a natural number k hereafter. Drop the index k and denote $\boldsymbol{\Phi}_\varepsilon = \boldsymbol{\Phi}_\varepsilon^{(k)}$, $\widetilde{\boldsymbol{\Phi}}_\varepsilon = \widetilde{\boldsymbol{\Phi}}_\varepsilon^{(k)}$, $\lambda(\varepsilon) = \lambda_k(\varepsilon)$. Note that $\widetilde{\boldsymbol{\Phi}}_0 = \boldsymbol{\Phi}_0$ because γ_0 is the identity map. Assume that $\widetilde{\boldsymbol{\Phi}}_\varepsilon$, $\lambda(\varepsilon)$ is differentiable in ε at 0 and put

$$\boldsymbol{\Psi}(y) = \big(\Psi_1(y), \Psi_2(y), \Psi_3(y)\big)^t = \big(\partial\widetilde{\boldsymbol{\Phi}}_\varepsilon^{(k)}/\partial\varepsilon\big)_{\varepsilon=0}, \qquad \boldsymbol{\kappa} = \big(d\lambda_k(\varepsilon)/d\varepsilon\big)(0). \tag{36}$$

We seek for the relation which $\boldsymbol{\Psi}$ and $\boldsymbol{\kappa}$ should satisfy if they exist. Take the derivative of (17), (18), (19), (20), (21) and put $\varepsilon = 0$ and calculate by the formula (34) and (35) and substitute $\varepsilon = 0$, we get

$$
\begin{aligned}
&\operatorname{div}(\nabla \Psi_i) + \operatorname{div}_y\big((\operatorname{div} g)\nabla_y \Phi_{0i}\big) - \operatorname{div}\left(\nabla \Phi_{0i}\left(\frac{\partial g}{\partial y} + \left(\frac{\partial g}{\partial y}\right)^t\right)\right) \\
&\quad + \kappa \Phi_{0i} + \lambda(0)(\operatorname{div} g)\Phi_{0i} + \lambda(0)\Psi_i = 0 \quad (y \in \Omega, i = 1,2,3), \qquad (37)
\end{aligned}
$$

$$
\operatorname{div} \boldsymbol{\Psi} = \sum_{i=1}^{3}\sum_{\ell=1}^{3} \frac{\partial \Phi_{0i}}{\partial y_\ell}\frac{\partial g_\ell}{\partial y_i} \quad \text{in } \Omega. \qquad (38)
$$

From (19), (20), (21), we have the boundary condition for $\boldsymbol{\Psi}$ which gives the values of $\nu \times \Psi$ on $\partial\Omega$,

$$
\begin{aligned}
\Psi_2\nu_1 - \Psi_1\nu_2 &= \Phi_{01}\left(\nu_2\frac{\partial g_1}{\partial y_1} - \nu_1\frac{\partial g_1}{\partial y_2}\right) + \Phi_{02}\left(\nu_2\frac{\partial g_2}{\partial y_1} - \nu_1\frac{\partial g_2}{\partial y_2}\right) \\
&\quad + \Phi_{03}\left(\nu_2\frac{\partial g_3}{\partial y_1} - \nu_1\frac{\partial g_3}{\partial y_2}\right), \qquad (39)
\end{aligned}
$$

$$
\begin{aligned}
\Psi_1\nu_3 - \Psi_3\nu_1 &= \Phi_{01}\left(\nu_1\frac{\partial g_1}{\partial y_3} - \nu_3\frac{\partial g_1}{\partial y_1}\right) + \Phi_{02}\left(\nu_1\frac{\partial g_2}{\partial y_3} - \nu_3\frac{\partial g_2}{\partial y_1}\right) \\
&\quad + \Phi_{03}\left(\nu_1\frac{\partial g_3}{\partial y_3} - \nu_3\frac{\partial g_3}{\partial y_1}\right), \qquad (40)
\end{aligned}
$$

$$
\begin{aligned}
\Psi_3\nu_2 - \Psi_2\nu_3 &= \Phi_{01}\left(\nu_3\frac{\partial g_1}{\partial y_2} - \nu_2\frac{\partial g_1}{\partial y_3}\right) + \Phi_{02}\left(\nu_3\frac{\partial g_2}{\partial y_2} - \nu_2\frac{\partial g_2}{\partial y_3}\right) \\
&\quad + \Phi_{03}\left(\nu_3\frac{\partial g_3}{\partial y_2} - \nu_2\frac{\partial g_3}{\partial y_3}\right). \qquad (41)
\end{aligned}
$$

For the domain derivative of solution of Poisson equations, we can learn a lot of things in [14, 15]. For convenience we define the vector field $\boldsymbol{\psi}_0$ by

$$
\boldsymbol{\psi}_0 = -\left(\frac{\partial g}{\partial y}\right)^t \boldsymbol{\Phi}_0 \quad \text{in } \Omega.
$$

Using $\boldsymbol{\psi}_0$, the boundary condition for $\boldsymbol{\Psi}$ (i.e. (39), (40), (41)) is written by

$$
\boldsymbol{\Psi} \times \nu = \boldsymbol{\psi}_0 \times \nu \quad \text{on } \partial\Omega. \qquad (42)
$$

We multiply both sides of Eq. (37) by Φ_{0i} and sum for $i = 1, 2, 3$.

$$
\begin{aligned}
&\sum_{i=1}^{3}\int_\Omega \left\{\Phi_{0i}\Delta\Psi_i + \Phi_{0i}\operatorname{div}\big((\operatorname{div} g)\nabla\Phi_{0i}\big) - \Phi_{0i}\operatorname{div}\left(\nabla\Phi_{0i}\left(\frac{\partial g}{\partial y} + \left(\frac{\partial g}{\partial y}\right)^t\right)\right)\right\} dy \\
&\quad + \sum_{i=1}^{3}\int_\Omega \big(\kappa\Phi_{0i}^2 + \lambda(0)(\operatorname{div} g)\Phi_{0i}^2 + \lambda(0)\Psi_i\Phi_{0i}\big)\, dy = 0.
\end{aligned}
$$

Denote the left hand side by J. Substitute $\Delta\boldsymbol{\Psi} = -\operatorname{rot}\operatorname{rot}\boldsymbol{\Psi} + \nabla\operatorname{div}\boldsymbol{\Psi}$ into J with (38) and integrate by parts, we get

$$J = \int_\Omega \left\langle \boldsymbol{\Phi}_0, \nabla\left(\sum_{i,\ell=1}^{3} \frac{\partial \Phi_{0i}}{\partial y_\ell}\frac{\partial g_\ell}{\partial y_i}\right)\right\rangle dy + \int_{\partial\Omega} \langle \boldsymbol{\Phi}_0, (-\nu)\times \operatorname{rot}\boldsymbol{\Psi}\rangle\, dS$$

$$- \int_\Omega \langle \operatorname{rot}\boldsymbol{\Phi}_0, \operatorname{rot}\boldsymbol{\Psi}\rangle\, dy$$

$$+ \sum_{i=1}^{3} \int_\Omega \left\{ \Phi_{0i} \operatorname{div}\big((\operatorname{div} g)\nabla\Phi_{0i}\big) - \Phi_{0i}\operatorname{div}\left(\nabla\Phi_{0i}\left(\frac{\partial g}{\partial y} + \left(\frac{\partial g}{\partial y}\right)^t\right)\right)\right\} dy$$

$$+ \sum_{i=1}^{3} \int_\Omega \big(\kappa\Phi_{0i}^2 + \lambda(0)(\operatorname{div} g)\Phi_{0i}^2 + \lambda(0)\Psi_i\Phi_{0i}\big)\, dy$$

$$= \int_{\partial\Omega} \langle \boldsymbol{\Phi}_0, \nu\rangle \left(\sum_{i,\ell=1}^{3} \frac{\partial \Phi_{0i}}{\partial y_\ell}\frac{\partial g_\ell}{\partial y_i}\right) dS - \int_\Omega (\operatorname{div}\boldsymbol{\Phi}_0)\left(\sum_{i,\ell=1}^{3} \frac{\partial \Phi_{0i}}{\partial y_\ell}\frac{\partial g_\ell}{\partial y_i}\right) dy$$

$$- \int_{\partial\Omega} \langle \operatorname{rot}\boldsymbol{\Psi}, \boldsymbol{\Phi}_0\times\nu\rangle\, dS - \int_{\partial\Omega} \langle \operatorname{rot}\boldsymbol{\Phi}_0, \nu\times\boldsymbol{\Psi}\rangle\, dS - \int_\Omega \langle \operatorname{rot}\operatorname{rot}\boldsymbol{\Phi}_0, \boldsymbol{\Psi}\rangle\, dy$$

$$+ \sum_{i=1}^{3} \int_{\partial\Omega} \Phi_{0i}(\operatorname{div} g)\langle \nu, \nabla\Phi_{0i}\rangle\, dS - \sum_{i=1}^{3} \int_\Omega (\operatorname{div} g)|\nabla\Phi_{0i}|^2\, dy$$

$$- \sum_{i=1}^{3} \int_{\partial\Omega} \Phi_{0i}\left\langle \nu, \nabla\Phi_{0i}\left(\frac{\partial g}{\partial y} + \left(\frac{\partial g}{\partial y}\right)^t\right)\right\rangle dS$$

$$+ \sum_{i=1}^{3} \int_\Omega \left\langle \nabla\Phi_{0i}, \nabla\Phi_{0i}\left(\frac{\partial g}{\partial y} + \left(\frac{\partial g}{\partial y}\right)^t\right)\right\rangle dy$$

$$+ \sum_{i=1}^{3} \int_\Omega \big(\kappa\Phi_{0i}^2 + \lambda(0)(\operatorname{div} g)\Phi_{0i}^2 + \lambda(0)\Psi_i\Phi_{0i}\big)\, dy.$$

Using $\boldsymbol{\Phi}_0 \times \nu = \mathbf{0}$ on $\partial\Omega$ and $\operatorname{rot}\operatorname{rot}\boldsymbol{\Phi}_0 - \lambda(0)\boldsymbol{\Phi}_0 = \mathbf{0}$ and $\operatorname{div}\boldsymbol{\Phi}_0 = 0$ in Ω, we can simplify this expression and get

$$J = \int_{\partial\Omega} \langle \boldsymbol{\Phi}_0, \nu\rangle \left(\sum_{i,\ell=1}^{3} \frac{\partial \Phi_{0i}}{\partial y_\ell}\frac{\partial g_\ell}{\partial y_i}\right) dS - \int_{\partial\Omega} \langle \operatorname{rot}\boldsymbol{\Phi}_0, \nu\times\boldsymbol{\Psi}\rangle\, dS$$

$$+ \sum_{i=1}^{3} \int_{\partial\Omega} \Phi_{0i}(\operatorname{div} g)\langle \nu, \nabla\Phi_{0i}\rangle\, dS - \sum_{i=1}^{3} \int_\Omega (\operatorname{div} g)|\nabla\Phi_{0i}|^2\, dy$$

$$- \sum_{i=1}^{3} \int_{\partial\Omega} \Phi_{0i}\left\langle \nu, \nabla\Phi_{0i}\left(\frac{\partial g}{\partial y} + \left(\frac{\partial g}{\partial y}\right)^t\right)\right\rangle dS$$

$$+\sum_{i=1}^{3}\int_{\Omega}\left\langle \nabla\Phi_{0i}, \nabla\Phi_{0i}\left(\frac{\partial g}{\partial y}+\left(\frac{\partial g}{\partial y}\right)^{t}\right)\right\rangle dy$$

$$+\sum_{i=1}^{3}\int_{\Omega}\left(\kappa\Phi_{0i}^{2}+\lambda(0)(\operatorname{div} g)\Phi_{0i}^{2}\right)dy,$$

$$J=-\int_{\partial\Omega}A\,dS+\int_{\partial\Omega}B\,dS-\sum_{i=1}^{3}\int_{\partial\Omega}\langle g,\nu\rangle|\nabla\Phi_{0i}|^{2}\,dS$$

$$+2\sum_{i=1}^{3}\int_{\partial\Omega}\langle g,\nabla\Phi_{0i}\rangle\langle\nu,\nabla\Phi_{0i}\rangle\,dS$$

$$+\lambda(0)\sum_{i=1}^{3}\int_{\partial\Omega}\langle g,\nu\rangle\Phi_{0i}^{2}\,dS+\kappa\sum_{i=1}^{3}\int_{\Omega}\Phi_{0i}^{2}\,dy.$$

Here A, B are given as follows. Note that the expression of $\nu\times\boldsymbol{\Psi}$ is substituted.

$$\begin{aligned}
A&=\langle \mathrm{rot}\boldsymbol{\Phi}_{0},\nu\times\boldsymbol{\Psi}\rangle\\
&=\left(\frac{\partial\Phi_{03}}{\partial y_2}-\frac{\partial\Phi_{02}}{\partial y_3}\right)\left[\Phi_{01}\left(\nu_3\frac{\partial g_1}{\partial y_2}-\nu_2\frac{\partial g_1}{\partial y_3}\right)+\Phi_{02}\left(\nu_3\frac{\partial g_2}{\partial y_2}-\nu_2\frac{\partial g_2}{\partial y_3}\right)\right.\\
&\quad\left.+\Phi_{03}\left(\nu_3\frac{\partial g_3}{\partial y_2}-\nu_2\frac{\partial g_3}{\partial y_3}\right)\right]\\
&\quad+\left(\frac{\partial\Phi_{01}}{\partial y_3}-\frac{\partial\Phi_{3}}{\partial y_1}\right)\left[\Phi_{01}\left(\nu_1\frac{\partial g_1}{\partial y_3}-\nu_3\frac{\partial g_1}{\partial y_1}\right)+\Phi_{02}\left(\nu_1\frac{\partial g_2}{\partial y_3}-\nu_3\frac{\partial g_2}{\partial y_1}\right)\right.\\
&\quad\left.+\Phi_{03}\left(\nu_1\frac{\partial g_3}{\partial y_3}-\nu_3\frac{\partial g_3}{\partial y_1}\right)\right]\\
&\quad+\left(\frac{\partial\Phi_{02}}{\partial y_1}-\frac{\partial\Phi_{1}}{\partial y_2}\right)\left[\Phi_{01}\left(\nu_2\frac{\partial g_1}{\partial y_1}-\nu_1\frac{\partial g_1}{\partial y_2}\right)+\Phi_{02}\left(\nu_2\frac{\partial g_2}{\partial y_1}-\nu_1\frac{\partial g_2}{\partial y_2}\right)\right.\\
&\quad\left.+\Phi_{03}\left(\nu_2\frac{\partial g_3}{\partial y_1}-\nu_1\frac{\partial g_3}{\partial y_2}\right)\right],
\end{aligned}$$

$$B=\langle\boldsymbol{\Phi}_0,\nu\rangle\sum_{i,\ell=1}^{3}\frac{\partial\Phi_{0i}}{\partial y_\ell}\frac{\partial g_\ell}{\partial y_i}+\sum_{i=1}^{3}(\operatorname{div} g)\Phi_{0i}\frac{\partial\Phi_{0i}}{\partial\nu}$$

$$-\sum_{i,j,\ell=1}^{3}\nu_\ell\left(\frac{\partial g_\ell}{\partial y_j}+\frac{\partial g_j}{\partial y_\ell}\right)\Phi_{0i}\frac{\partial\Phi_{0i}}{\partial y_j}.$$

We mention some useful property for the boundary condition of $\mathrm{rot}\,\Phi_0$.

Lemma 8 *We have* $\langle\mathrm{rot}\,\boldsymbol{\Phi}_0,\nu\rangle=0$ *on* $\partial\Omega$.

Proof From the direct calculation near $\partial\Omega$, the boundary condition $\boldsymbol{\Phi}_0\times\nu=\mathbf{0}$ on $\partial\Omega$ gives this property of $\mathrm{rot}\,\boldsymbol{\Phi}_0$. □

Evaluation of A, B We see the values A and B in terms of Ω, $\boldsymbol{\Phi}_0$, ρ. For that purpose, we take an arbitrary point of $\partial\Omega$ and a special coordinate around the point to calculate A and B. Take any point $O \in \partial\Omega$ and take the orthogonal coordinate $y=(y_1, y_2, y_3)$ centered at O such that $\nu(O)=(1,0,0)$. We express $\partial\Omega$ by a graph $y_1 = h(y_2, y_3)$ near O. There exists a $\delta > 0$ and C^2 function such that

$$\Omega \cap U(O,\delta) = \big\{(y_1, y_2, y_3) \in \mathbb{R}^3 \,\big|\, |y| < \delta,\ y_1 < h(y_2, y_3)\big\}.$$

It holds that $(\partial h/\partial y_2)(0,0)=0$, $(\partial h/\partial y_2)(0,0)=0$. We can assume that two vectors $(0,1,0)$ and $(0,0,1)$ are principal directions in the tangent space of $\partial\Omega$ at O. In this case

$$\frac{\partial \nu}{\partial y_2}(O) = \alpha(0,1,0), \qquad \frac{\partial \nu}{\partial y_3}(O) = \beta(0,0,1),$$

where α and β are the principal curvatures of $\partial\Omega$ at O. Put $\phi(y) = \langle \boldsymbol{\Phi}_0(y), \nu(y)\rangle$ for $y \in \partial\Omega$ for simplicity. We note that

$$\nu_1(O)=1, \qquad \nu_2(O)=0, \qquad \nu_3(O)=0, \qquad \Phi_{01}(O) = \big\langle \boldsymbol{\Phi}_0(O), \nu(O)\big\rangle,$$
$$\Phi_{02}(O)=0, \qquad \Phi_{03}(O)=0, \qquad \frac{\partial g_1}{\partial y_1}(O)=0, \qquad \frac{\partial g_2}{\partial y_1}(O)=0,$$
$$\frac{\partial g_3}{\partial y_1}(O)=0, \qquad \frac{\partial g_1}{\partial y_2}(O) = \frac{\partial \rho}{\partial y_2}(O),$$
$$\frac{\partial g_1}{\partial y_3}(O) = \frac{\partial \rho}{\partial y_3}(O), \qquad \frac{\partial g_2}{\partial y_2}(O) = \rho(O)\frac{\partial \nu_2}{\partial y_2}(O) = \rho(O)\alpha,$$
$$\frac{\partial g_2}{\partial y_3}(O) = \frac{\partial g_3}{\partial y_2}(O) = 0, \qquad \frac{\partial g_3}{\partial y_3}(O) = \rho(O)\frac{\partial \nu_3}{\partial y_3}(O) = \rho(O)\beta.$$

From the condition $\boldsymbol{\Phi}_0 \times \nu = 0$ on the boundary, we have

$$\Phi_{0i}(\xi)\nu_j(\xi) - \Phi_{0j}(\xi)\nu_i(\xi) = 0 \quad (\xi \in \partial\Omega,\ 1 \leqq i, j \leqq 3).$$

We can operate $\partial/\partial y_2$, $\partial/\partial y_3$ (tangential derivative) on the above equations at O and get the following properties,

$$\frac{\partial \Phi_{02}}{\partial y_2}(O) = \alpha\Phi_{01}(O), \qquad \frac{\partial \Phi_{03}}{\partial y_3}(O) = \beta\Phi_{01}(O),$$
$$\frac{\partial \Phi_{01}}{\partial y_1}(O) = -(\alpha+\beta)\Phi_{01}(O), \qquad \frac{\partial \Phi_{02}}{\partial y_3}(O) = \frac{\partial \Phi_{03}}{\partial y_2}(O) = 0.$$

Substituting these quantities into A and B, we have

$$\begin{aligned} A(O) &= \left(\frac{\partial \Phi_{01}}{\partial y_3} - \frac{\partial \Phi_{03}}{\partial y_1}\right)\phi(O)\frac{\partial \rho}{\partial y_3}(O) + \left(\frac{\partial \Phi_{02}}{\partial y_1} - \frac{\partial \Phi_{01}}{\partial y_2}\right)\phi(O)(-1)\frac{\partial \rho}{\partial y_2}(O) \\ &= \langle \operatorname{rot}\boldsymbol{\Phi}_0 \times \nabla\rho, \nu\rangle\langle \boldsymbol{\Phi}_0, \nu\rangle, \\ B(O) &= \frac{\partial \rho}{\partial y_2}(O)\phi(O)\left(\frac{\partial \Phi_{02}}{\partial y_1} - \frac{\partial \Phi_{01}}{\partial y_2}\right) + \frac{\partial \rho}{\partial y_3}(O)\phi(O)\left(\frac{\partial \Phi_{03}}{\partial y_1} - \frac{\partial \Phi_{01}}{\partial y_3}\right) \\ &\quad + \alpha^2\phi(O)^2\rho(O) + \beta^2\phi(O)^2\rho(O) - \rho(O)\phi(O)^2(\alpha+\beta)^2 \\ &= \phi(O)\langle \nabla\rho \times \operatorname{rot}\boldsymbol{\Phi}_0, \nu\rangle - 2K(O)\rho(O)\phi(O)^2. \end{aligned}$$

Note that $K(O) = \alpha\beta$ is the Gaussian curvature of $\partial\Omega$ at O. Summing up these quantities $A(O)$, $B(O)$ and put them into $J = 0$ (recall $\boldsymbol{\Phi}_0(y) = \boldsymbol{\Phi}_0^{(k)}(y)$), we get

$$\begin{aligned}
&\kappa \int_{\Omega} |\boldsymbol{\Phi}_0^{(k)}|^2 dx \\
&= \int_{\partial\Omega} \left(|\nabla \boldsymbol{\Phi}_0^{(k)}|^2 - 2\left|\frac{\partial \boldsymbol{\Phi}_0^{(k)}}{\partial \nu}\right|^2 + \big(2K(x) - \lambda_k(0)\big)|\boldsymbol{\Phi}_0^{(k)}(x)|^2 \right) \rho \, dS \\
&\quad + 2\int_{\partial\Omega} \langle \boldsymbol{\Phi}_0^{(k)}, \nu\rangle \langle \mathrm{rot}\, \boldsymbol{\Phi}_0^{(k)} \times \nabla\rho, \nu\rangle dS.
\end{aligned} \tag{43}$$

Thus we have obtained the candidate of $(d\lambda_k(\varepsilon)/d\varepsilon)(0)$ which is the value $\boldsymbol{\kappa}$.

5 Justification of the Formula

In this section, we justifies the formula $\lim_{\varepsilon\to 0}(\lambda_k(\varepsilon) - \lambda_k(0))/\varepsilon = \boldsymbol{\kappa}$. First we prepare some perturbation properties of γ_ε for the later calculation.

Lemma 9

$$\frac{\partial \gamma_\varepsilon}{\partial y}(y) = I + \frac{\partial g}{\partial y}(y)\varepsilon + O(\varepsilon^2), \qquad \left(\frac{\partial \gamma_\varepsilon}{\partial y}(y)\right)^{-1} = I - \frac{\partial g}{\partial y}(y)\varepsilon + O(\varepsilon^2), \tag{44}$$

$$\det\left(\frac{\partial \gamma_\varepsilon}{\partial y}\right) = 1 + (\mathrm{div}\, g)\varepsilon + O(\varepsilon^2). \tag{45}$$

Proof These formulas are just the Taylor expansion in ε. □

Lemma 10 *The map $\gamma_\varepsilon : \Omega \longrightarrow \Omega(\varepsilon)$ (sufficiently smooth up to $\partial\Omega$) induces the following transformation of the surface element and the unit outward normal vector field as follows,*

$$\begin{aligned}
&dS_\varepsilon = \big(1 + \varepsilon\rho H + O(\varepsilon^2)\big)\, dS, \\
&\nu_\varepsilon(x) = \left(\left[\frac{\partial \gamma_\varepsilon}{\partial y}\right]^{-1}\right)^t \nu(y) \Big/ \left|\left(\left[\frac{\partial \gamma_\varepsilon}{\partial y}\right]^{-1}\right)^t \nu(y)\right| \\
&\big(\text{for } x = \gamma_\varepsilon(y),\ y \in \partial\Omega\big).
\end{aligned} \tag{46}$$

See [14, 15, 19] for the proofs of these formulas.

Proof of the main result We consider the variational equation (37)–(41). First define the value $\boldsymbol{\kappa}$ by the relation (43). We use an approximate eigenfunction in the problem (17)–(21) in the following form

$$\widetilde{\boldsymbol{\Phi}}_\varepsilon^*(y) = \boldsymbol{\Phi}_0(y) + \varepsilon \boldsymbol{\Psi}_0(y) \quad \text{in } \Omega \tag{47}$$

where $\boldsymbol{\Psi}_0$ is a solution of (37)–(41) (the variational equation). Existence of $\boldsymbol{\Psi}_0$ is seen from the following arguments. To obtain $\boldsymbol{\Psi}_0$, we consider the new unknown variable $\boldsymbol{\Gamma}$. For that purpose we prepare the followings.

Define the vector valued functions $\boldsymbol{\psi}_0(y)$, $\boldsymbol{\psi}_1(y)$ by

$$\boldsymbol{\psi}_0(y) = -\left(\frac{\partial g}{\partial y}\right)^t \boldsymbol{\Phi}_0(y), \qquad \boldsymbol{\psi}_1(y) = \left(\sum_{\ell=1}^{3} \frac{\partial \Phi_{0i}}{\partial y_\ell} g_\ell\right)_{1 \leqq i \leqq 3}.$$

Then we have the following result.

Lemma 11

$$\operatorname{div} \boldsymbol{\psi}_1(y) = \sum_{\ell=1}^{3} \sum_{i=1}^{3} \frac{\partial \Phi_{0i}}{\partial y_\ell} \frac{\partial g_\ell}{\partial y_i} \quad \text{in } \Omega.$$

Proof This is deduced by a direct calculation. □

Lemma 12 *There exists a solution $\boldsymbol{\psi}_2(y)$ to the equation,*

$$\operatorname{div} \boldsymbol{\psi}_2 = -\operatorname{div} \boldsymbol{\psi}_0 \quad \text{in } \Omega, \qquad \boldsymbol{\psi}_2(y) = \boldsymbol{\psi}_1(y) - \langle \boldsymbol{\psi}_1(y), \nu(y) \rangle \nu(y) \quad \text{on } \partial\Omega. \tag{48}$$

Proof Existence of $\boldsymbol{\psi}_2$ in (48) follows from the property

$$\int_\Omega (-1) \operatorname{div} \boldsymbol{\psi}_0 \, dy = \int_{\partial\Omega} \left\langle \left(\frac{\partial g}{\partial y}\right)^t \boldsymbol{\Phi}_0(y), \nu \right\rangle dS = \int_{\partial\Omega} \left\langle \boldsymbol{\Phi}_0(y), \frac{\partial g}{\partial y} \nu \right\rangle dS = 0$$

(cf. Lemma 4) and the boundary value of $\boldsymbol{\psi}_2$ is imposed to be a tangential field. So we can apply the existence theorem in [21]. □

Here we give a new unknown variable as follows:

$$\boldsymbol{\Gamma}(y) = \boldsymbol{\Psi}(y) - \boldsymbol{\psi}_0(y) - \boldsymbol{\psi}_1(y) - \boldsymbol{\psi}_2(y). \tag{49}$$

Through this change of variable from $\boldsymbol{\Psi}$ to $\boldsymbol{\Gamma}$ with Lemma 11, Lemma 12, we have following equation which is equivalent to (37)–(41).

$$\operatorname{rot} \operatorname{rot} \boldsymbol{\Gamma} - \lambda(0) \boldsymbol{\Gamma} = \boldsymbol{F}, \qquad \operatorname{div} \boldsymbol{\Gamma} = 0 \quad \text{in } \Omega, \qquad \boldsymbol{\Gamma} \times \nu = \boldsymbol{0} \quad \text{on } \partial\Omega, \tag{50}$$

where

$$\begin{aligned} \boldsymbol{F} = {} & \kappa \boldsymbol{\Phi}_0 - \operatorname{div}\left(\nabla \boldsymbol{\Phi}_0 \left(\frac{\partial g}{\partial y} + \left(\frac{\partial g}{\partial y}\right)^t \right) \right) + \nabla(\operatorname{div} g) \cdot \nabla \boldsymbol{\Phi}_0 \\ & + \Delta(\boldsymbol{\psi}_0 + \boldsymbol{\psi}_1 + \boldsymbol{\psi}_2) + \lambda(0)(\boldsymbol{\psi}_0 + \boldsymbol{\psi}_1 + \boldsymbol{\psi}_2). \end{aligned} \tag{51}$$

For the proof of the existence of a solution to (50), it suffices to show that $\boldsymbol{F}$ is orthogonal to the kernel of the operator rot rot (with the conditions: "div-free" and the boundary condition). This is justified by an alternative theorem for a linear inhomogeneous equation including a self-adjoint operator. So it is necessary to show $\boldsymbol{F} \perp \boldsymbol{\Phi}_0$ because that kernel is spanned by $\boldsymbol{\Phi}_0$ from the assumption of the theorem.

However this condition holds due to the calculation in the definition of $\boldsymbol{\kappa}$ and we already justified this condition. So we know the existence of $\boldsymbol{\Psi}_0$.

Recall $\boldsymbol{\Phi}_\varepsilon = \boldsymbol{\Phi}_\varepsilon^{(k)}(x)$ is the k-th eigenfunction of (12) and $\boldsymbol{\Psi}_0$ is the function constructed above. For later arguments, we define

$$\widehat{\boldsymbol{\Phi}}_\varepsilon(x) = \boldsymbol{\Phi}_0\big(\gamma_\varepsilon^{-1}(x)\big), \qquad \widehat{\boldsymbol{\Psi}}_\varepsilon(x) = \boldsymbol{\Psi}_0\big(\gamma_\varepsilon^{-1}(x)\big).$$

As the assumption that the eigenvalue $\lambda(0) = \lambda_k(0)$ is simple, we can assume (without loss of generality) that $\widetilde{\boldsymbol{\Phi}}_\varepsilon(y) = \boldsymbol{\Phi}_\varepsilon(\gamma_\varepsilon(y))$ approaches $\boldsymbol{\Phi}_0$ for $\varepsilon \to 0$ (strongly in $L^2(\Omega;\mathbb{R}^3)$ and weakly in $H^1(\Omega;\mathbb{R}^3)$) because we can multiply $\widetilde{\boldsymbol{\Phi}}_\varepsilon$ by -1 for each ε if necessary. As in the standard regularity argument of elliptic equation theory (cf. [13]), we can prove the convergence in higher norm (like C^m). Multiply both sides of the eigenvalue equation (12) by $\widehat{\boldsymbol{\Phi}}_\varepsilon(x) + \varepsilon\widehat{\boldsymbol{\Psi}}_\varepsilon(x)$ and integrate, we have

$$\int_{\Omega(\varepsilon)} \big\langle \mathrm{rot}\,\mathrm{rot}\,\boldsymbol{\Phi}_\varepsilon - \lambda(\varepsilon)\boldsymbol{\Phi}_\varepsilon, (\widehat{\boldsymbol{\Phi}}_\varepsilon + \varepsilon\widehat{\boldsymbol{\Psi}})\big\rangle\, dx = 0. \tag{52}$$

By the partial integration, we have

$$\int_{\partial\Omega(\varepsilon)} \langle \nu_\varepsilon \times \mathrm{rot}\,\boldsymbol{\Phi}_\varepsilon, \widehat{\boldsymbol{\Phi}}_\varepsilon + \varepsilon\widehat{\boldsymbol{\Psi}}_\varepsilon\rangle\, dS_\varepsilon + \int_{\partial\Omega(\varepsilon)} \big\langle \nu_\varepsilon \times \boldsymbol{\Phi}_\varepsilon, \mathrm{rot}(\widehat{\boldsymbol{\Phi}}_\varepsilon + \varepsilon\,\widehat{\boldsymbol{\Psi}}_\varepsilon)\big\rangle\, dS_\varepsilon$$
$$+ \int_{\Omega(\varepsilon)} \big\langle \boldsymbol{\Phi}_\varepsilon, \mathrm{rot}\,\mathrm{rot}(\widehat{\boldsymbol{\Phi}}_\varepsilon + \varepsilon\widehat{\boldsymbol{\Psi}}_\varepsilon)\big\rangle\, dx - \lambda(\varepsilon)\int_{\Omega(\varepsilon)} \big\langle \boldsymbol{\Phi}_\varepsilon, (\widehat{\boldsymbol{\Phi}}_\varepsilon + \varepsilon\widehat{\boldsymbol{\Psi}}_\varepsilon)\big\rangle\, dx = 0. \tag{53}$$

We use the boundary condition $\boldsymbol{\Phi}_\varepsilon \times \nu_\varepsilon = \mathbf{0}$ on $\partial\Omega(\varepsilon)$ and see the second term vanishes. Using $\mathrm{rot}\,\mathrm{rot} = \mathrm{grad}\,\mathrm{div} - \Delta$, we have

$$J_1(\varepsilon) + J_2(\varepsilon) - J_4(\varepsilon) - \lambda(\varepsilon)J_3(\varepsilon) = 0, \tag{54}$$

where

$$J_1(\varepsilon) = \int_{\partial\Omega(\varepsilon)} \big\langle \mathrm{rot}\,\boldsymbol{\Phi}_\varepsilon, (\widehat{\boldsymbol{\Phi}}_\varepsilon + \varepsilon\widehat{\boldsymbol{\Psi}}_\varepsilon) \times \nu_\varepsilon\big\rangle\, dS_\varepsilon,$$

$$J_2(\varepsilon) = \int_{\Omega(\varepsilon)} \big\langle \boldsymbol{\Phi}_\varepsilon, \nabla\,\mathrm{div}(\widehat{\boldsymbol{\Phi}}_\varepsilon + \varepsilon\widehat{\boldsymbol{\Psi}}_\varepsilon)\big\rangle\, dx,$$

$$J_4(\varepsilon) = \int_{\Omega(\varepsilon)} \big\langle \boldsymbol{\Phi}_\varepsilon, \Delta(\widehat{\boldsymbol{\Phi}}_\varepsilon + \varepsilon\widehat{\boldsymbol{\Psi}}_\varepsilon)\big\rangle\, dx, \qquad J_3(\varepsilon) = \int_{\Omega(\varepsilon)} \langle \boldsymbol{\Phi}_\varepsilon, \widehat{\boldsymbol{\Phi}}_\varepsilon + \varepsilon\,\widehat{\boldsymbol{\Psi}}_\varepsilon\rangle\, dx.$$

We change the variable from x to y by $x = \gamma_\varepsilon(y)$ and express $J_j(\varepsilon)$ $(j = 1, 2, 3, 4)$ in the form of integration in the domain Ω and calculate the ε-expansion to the first order. Later we use the Landau's symbols. As well as the standard cases, if it is the case for a function h_ε like $h_\varepsilon = O(\varepsilon^p)$, it implies $\|h_\varepsilon\|_{C^2(\overline{\Omega})} = O(\varepsilon^p)$. To deal with $J_3(\varepsilon)$, $J_4(\varepsilon)$, $J_2(\varepsilon)$, we will use Lemma 6 and Lemma 9 below.

$$
\begin{aligned}
J_3(\varepsilon) &= \int_\Omega \langle \widetilde{\boldsymbol{\Phi}}_\varepsilon, \boldsymbol{\Phi}_0 + \varepsilon \boldsymbol{\Psi}_0 \rangle \det\left(\frac{\partial \gamma_\varepsilon}{\partial y} \right) dy \\
&= \int_\Omega \langle \widetilde{\boldsymbol{\Phi}}_\varepsilon, \boldsymbol{\Phi}_0 + \varepsilon \boldsymbol{\Psi}_0 \rangle \big(1 + (\operatorname{div} g)\varepsilon + O(\varepsilon^2)\big)\, dy \\
&= \int_\Omega \langle \widetilde{\boldsymbol{\Phi}}_\varepsilon, \boldsymbol{\Phi}_0 \rangle\, dy + \varepsilon \int_\Omega \big(\langle \widetilde{\boldsymbol{\Phi}}_\varepsilon, \boldsymbol{\Psi}_0 \rangle + \langle \widetilde{\boldsymbol{\Phi}}_\varepsilon, \boldsymbol{\Phi}_0 \rangle (\operatorname{div} g) \big)\, dy + O(\varepsilon^2), \\
J_4(\varepsilon) &= \int_\Omega \left\langle \widetilde{\boldsymbol{\Phi}}_\varepsilon, \operatorname{div}\left(\nabla(\boldsymbol{\Phi}_0 + \varepsilon \boldsymbol{\Psi}_0) \left(\frac{\partial \gamma_\varepsilon}{\partial y} \right)^{-1} \left[\left(\frac{\partial \gamma_\varepsilon}{\partial y} \right)^{-1} \right]^t \det\left(\frac{\partial \gamma_\varepsilon}{\partial y} \right) \right) \right\rangle dy \\
&= \int_\Omega \left\langle \widetilde{\boldsymbol{\Phi}}_\varepsilon, \operatorname{div}\left[\nabla(\boldsymbol{\Phi}_0 + \varepsilon \boldsymbol{\Psi}_0) \left(I - \frac{\partial g}{\partial y} \varepsilon \right) \left(I - \left(\frac{\partial g}{\partial y} \right)^t \varepsilon \right) \right. \right. \\
&\quad \left. \left. \times \big(1 + (\operatorname{div} g)\varepsilon\big) \right] \right\rangle dy + O(\varepsilon^2) \\
&= \int_\Omega \langle \widetilde{\boldsymbol{\Phi}}_\varepsilon, \Delta \boldsymbol{\Phi}_0 \rangle\, dy \\
&\quad + \varepsilon \int_\Omega \left\{ \langle \widetilde{\boldsymbol{\Phi}}_\varepsilon, \Delta \boldsymbol{\Psi}_0 \rangle - \left\langle \widetilde{\boldsymbol{\Phi}}_\varepsilon, \operatorname{div}\left[\nabla \boldsymbol{\Phi}_0 \left(\frac{\partial g}{\partial y} + \left(\frac{\partial g}{\partial y} \right)^t \right) \right] \right\rangle \right\} dy \\
&\quad + \varepsilon \int_\Omega \big\langle \widetilde{\boldsymbol{\Phi}}_\varepsilon, \operatorname{div}\big(\nabla \boldsymbol{\Phi}_0 (\operatorname{div} g) \big) \big\rangle\, dy + O(\varepsilon^2), \\
J_2(\varepsilon) &= \sum_{\ell=1}^3 \int_\Omega \widetilde{\Phi}_{\varepsilon,\ell} \sum_{p,j,i=1}^3 \frac{\partial}{\partial y_p} \left(\left(\frac{\partial \Phi_{0i}}{\partial y_j} + \varepsilon \frac{\partial \Psi_{0i}}{\partial y_j} \right) \left[\left(\frac{\partial \gamma_\varepsilon}{\partial y} \right)^{-1} \right]_{ji} \right) \\
&\quad \times \left[\left(\frac{\partial \gamma_\varepsilon}{\partial y} \right)^{-1} \right]_{pl} dy \\
&= \sum_{\ell=1}^3 \int_\Omega \widetilde{\Phi}_{\varepsilon,\ell} \sum_{p,j,i=1}^3 \frac{\partial}{\partial y_p} \left(\left(\frac{\partial \Phi_{0i}}{\partial y_j} + \varepsilon \frac{\partial \Psi_{0i}}{\partial y_j} \right) \left(\delta(j,i) - \frac{\partial g_j}{\partial y_i} \varepsilon \right) \right) \\
&\quad \times \left(\delta(p,\ell) - \frac{\partial g_p}{\partial y_\ell} \varepsilon \right) dy \\
&\quad + O(\varepsilon^2) = \int_\Omega \langle \widetilde{\boldsymbol{\Phi}}_\varepsilon, \nabla \operatorname{div} \boldsymbol{\Phi}_0 \rangle\, dy \\
&\quad + \varepsilon \int_\Omega \left\langle \widetilde{\boldsymbol{\Phi}}_\varepsilon, \nabla \left(\operatorname{div} \boldsymbol{\Psi}_0 - \sum_{j,i=1}^3 \frac{\partial \Phi_{0i}}{\partial y_j} \frac{\partial g_j}{\partial y_i} \right) \right\rangle dy \\
&\quad + \varepsilon \int_\Omega \sum_{\ell,r=1}^3 \widetilde{\Phi}_{\varepsilon\ell} \frac{\partial (\operatorname{div} \Phi_0)}{\partial y_r} \left(\delta(r,\ell)(\operatorname{div} g) - \frac{\partial g_r}{\partial y_\ell} \right) dy + O(\varepsilon^2).
\end{aligned}
$$

We used the condition $\operatorname{div} \boldsymbol{\Phi}_0 = 0$ in Ω and the expression for $\operatorname{div} \boldsymbol{\Psi}_0$. We have $J_2(\varepsilon) = O(\varepsilon^2)$. Next we deal with $J_1(\varepsilon)$.

$$
J_1(\varepsilon) = \int_{\partial\Omega} \big\langle \Upsilon_\varepsilon(y), (\boldsymbol{\Phi}_0 + \varepsilon \boldsymbol{\Psi}_0) \times \nu_\varepsilon\big(\gamma_\varepsilon(y)\big) \big\rangle \big(1 + H\rho\varepsilon + O(\varepsilon^2)\big)\, dS
$$

where $\nu_\varepsilon(\gamma_\varepsilon(y))$ and $\Upsilon_\varepsilon(y)$ are given as follows:

$$\begin{aligned}\nu_\varepsilon\big(\gamma_\varepsilon(y)\big) &= \left[\left(\frac{\partial \gamma_\varepsilon}{\partial y}\right)^{-1}\right]^t \nu(y) \Big/ \left|\left[\left(\frac{\partial \gamma_\varepsilon}{\partial y}\right)^{-1}\right]^t \nu(y)\right| \\ &= \nu + \varepsilon\left[-\left(\frac{\partial g}{\partial y}\right)^t \nu + \left\langle\left(\frac{\partial g}{\partial y}\right)^t \nu, \nu\right\rangle \nu\right] + O\big(\varepsilon^2\big)\end{aligned}$$

$$\Upsilon_\varepsilon(y) = (\mathrm{rot}\,\widetilde{\boldsymbol{\Phi}}_\varepsilon)\big(\gamma_\varepsilon(y)\big) = \begin{pmatrix} \sum_{j=1}^3 \frac{\partial \widetilde{\Phi}_{\varepsilon 3}}{\partial y_j}[(\frac{\partial \gamma_\varepsilon}{\partial y})^{-1}]_{j2} - \sum_{j=1}^3 \frac{\partial \widetilde{\Phi}_{\varepsilon 2}}{\partial y_j}[(\frac{\partial \gamma_\varepsilon}{\partial y})^{-1}]_{j3} \\ \sum_{j=1}^3 \frac{\partial \widetilde{\Phi}_{\varepsilon 1}}{\partial y_j}[(\frac{\partial \gamma_\varepsilon}{\partial y})^{-1}]_{j3} - \sum_{j=1}^3 \frac{\partial \widetilde{\Phi}_{\varepsilon 3}}{\partial y_j}[(\frac{\partial \gamma_\varepsilon}{\partial y})^{-1}]_{j1} \\ \sum_{j=1}^3 \frac{\partial \widetilde{\Phi}_{\varepsilon 2}}{\partial y_j}[(\frac{\partial \gamma_\varepsilon}{\partial y})^{-1}]_{j1} - \sum_{j=1}^3 \frac{\partial \widetilde{\Phi}_{\varepsilon 1}}{\partial y_j}[(\frac{\partial \gamma_\varepsilon}{\partial y})^{-1}]_{j2} \end{pmatrix}$$

$$\begin{aligned}&(\boldsymbol{\Phi}_0 + \varepsilon \boldsymbol{\Psi}_0) \times \nu_\varepsilon\big(\gamma_\varepsilon(y)\big) \\ &= \boldsymbol{\Phi}_0 \times \nu + \varepsilon(\boldsymbol{\Psi}_0 \times \nu) + \varepsilon\left(\boldsymbol{\Phi}_0 \times \left[-\left(\frac{\partial g}{\partial y}\right)^t \nu + \left\langle\left(\frac{\partial g}{\partial y}\right)^t \nu, \nu\right\rangle \nu\right]\right) \\ &\quad + O\big(\varepsilon^2\big) = \varepsilon(\boldsymbol{\Psi}_0 \times \nu) - \varepsilon\left(\boldsymbol{\Phi}_0 \times \left(\frac{\partial g}{\partial y}\right)^t \nu\right) + O\big(\varepsilon^2\big).\end{aligned} \tag{55}$$

We used the boundary condition $\boldsymbol{\Phi}_0 \times \nu = \mathbf{0}$ on $\partial\Omega$ above. We deal with the first and second term. Again, from the same boundary condition, we have

$$\boldsymbol{\Phi}_0(y) = \big\langle \boldsymbol{\Phi}_0(y), \nu(y)\big\rangle \nu(y) \quad (y \in \partial\Omega).$$

We use (42) and calculate on $\partial\Omega$ that

$$\begin{aligned}&\boldsymbol{\Psi}_0(y) \times \nu(y) - \boldsymbol{\Phi}_0(y) \times \left(\frac{\partial g}{\partial y}\right)^t \nu(y) \\ &= \boldsymbol{\psi}_0(y) \times \nu(y) - \big\langle \boldsymbol{\Phi}_0(y), \nu(y)\big\rangle \left(\nu(y) \times \left(\frac{\partial g}{\partial y}\right)^t \nu(y)\right) \\ &= \boldsymbol{\psi}_0(y) \times \nu - \nu(y) \times \left(\frac{\partial g}{\partial y}\right)^t \big\langle \boldsymbol{\Phi}_0(y), \nu(y)\big\rangle \nu(y) \\ &= \boldsymbol{\psi}_0(y) \times \nu(y) + \nu(y) \times \left[-\left(\frac{\partial g}{\partial y}\right)^t \boldsymbol{\Phi}_0(y)\right] \\ &= \boldsymbol{\psi}_0(y) \times \nu(y) + \nu(y) \times \boldsymbol{\psi}_0(y) = \mathbf{0}.\end{aligned}$$

Using this property in (55), we substitute it into the right hand of $J_1(\varepsilon)$, we have

$$J_1(\varepsilon) = O\big(\varepsilon^2\big). \tag{56}$$

Now we calculate $J_4(\varepsilon) + \lambda(\varepsilon) J_3(\varepsilon)$ with the equations for $\boldsymbol{\Psi}_0$ and $\boldsymbol{\Phi}_0$ and get the

$$\begin{aligned}J_4(\varepsilon) + \lambda(\varepsilon) J_3(\varepsilon) &= \big(\lambda(\varepsilon) - \lambda(0)\big) \int_\Omega \langle \widetilde{\boldsymbol{\Phi}}_\varepsilon, \boldsymbol{\Phi}_0\rangle\, dy - \kappa\varepsilon \int_\Omega \langle \widetilde{\boldsymbol{\Phi}}_\varepsilon, \boldsymbol{\Phi}_0\rangle\, dy \\ &\quad + \big(\lambda(\varepsilon) - \lambda(0)\big)\varepsilon \int_\Omega \big(\langle \widetilde{\boldsymbol{\Phi}}_\varepsilon, \boldsymbol{\Psi}_0\rangle + \langle \widetilde{\boldsymbol{\Phi}}_\varepsilon, \boldsymbol{\Phi}_0\rangle (\mathrm{div}\, g)\big)\, dy \\ &\quad + O\big(\varepsilon^2\big).\end{aligned} \tag{57}$$

We use $J_1(\varepsilon) = O(\varepsilon^2)$, $J_2(\varepsilon) = O(\varepsilon^2)$ in (54) and get $J_4(\varepsilon) + \lambda(\varepsilon)J_3(\varepsilon) = O(\varepsilon^2)$. We divide (57) by ε and take the limit ε and we get

$$\lim_{\varepsilon\to 0} \frac{\lambda(\varepsilon) - \lambda(0)}{\varepsilon} = \boldsymbol{\kappa}.$$

This completes the proof of the main result (Theorem 1).

References

1. Courant, R., Hilbert, D.: Method of Mathematical Physics, I. Wiley-Interscience, New York (1953)
2. Davies, E.B.: Spectral Theory and Differential Operators. Cambridge University Press, Cambridge (1995)
3. Edmunds, D.E., Evans, W.D.: Spectral Theory and Differential Operators. Oxford Mathematical Monographs. Oxford University Press, London (1987)
4. Garabedian, P., Schiffer, M.: Convexity of domain functionals. J. Anal. Math. **2**, 281–368 (1952–1953)
5. Gilbarg, D., Trudinger, N.: Elliptic Partial Differential Equations of Second Order. Springer, Berlin (1977)
6. Hadamard, J.: Memoire sur le problème d'analyse relatif a equilibre des plaques elastiques encastrees. Memoires des Savants Etrangers, vol. 33 (1908) (cf. Oeuvres De Jacqes Hadamard Tom II, Centre National de la Recherche Scientifiques 1968, pp. 515–631)
7. Hirakawa, K.: Denki Rikigaku. Baifukan, Tokyo (1973) (in Japanese)
8. Jimbo, S., Kosugi, S.: Spectra of domains with partial degeneration. J. Math. Sci. Univ. Tokyo **16**, 269–414 (2009)
9. Kato, T.: Perturbation Theory for Linear Operators. Springer, Berlin (1976)
10. Maz'ya, V., Nazarov, S., Plamenevskij, B.: Asymptotic theory of elliptic boundary value problems in singularly perturbed domains, I. Operator Theory Advances and Applications, vol. 111. Birkhäuser, Basel (2000)
11. Maz'ya, V., Nazarov, S., Plamenevskij, B.: Asymptotic theory of elliptic boundary value problems in singularly perturbed domains, II. Operator Theory Advances and Applications, vol. 112. Birkhäuser, Basel (2000)
12. Mizohata, S.: The Theory of Partial Differential Equations. Cambridge University Press, Cambridge (1979)
13. Morrey, C.: Multiple Integrals in the Calculus of Variations. Springer, Berlin (1966)
14. Murat, F., Simon, J.: Quelques resultats sur le controle par un domaine geometrique. Rapport L. A., vol. 189. Universite Paris VI, Paris (1974)
15. Murat, F., Simon, J.: Sur le controle par un domaine geometrique. Laboratoire Analyse Numerique, L. A., vol. 189. Universite Paris VI, Paris (1976)
16. Ozawa, S.: Ryouiki no Setsudo to Koyuti Mondai. Sugaku **33**, 248–261 (1981) (in Japanese)
17. Reed, M., Simon, B.: Methods of Modern Mathemtical Physics IV: Analysis of Operators. Academic Press, San Diego (1978)
18. Shimakura, N.: La première valeur propre du laplacien pour le problème de Dirichlet. J. Math. Pures Appl. **62**, 129–152 (1983)
19. Simon, J.: Differentiation with respect to the domain in boundary value problems. Numer. Funct. Anal. Optim. **2**, 649–687 (1980)
20. Sokolowski, J., Zolesio, J.P.: Introduction to Shape Optimization, Shape Sensitivity Analysis. Springer Series in Computational Mathematics, vol. 16. Springer, Berlin (1992)
21. Temam, R.: Navier-Stokes Equations. AMS, Providence (1984)

Global Structure of the Solution Set for a Semilinear Elliptic Problem Related to the Liouville Equation on an Annulus

Toru Kan

Abstract A semilinear elliptic problem related to the Liouville equation on a two-dimensional annulus is studied. The problem appears as the limiting problem of the Liouville equation as the inside radius of the annulus tends to 0, and is derived by the method of matched asymptotic expansions. Our concern is the solution set of the problem in the bifurcation diagram. We find explicit solutions including non-radially symmetric solutions and determine the connected component containing the solutions. As a consequence, we provide a suggestive evidence for the global structure of the solution set of the Liouville equation.

Keywords Semilinear elliptic equation · Liouville equation · Bifurcation

1 Introduction

In this paper we study the solution set of the following semilinear elliptic equation:

$$\Delta u + Ae^{u} = 0 \quad \text{in } \mathbb{R}^2 \setminus \{0\}, \tag{1}$$

$$u(x) = \begin{cases} (B-2)\log|x| + o(1) & \text{as } |x| \to 0, \\ -(B+2)\log|x| + o(1) & \text{as } |x| \to \infty. \end{cases} \tag{2}$$

Here $A > 0$ and $B \geq 2$ are (bifurcation) parameters. Our motivation originally comes from the Dirichlet problem for the Liouville equation

$$\begin{cases} \Delta v + \lambda e^{v} = 0 & \text{in } \Omega, \\ v = 0 & \text{on } \partial\Omega, \end{cases} \tag{3}$$

where $\lambda > 0$ is a parameter and Ω is a bounded domain in $\mathbb{R}^2$. Equation (1)–(2) is strongly related to (3) for the case of an annular domain defined by $\Omega = \Omega_\varepsilon := \{x \in \mathbb{R}^2; \varepsilon < |x| < 1\}$, where $0 < \varepsilon < 1$. In fact, in the next section, we will show that (1)–(2) appears as the limit of (3) as $\varepsilon \to 0$.

T. Kan (✉)
Department of Mathematics, Tokyo Institute of Technology, 2-12-1 Ookayama, Meguro-ku, Tokyo 152-8551, Japan
e-mail: kan@math.titech.ac.jp

R. Magnanini et al. (eds.), *Geometric Properties for Parabolic and Elliptic PDE's*, Springer INdAM Series 2, DOI 10.1007/978-88-470-2841-8_13,

If the domain Ω is a disk, from the well known result obtained by Gidas, Ni and Nirenberg [4], there is no non-radially symmetric solution in (3). On the other hand in the case of an annulus, the existence of non-radially symmetric solutions is revealed by Lin [7] and Nagasaki and Suzuki [12]. More precisely, Lin showed that non-radially symmetric solutions appear through a bifurcation from radially symmetric solutions and Nagasaki and Suzuki proved that for any $k \in \mathbb{N}$, there exists a k-mode solution such that $\int_\Omega e^v \, dx$ is large. Here, by k-mode solution, we mean a solution which is invariant under the rotation of $2\pi/k$, and is not invariant under the rotation of $2\pi/m$ for $m > k$. From the subsequent work of Dancer [1], the set of the bifurcating non-radially symmetric solutions is unbounded in (λ, v) plane. Additionally, for a general non-simply connected domain, del Pino, Kowalczyk and Musso [3] obtained a solution which blows up at k different points as $\lambda \to 0$.

From these results, it is expected that the bifurcating solutions connect to the large solutions obtained in [3, 12]. The aim of this paper is to provide a suggestive evidence for this expectation. To accomplish this, we consider the limiting case as the inside radius ε tends to 0.

First we have to derive an appropriate limiting equation for (3) as $\varepsilon \to 0$. Since (3) has no non-radially symmetric solution when Ω is a disk, it is not suitable for our purpose to regard a disk as the limit. This indicates that the information on non-radially symmetric solutions can be obtained only under appropriate scaling. This problem is discussed in the next section.

Next we need to study the solution set of the limiting equation (1)–(2) in (A, B, u) space. One of our main results is concerned with the connected component of the solution set containing all radially symmetric solutions. We find non-radially symmetric solutions bifurcating from radially symmetric solutions explicitly, and prove that the union of these solutions and all radially symmetric solutions is the maximal connected subset of the solution set. Furthermore, to prove this result, we also study the linearized operator around the solution. More precisely, we determine the null space and find a fundamental solution for the linearized operator. See Theorems 1 through 3 in Sect. 3 for precise statements of these results.

The solution set in a bifurcation diagram was also considered in other equations. Holzmann and Kielhöfer [5] studied positive solutions of $\Delta u + \lambda f(u) = 0$ under the Dirichlet boundary condition on a two-dimensional bounded domain which is axially symmetric with respect to x-axis, y-axis and is partially convex. They considered f such that $f(u) \geq 0$ for $u \geq 0$, and proved that the set of positive solutions are always parametrized by the amplitude of a solution. See also [6]. Furthermore, for a general class of f, Miyamoto [9, 10] studied the same equation under the Neumann boundary condition on a disk and obtained the global branch emanating from the second eigenvalue of the Laplacian. In [11], the case of a more general domain is also considered for a certain class of f.

This paper is organized as follows. In Sect. 2, by a formal argument, we observe that Eqs. (1)–(2) is derived as the limit of (3). In Sect. 3 we state the main results of this paper. Section 4 is devoted to the proofs of the main results and Appendix presents a lemma often used in the proofs.

Fig. 1 The outer region and the inner region

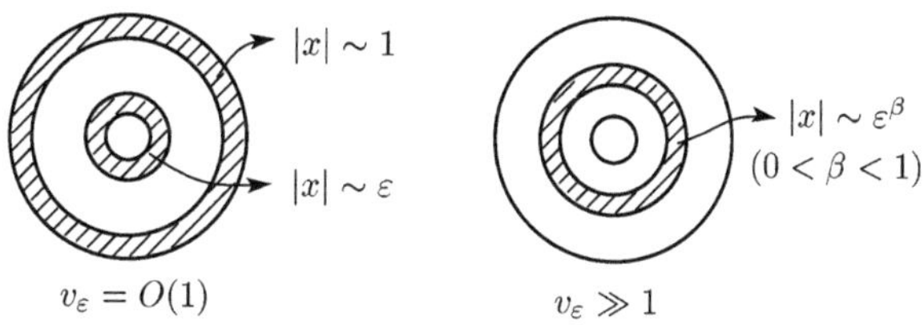

2 Derivation of the Limiting Equation

We begin this section with some important observation before deriving the limiting equation (1)–(2). Let us consider a bifurcation point $(\tilde{\lambda}_\varepsilon, \tilde{v}_\varepsilon)$ of (3) from which non-radially symmetric solutions bifurcate. Then, by using the explicit representation of this point obtained by Lin [7], we can show that $\tilde{\lambda}_\varepsilon \to 0$, $\|\tilde{v}_\varepsilon\|_{L^\infty} \to \infty$ as $\varepsilon \to 0$. This fact indicates that if we wish to obtain the informations of non-radially symmetric solutions in the limit, we have to treat the situation where $\lambda \ll 1$ and $\|v\|_{L^\infty} \gg 1$.

Based on this observation, we derive (1)–(2) by using the method of matched asymptotic expansions. Let $(\lambda_\varepsilon, v_\varepsilon)$ be a solution of (3) and assume that

$$\lambda_\varepsilon = A\varepsilon^\alpha\big(1 + o(1)\big) \tag{4}$$

for some $A > 0$ and $\alpha > 0$ as $\varepsilon \to 0$. We separate the domain Ω_ε into the outer region ($|x| \sim 1$ or $|x| \sim \varepsilon$) and the inner region ($|x| \sim \varepsilon^\beta$), where $0 < \beta < 1$, and assume that $v_\varepsilon = O(1)$ in the outer region and $v_\varepsilon \gg 1$ in the inner region (see Fig. 1).

Now we find the expansion of v_ε in the outer region. First we consider the region near the outside boundary. In this case from the assumption (4) and $v_\varepsilon = O(1)$, the second term of the left hand side of (3) is small. Hence we infer that the limiting equation in this region is

$$\begin{cases} -\Delta v = 0, & 0 < |x| < 1, \\ v = 0, & |x| = 1. \end{cases}$$

Since the maximum principle implies that v_ε is a positive function, the limiting function (the solution of the above equation) must be nonnegative. From this, the limiting function is expressed as $v(x) = C_1 \log 1/|x|$ for some nonnegative constant C_1, and the expansion in this region is given by

$$v_\varepsilon(x) = C_1 \log \frac{1}{|x|} + o(1). \tag{5}$$

This expansion is expected to be valid for $\delta_1 \le |x| \le 1$, where $\delta_1 > 0$ is an arbitrary fixed constant. A similar manner allows us to deduce that the expansion near the inside boundary is

$$v_\varepsilon(x) = C_2 \log \frac{|x|}{\varepsilon} + o(1) \tag{6}$$

for some nonnegative undetermined constant C_2. This expansion will be valid for $\varepsilon \le |x| \le \delta_2\varepsilon$, where $\delta_2 > 1$ is an arbitrary fixed constant.

Next we find the expansion in the inner region. Since the inner region is $|x| \sim \varepsilon^{\beta}$, it is convenient to perform the change of variables $x \mapsto \varepsilon^{\beta} x$. Then we have

$$\Delta v_{\varepsilon} + \varepsilon^{2\beta} \lambda_{\varepsilon} e^{v_{\varepsilon}} = 0, \quad \varepsilon^{1-\beta} < |x| < \varepsilon^{-\beta}.$$

Notice that $\lambda_{\varepsilon} \varepsilon^{2\beta} e^{v_{\varepsilon}} = A(1+o(1)) e^{v_{\varepsilon} - (\alpha + 2\beta) \log 1/\varepsilon}$. From this, putting $u_{\varepsilon} := v_{\varepsilon} - (\alpha + 2\beta) \log 1/\varepsilon$, we see that u_{ε} satisfies

$$\Delta u_{\varepsilon} + A\bigl(1+o(1)\bigr) e^{u_{\varepsilon}} = 0, \quad \varepsilon^{1-\beta} < |x| < \varepsilon^{-\beta}.$$

Therefore it is expected that the limiting equation for this region is (1) and v_{ε} is expanded as

$$v_{\varepsilon}(x) = (\alpha + 2\beta) \log \frac{1}{\varepsilon} + u\bigl(\varepsilon^{-\beta} x\bigr) + o(1), \tag{7}$$

where u is a solution of (1). For arbitrary fixed $0 < \delta_3 < \delta_4$, this expansion will be valid for $\delta_3 \varepsilon^{\beta} \le |x| \le \delta_4 \varepsilon^{\beta}$.

Finally we determine the undetermined constants and the boundary condition of (1) by matching the inner expansion and the outer expansion. We first consider the matching condition for the region between the outside boundary and the inner region. Substituting $x = \varepsilon^{\beta/2} y$ in (5) and (7), and comparing the expansions, we have

$$(\alpha + 2\beta) \log \frac{1}{\varepsilon} + u\bigl(\varepsilon^{-\beta/2} y\bigr) \sim C_1 \log \frac{1}{\varepsilon^{\beta/2} |y|}.$$

Hence

$$u\bigl(\varepsilon^{-\beta/2} y\bigr) \sim -(\alpha + 2\beta - C_1 \beta) \log \frac{1}{\varepsilon} + C_1 \log \frac{1}{\varepsilon^{-\beta/2} |y|}.$$

This implies that the constants α, β, C_1 and the function u satisfy

$$\alpha + 2\beta - C_1 \beta = 0, \tag{8}$$

$$u(x) = C_1 \log \frac{1}{|x|} + o(1) \quad \text{as } |x| \to \infty. \tag{9}$$

Similarly, by substituting $x = \varepsilon^{(1+\beta)/2} y$ in (6) and (7), we have

$$u\bigl(\varepsilon^{(1-\beta)/2} y\bigr) \sim -\bigl\{\alpha + 2\beta - C_2(1-\beta)\bigr\} \log \frac{1}{\varepsilon} + C_2 \log \varepsilon^{(1-\beta)/2} |y|,$$

which gives

$$\alpha + 2\beta - C_2(1-\beta) = 0, \tag{10}$$

$$u(x) = C_2 \log |x| + o(1) \quad \text{as } |x| \to 0. \tag{11}$$

Now we observe that constants C_1 and C_2 are not independent. More precisely, we can show that if a solution u of (1), (9), (11) exists, constants C_1, C_2 must satisfy $C_1 - 2 = C_2 + 2$. Indeed, this fact is shown as follows. Let u be a solution of (1), (9), (11) and put $w(s,\theta) = u(x) + 2s$, $x = (e^s \cos\theta, e^s \sin\theta)$. Then w satisfies

$$\Delta_{(s,\theta)} w + Ae^w = 0, \quad (s,\theta) \in \mathbb{R} \times S^1, \tag{12}$$

$$w(s,\theta) = \begin{cases} B_2 s + o(1) & \text{as } s \to -\infty, \\ -B_1 s + o(1) & \text{as } s \to \infty, \end{cases} \tag{13}$$

where $\Delta_{(s,\theta)} = \partial^2/\partial s^2 + \partial^2/\partial\theta^2$, $S^1 =$ (the unit circle) $\cong \mathbb{R}/2\pi\mathbb{Z}$, $B_1 = C_1 - 2$ and $B_2 = C_2 + 2$. We multiply w_s by (12) and integrate over $\mathbb{R} \times (0, 2\pi)$. Then, each term is formally calculated as

$$\int_{-\infty}^{\infty}\int_0^{2\pi} w_{ss} w_s \, d\theta \, ds = \frac{1}{2}\int_{-\infty}^{\infty}\int_0^{2\pi} \frac{\partial}{\partial s} w_s^2 \, d\theta \, ds = \pi\left(B_1^2 - B_2^2\right),$$

$$\begin{aligned} \int_{-\infty}^{\infty}\int_0^{2\pi} w_{\theta\theta} w_s \, d\theta \, ds &= -\int_{-\infty}^{\infty}\int_0^{2\pi} w_\theta w_{s\theta} \, d\theta \, ds \\ &= -\frac{1}{2}\int_{-\infty}^{\infty}\int_0^{2\pi} \frac{\partial}{\partial s} w_\theta^2 \, d\theta \, ds = 0, \end{aligned}$$

$$\int_{-\infty}^{\infty}\int_0^{2\pi} Ae^w w_s \, d\theta \, ds = A\int_{-\infty}^{\infty}\int_0^{2\pi} \frac{\partial}{\partial s} e^w \, d\theta \, ds = 0.$$

Hence $\pi(B_2^2 - B_1^2) = 0$, which gives $B_1 = B_2$. The above formal calculation is verified by Lemma 1 in Appendix.

Consequently, by putting $B := C_1 - 2 = C_2 + 2$ (≥ 2), we obtain the condition (2). Furthermore, from (8) and (10), if we find a solution of the problem (1)–(2), α and β are determined by

$$\alpha = \frac{1}{2}(B-2), \qquad \beta = \frac{1}{2}\left(1 - \frac{2}{B}\right).$$

When $B = 2$, the formal expansion fails since in this case $\alpha = \beta = 0$. This is because the assumption (4) is incorrect and the range which the inner expansion is valid is narrow. However, since (1) should have an information of (3) even for $B = 2$, we also deal with this case.

3 Main Results

In this section, we consider the limiting equation (1)–(2) and state our main results. The first half discusses the solution set in (A, B, u) space and the second half deals with the linearized equations. Hereafter, instead of (1)–(2), we consider the following transformed equation:

$$\begin{cases} \Delta_{(s,\theta)} u + Ae^u = 0, & (s,\theta) \in \mathbb{R} \times S^1, \\ u(s,\theta) = -B|s| + o(1) & \text{as } |s| \to \infty. \end{cases} \tag{14}$$

This equation is obtained by the same change of variables as in (12)–(13).

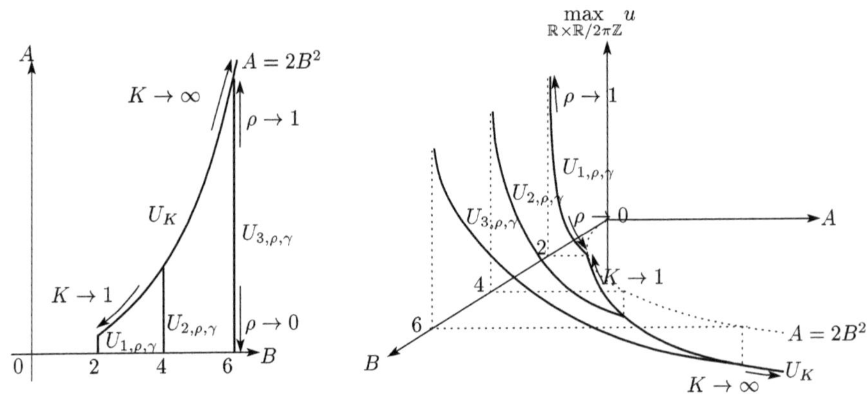

Fig. 2 Bifurcation diagram of (14)

3.1 Solution Set of (14)

We first consider radially symmetric solutions. In this case, the problem is reduced to an autonomous ordinary differential equation, and all radially symmetric solutions can be found. With a parameter $K \geq 1$, they are given by

$$\begin{aligned}(A, B, u) &= \big(8K^2, 2K, U_K\big),\\ U_K(s) &= \log\frac{1}{(e^{Ks}+e^{-Ks})^2} = \log\frac{1}{4\cosh^2 Ks},\end{aligned} \tag{15}$$

and there is no other radially symmetric solution.

Next we consider non-radially symmetric solutions. Fortunately, explicit solutions can be found. In fact, it can be checked that the following functions are solutions of (14):

$$\begin{aligned}(A, B, u) &= \big(8k^2(1-\rho^2), 2k, U_{k,\rho,\gamma}\big),\\ U_{k,\rho,\gamma}(s,\theta) &= \log\frac{1}{\{e^{ks}+e^{-ks}-2\rho\cos(k\theta+\gamma)\}^2}\\ &= \log\frac{1}{4\{\cosh ks - \rho\cos(k\theta+\gamma)\}^2}.\end{aligned} \tag{16}$$

Here $k \in \mathbb{N}$, $\rho \in (0, 1)$ and $\gamma \in S^1$. The parameters ρ and γ represent a dilation and a rotation respectively.

The bifurcation diagram for the above solutions are drawn in Fig. 2. As is seen from the figure, non-radially symmetric solutions bifurcate from radially symmetric solutions when $(A, B) = (8k^2, 2k)$, $k = 1, 2, 3, \ldots$, and they blow up at the vertices of k-sided regular polygons as the dilation parameter ρ tends to 1. This suggests that the bifurcating non-radially symmetric solutions of (3) connect to the solutions obtained in [12] and [3], as we expected in Sect. 1.

A natural way to get a deeper understanding of the global structure of the solution set of (14) is to determine the connected component containing the solutions obtained above. Our result gives an answer to this question.

Before stating the result, we introduce some notation. What we need to take care of is the topology of the space of solutions since solutions of (14) diverge as $|s| \to \infty$ because of the boundary condition. The growth rate which solutions of (14) have is linear growth, and therefore it is appropriate to take the function $1/(1+|s|)$ as a weight function. Based on this consideration, we define the function space X by

$$X := \left\{ u \in C^2(\mathbb{R} \times S^1);\quad \sup_{(s,\theta)\in\mathbb{R}\times S^1} \frac{|u(s,\theta)|}{1+|s|} < \infty \right\}$$

and the norm as

$$\|u\|_X := \sup_{(s,\theta)\in\mathbb{R}\times S^1} \frac{|u(s,\theta)|}{1+|s|}.$$

(We note that X is not complete.) Let $\mathscr{S} \subset (0,\infty) \times [2,\infty) \times X$ be the set of all solutions of (14) and $\mathscr{C} \subset \mathscr{S}$ be the union of the set of solutions (15) and (16), that is,

$$\mathscr{C} := \{(8K^2, 2K, U_K)\}_{K\ge 1} \cup \{(8k^2(1-\rho^2), 2k, U_{k,\rho,\gamma})\}_{k\in\mathbb{N},\, \rho\in(0,1),\, \gamma\in S^1}.$$

Then it is easily seen that $\mathscr{C}$ is connected in $(0,\infty) \times [2,\infty) \times X$.

We are now in a position to state the main result in this subsection. The following theorem shows that there is no bifurcation point on $\mathscr{C}$ except for $\{(8k^2, 2k, U_k)\}_{k\in\mathbb{N}}$. Hence, in particular, it can be concluded that the connected component of $\mathscr{S}$ containing $\mathscr{C}$ is $\mathscr{C}$ itself.

Theorem 1 *There do not exist $(A_0, B_0, u_0) \in \mathscr{C}$ and $\{(A_n, B_n, u_n)\}_{n=1}^{\infty} \subset \mathscr{S} \setminus \mathscr{C}$ such that $(A_n, B_n, u_n) \to (A_0, B_0, u_0)$ in $\mathbb{R} \times \mathbb{R} \times X$ as $n \to \infty$.*

The main key and also the main difficulty in proving Theorem 1 is to determine the null space of the linearized operator $\Delta_{(s,\theta)} + Ae^u$ in X. This is because if (A, B, u) is a bifurcation point, there is a nontrivial function Φ such that $\Delta_{(s,\theta)}\Phi + Ae^u\Phi = 0$ and $\Phi(s,\theta) = b|s| + o(1)$ for some $b \in \mathbb{R}$ as $|s| \to \infty$. We discuss this problem in the next subsection.

3.2 Linearized Equations

We consider the following linear homogeneous equation:

$$\Delta_{(s,\theta)}\Phi + Ae^u\Phi = 0, \quad (s,\theta) \in \mathbb{R} \times S^1. \tag{17}$$

Here (A, B, u) is a solution of (14). As mentioned in the previous subsection, nontrivial solutions of (17) play a crucial role to prove the nonexistence of a bifurcation point in (14).

When $(A, B, u) = (8K^2, 2K, U_K)$, it is easily seen that the functions

$$\Phi_{K,1}(s) = \tanh Ks, \qquad \Phi_{K,2}(s) = Ks \tanh Ks - 1$$

are solutions of (17) in X. Moreover, if $K = k \in \mathbb{N}$, (17) has the following solutions coming from the symmetry-breaking bifurcation:

$$\Phi_{k,3}(s,\theta) = \frac{\cos k\theta}{\cosh ks}, \qquad \Phi_{k,4}(s,\theta) = \frac{\sin k\theta}{\cosh ks}.$$

In the case where $(A, B, u) = (8k^2(1-\rho^2), 2k, U_{k,\rho,\gamma})$, we can find the following three solutions:

$$\Phi_{k,\rho,\gamma,1}(s,\theta) = \frac{\sinh ks}{\cosh ks - \rho\cos(k\theta+\gamma)},$$
$$\Phi_{k,\rho,\gamma,2}(s,\theta) = \frac{\cos(k\theta+\gamma) - \rho\cosh ks}{\cosh ks - \rho\cos(k\theta+\gamma)},$$
$$\Phi_{k,\rho,\gamma,3}(s,\theta) = \frac{\sin(k\theta+\gamma)}{\cosh ks - \rho\cos(k\theta+\gamma)}.$$

These solutions represent a translation, a dilation, and a rotation respectively.

The first result in this subsection shows that the solution set of (17) in X is spanned by the above functions.

Theorem 2 *Let $\Phi \in X$ be a solution of* (17). *Then Φ is a linear combination of the following functions*:

(i) $\Phi_{K,1}$ *and* $\Phi_{K,2}$ *if* $(A, B, u) = (8K^2, 2K, U_K)$, $K \notin \mathbb{N}$;
(ii) $\Phi_{k,1}$, $\Phi_{k,2}$, $\Phi_{k,3}$ *and* $\Phi_{k,4}$ *if* $(A, B, u) = (8k^2, 2k, U_k)$, $k \in \mathbb{N}$;
(iii) $\Phi_{k,\rho,\gamma,1}$, $\Phi_{k,\rho,\gamma,2}$ *and* $\Phi_{k,\rho,\gamma,3}$ *if* $(A, B, u) = (8k^2(1-\rho^2), 2k, U_{k,\rho,\gamma})$.

The most difficult case is (iii) since in this case we cannot apply the method of separation of variables. Our approach to handling this case is to consider the nonhomogeneous equation

$$L[\Psi] := \Delta_{(s,\theta)}\Psi + \frac{2k^2(1-\rho^2)}{\{\cosh ks - \rho\cos(k\theta+\gamma)\}^2}\Psi = f(s,\theta), \quad (s,\theta) \in \mathbb{R}\times S^1. \tag{18}$$

Fortunately, we can obtain a particular solution of this equation. Specifically, a fundamental solution for L is found.

Let Γ_j $(j = 0, 1, 2)$ be a function defined by

$$\Gamma_0(s,\theta,t,\tau) := \frac{1}{4\pi}\log 2(\cosh r - \cos\sigma),$$
$$\Gamma_1(s,\theta,t,\tau) := \left(1-\rho^2\right)\Phi_{k,\rho,\gamma,1}(s,\theta)\Phi_{k,\rho,\gamma,1}(t,\tau) + \Phi_{k,\rho,\gamma,2}(s,\theta)\Phi_{k,\rho,\gamma,2}(t,\tau)$$
$$+ \left(1-\rho^2\right)\Phi_{k,\rho,\gamma,3}(s,\theta)\Phi_{k,\rho,\gamma,3}(t,\tau),$$

$$\Gamma_2(s,\theta,t,\tau)$$
$$:= \frac{1}{4\pi}\Bigg[1 - \rho ks\Phi_{k,\rho,\gamma,2}(s,\theta)\Phi_{k,\rho,\gamma,1}(t,\tau)$$

$$- \frac{1-\rho^2}{\rho}\left(ks\Phi_{k,\rho,\gamma,1}(s,\theta) - \frac{1}{1-\rho^2}\right)\Phi_{k,\rho,\gamma,2}(t,\tau)$$
$$+ \cosh ks \sin(k\theta+\gamma)\Phi_{k,\rho,\gamma,3}(t,\tau)$$
$$\left. + \sum_{j=1}^{k-1}\left\{ \frac{k}{j}\frac{\cosh jr\cos(k-j)\sigma - \rho\cosh(kt+(k-j)r)\cos(k-j)\sigma}{\{\cosh ks - \rho\cos(k\theta+\gamma)\}\{\cosh kt - \rho\cos(k\tau+\gamma)\}}\right\}\right],$$

where $s, t \in \mathbb{R}, \theta, \tau \in S^1, r = s - t$ and $\sigma = \theta - \tau$. Then $\Gamma_0\Gamma_1 + \Gamma_2$ is a fundamental solution for L. In fact, the following theorem holds.

Theorem 3 *Suppose that $f \in C^\infty(\mathbb{R}\times S^1)$ has a compact support in $\mathbb{R}\times S^1$. Then the function Ψ defined by*

$$\Psi(s,\theta) := \int_0^{2\pi}\int_{-\infty}^{\infty}\big\{\Gamma_0(s,\theta,t,\tau)\Gamma_1(s,\theta,t,\tau) + \Gamma_2(s,\theta,t,\tau)\big\} f(t,\tau)\,dt\,d\tau$$

is a (classical) solution of (18). *Furthermore, if f is perpendicular to $\Phi_{k,\rho,\gamma,3}$ in $L^2(\mathbb{R}\times S^1)$, then $\Psi(s,\theta) = O(|s|)$, $\nabla_{(s,\theta)}\Psi(s,\theta) = O(1)$ as $|s|\to\infty$.*

This theorem gives us a way to show that a solution of (17) decaying exponentially as $|s|\to\infty$ must be parallel to $\Phi_{k,\rho,\gamma,3}$. For details, see the proof of Theorem 2 in the next section.

Note After Submission

After completing this work, the author learned that solutions of (14) has been already classified by Prajapat and Tarantello [13]. By their result, it can be concluded that $\mathscr{S} = \mathscr{C}$. Furthermore, the author also learned that the homogeneous equation (17) was considered by del Pino, Esposito and Musso [2]. They showed Theorem 2(iii) under the condition that $\Phi \in L^\infty(\mathbb{R}\times S^1)$.

In our study, a relation between Eqs. (1)–(2) and (3) is found from the view point of a singular limit of a domain. Moreover, we obtain a fundamental solution for the linearized operator L and give a proof of Theorem 2(iii) in a different way.

4 Proofs of Theorems

4.1 Proofs of Theorems 2 and 3

First we prove Theorem 3.

Proof of Theorem 3 For the sake of simplicity, we only deal with the case where $\gamma = 0$. Another case can be treated in the same way.

Put $x=(e^s\cos\theta, e^s\sin\theta)$ and $y=(e^t\cos\tau, e^t\sin\tau)$. Then we have $\Gamma_0(s,\theta,t,\tau) = \frac{1}{4\pi}\log\frac{|x-y|^2}{|x||y|}$, that is, Γ_0 corresponds to a fundamental solution of Laplace's equation in terms of the original variables. From this, we have

$$\begin{aligned}L[\Psi] &= \Gamma_1(s,\theta,s,\theta)f(s,\theta)\\&\quad+2\int_0^{2\pi}\int_{-\infty}^{\infty}\nabla_{(s,\theta)}\Gamma_0(s,\theta,t,\tau)\cdot\nabla_{(s,\theta)}\Gamma_1(s,\theta,t,\tau)f(t,\tau)\,dt\,d\tau\\&\quad+\int_0^{2\pi}\int_{-\infty}^{\infty}\Gamma_0(s,\theta,t,\tau)L\bigl[\Gamma_1(s,\theta,t,\tau)\bigr]f(t,\tau)\,dt\,d\tau\\&\quad+\int_0^{2\pi}\int_{-\infty}^{\infty}L\bigl[\Gamma_2(s,\theta,t,\tau)\bigr]f(t,\tau)\,dt\,d\tau.\end{aligned}$$

Now we calculate each term of the above equality. Since

$$\begin{aligned}\Gamma_1(s,\theta,s,\theta) &= \bigl(1-\rho^2\bigr)\Phi_1(s,\theta)^2+\Phi_2(s,\theta)^2+\bigl(1-\rho^2\bigr)\Phi_3(s,\theta)^2\\&=\frac{(1-\rho^2)\sinh^2 ks+(\cos k\theta-\rho\cosh ks)^2+(1-\rho^2)\sin^2 k\theta}{(\cosh ks-\rho\cos k\theta)^2}\\&=1,\end{aligned}$$

the first term equals to $f(s,\theta)$. The integrand of the second term is calculated as

$$\begin{aligned}&\nabla_{(s,\theta)}\Gamma_0(s,\theta,t,\tau)\cdot\nabla_{(s,\theta)}\Gamma_1(s,\theta,t,\tau)\\&=\frac{\sinh r}{4\pi(\cosh r-\cos\sigma)}\biggl\{\frac{(1-\rho^2)\sinh kt}{\cosh kt-\rho\cos k\tau}\cdot\frac{k(1-\rho\cosh ks\cos k\theta)}{(\cosh ks-\rho\cos k\theta)^2}\\&\quad-\frac{\cos k\tau-\rho\cosh kt}{\cosh kt-\rho\cos k\tau}\cdot\frac{k(1-\rho^2)\sinh ks\cos k\theta}{(\cosh ks-\rho\cos k\theta)^2}\\&\quad-\frac{(1-\rho^2)\sin k\tau}{\cosh kt-\rho\cos k\tau}\cdot\frac{k\sinh ks\sin k\theta}{(\cosh ks-\rho\cos k\theta)^2}\biggr\}\\&\quad+\frac{\sin\sigma}{4\pi(\cosh r-\cos\sigma)}\biggl\{\frac{(1-\rho^2)\sinh kt}{\cosh kt-\rho\cos k\tau}\cdot\frac{-k\rho\sinh ks\sin k\theta}{(\cosh ks-\rho\cos k\theta)^2}\\&\quad-\frac{\cos k\tau-\rho\cosh kt}{\cosh kt-\rho\cos k\tau}\cdot\frac{k(1-\rho^2)\cosh ks\sin k\theta}{(\cosh ks-\rho\cos k\theta)^2}\\&\quad+\frac{(1-\rho^2)\sin k\tau}{\cosh kt-\rho\cos k\tau}\cdot\frac{k(\cosh ks\cos k\theta-\rho)}{(\cosh ks-\rho\cos k\theta)^2}\biggr\}\\&=\frac{k(1-\rho^2)\sinh r}{4\pi(\cosh r-\cos\sigma)(\cosh kt-\rho\cos k\tau)(\cosh ks-\rho\cos k\theta)^2}\\&\quad\times\bigl\{\sinh kt(1-\rho\cosh ks\cos k\theta)\\&\quad-(\cos k\tau-\rho\cosh kt)\sinh ks\cos k\theta-\sin k\tau\sinh ks\sin k\theta\bigr\}\\&\quad+\frac{k(1-\rho^2)\sin\sigma}{4\pi(\cosh r-\cos\sigma)(\cosh kt-\rho\cos k\tau)(\cosh ks-\rho\cos k\theta)^2}\end{aligned}$$

$$
\begin{aligned}
&\times \big\{-\rho \sinh kt \sinh ks \sin k\theta \\
&- (\cos k\tau - \rho \cosh kt) \cosh ks \sin k\theta + \sin k\tau (\cosh ks \cos k\theta - \rho)\big\} \\
= &\frac{k(1-\rho^2)}{4\pi(\cosh r - \cos\sigma)(\cosh kt - \rho\cos k\tau)(\cosh ks - \rho\cos k\theta)^2} \\
&\times (\sinh r \sinh kt + \rho \sinh r \sinh kr \cos k\theta - \sinh r \sinh ks \cos k\sigma \\
&+ \rho \cosh kr \sin k\theta \sin\sigma - \cosh ks \sin k\sigma \sin\sigma - \rho \sin k\tau \sin\sigma).
\end{aligned}
$$

The third term vanishes since $\Phi_{k,\rho,\gamma,j}$ satisfies $L[\Phi_{k,\rho,\gamma,j}] = 0$ for $j = 1,2,3$. Finally we compute the last term. A straightforward computation yields

$$
\begin{aligned}
L\big[s\Phi_2(s,\theta)\big] &= 2(\Phi_2)_s(s,\theta) = \frac{-2k(1-\rho^2)\sinh ks \cos k\theta}{(\cosh ks - \rho\cos k\theta)^2}, \\
L\left[ks\Phi_1(s,\theta) - \frac{1}{1-\rho^2}\right] &= 2k(\Phi_1)_s(s,\theta) - \frac{2k^2}{(\cosh ks - \rho\cos k\theta)^2} \\
&= \frac{-2k^2\rho\cosh ks \cos k\theta}{(\cosh ks - \rho\cos k\theta)^2}.
\end{aligned}
$$

Furthermore,

$$
\begin{aligned}
&L\left[\frac{\cosh jr \cos(k-j)\sigma}{\cosh ks - \rho\cos k\theta}\right] \\
&= \Delta_{(s,\theta)}\big\{\cosh jr \cos(k-j)\sigma\big\} \cdot \frac{1}{\cosh ks - \rho\cos k\theta} \\
&\quad + 2\nabla_{(s,\theta)}\big\{\cosh jr \cos(k-j)\sigma\big\} \cdot \nabla_{(s,\theta)}\left\{\frac{1}{\cosh ks - \rho\cos k\theta}\right\} \\
&\quad + \cosh jr \cos(k-j)\sigma \cdot L\left\{\frac{1}{\cosh ks - \rho\cos k\theta}\right\} \\
&= \frac{(-k^2 + 2kj)\cosh jr \cos(k-j)\sigma}{\cosh ks - \rho\cos k\theta} \\
&\quad + \frac{2\rho k(k-j)\cosh jr \sin k\theta \sin(k-j)\sigma - 2kj \sinh jr \sinh ks \cos(k-j)\sigma}{(\cosh ks - \rho\cos k\theta)^2} \\
&\quad + \frac{k^2\cosh jr \cos(k-j)\sigma(\cosh ks + \rho\cos k\theta)}{(\cosh ks - \rho\cos k\theta)^2} \\
&= \frac{2\rho k(k-j)\cosh jr \cos(k\tau + j\sigma) + 2kj\cosh\{kt + (k-j)r\}\cos(k-j)\sigma}{(\cosh ks - \rho\cos k\theta)^2},
\end{aligned}
$$

and in a similar manner,

$$
\begin{aligned}
&L\left[\frac{\cosh\{kt + (k-j)r\}\cos(k\tau + j\sigma)}{\cosh ks - \rho\cos k\theta}\right] \\
&= \frac{2k(k-j)\cosh jr \cos(k\tau + j\sigma) + 2\rho kj\cosh\{kt + (k-j)r\}\cos(k-j)\sigma}{(\cosh ks - \rho\cos k\theta)^2}.
\end{aligned}
$$

Hence we have

$$
\begin{aligned}
&L\big[\Gamma_2(s,\theta,t,\tau)\big]\\
&\quad=\frac{2k^2(1-\rho^2)}{(\cosh ks-\rho\cos k\theta)^2}+\frac{2k^2\rho(1-\rho^2)\sinh ks\cos k\theta}{(\cosh ks-\rho\cos k\theta)^2}\cdot\frac{\sinh kt}{\cosh kt-\rho\cos k\tau}\\
&\qquad+\frac{2k^2(1-\rho^2)\cosh ks\cos k\theta}{(\cosh ks-\rho\cos k\theta)^2}\cdot\frac{\cos k\tau-\rho\cosh kt}{\cosh kt-\rho\cos k\tau}\\
&\qquad+\frac{2k^2(1-\rho^2)}{(\cosh ks-\rho\cos k\theta)^2(\cosh kt-\rho\cos k\tau)}\\
&\qquad\times\sum_{j=1}^{k-1}\cosh\big\{kt+(k-j)r\big\}\cos(k-j)\sigma\\
&\qquad-\frac{2k^2\rho(1-\rho^2)}{(\cosh ks-\rho\cos k\theta)^2(\cosh kt-\rho\cos k\tau)}\sum_{j=1}^{k-1}\cosh jr\cos(k\tau+j\sigma)\\
&\quad=\frac{2k^2(1-\rho^2)}{(\cosh ks-\rho\cos k\theta)^2(\cosh kt-\rho\cos k\tau)}\big\{S_1(s,t,\theta,\tau)-\rho S_2(s,t,\theta,\tau)\big\},
\end{aligned}
$$

where

$$S_1(s,t,\theta,\tau):=\cosh kt+\cosh ks\cos k\sigma+2\sum_{j=1}^{k-1}\cosh(kt+jr)\cos j\sigma,$$

$$S_2(s,t,\theta,\tau):=\cos k\tau+\cosh kr\cos k\theta+2\sum_{j=1}^{k-1}\cosh jr\cos(k\tau+j\sigma).$$

Therefore if we prove the equalities

$$S_1(s,t,\theta,\tau)=\frac{\cosh ks\sin k\sigma\sin\sigma+\sinh r\sinh ks\cos k\sigma-\sinh r\sinh kt}{\cosh r-\cos\sigma}, \tag{19}$$

$$S_2(s,t,\theta,\tau)=\frac{\sinh r\sinh kr\cos k\theta+\cosh kr\sin k\theta\sin\sigma-\sinh k\tau\sin\sigma}{\cosh r-\cos\sigma}, \tag{20}$$

the assertion follows.

(19) and (20) are shown in the following way. From the formula $\sum_{j=0}^{k}e^{jz}=(e^{(k+1)z}-1)/(e^z-1)$, $z\in\mathbb{C}$, we have

$$
\begin{aligned}
2\sum_{j=0}^{k}\cosh jr\cos j\sigma&=\mathrm{Re}\left(\sum_{j=0}^{k}e^{jz}+\sum_{j=0}^{k}e^{-jz}\right)\Bigg|_{z=r+\sqrt{-1}\sigma}\\
&=\mathrm{Re}\left(\frac{e^{(k+1)z}-1}{e^z-1}+\frac{e^{-(k+1)z}-1}{e^{-z}-1}\right)\Bigg|_{z=r+\sqrt{-1}\sigma}\\
&=\mathrm{Re}\left(\frac{e^{(k+1)z}-e^{-kz}}{e^z-1}+1\right)\Bigg|_{z=r+\sqrt{-1}\sigma}
\end{aligned}
$$

$$
\begin{aligned}
&= \frac{(e^{(k+1)r}\cos(k+1)\sigma - e^{-kr}\cos k\sigma)(e^r\cos\sigma - 1)}{e^{2r}+1-2e^r\cos\sigma} \\
&\quad + \frac{(e^{(k+1)r}\sin(k+1)\sigma + e^{-kr}\sin k\sigma)e^r\sin\sigma}{e^{2r}+1-2e^r\cos\sigma} + 1 \\
&= \frac{\sinh kr\sinh r\cos k\sigma + \cosh kr\sin k\sigma\sin\sigma}{\cosh r - \cos\sigma} \\
&\quad + \cosh kr\cos k\sigma + 1.
\end{aligned}
$$

A similar calculation gives

$$
\begin{aligned}
&2\sum_{j=0}^{k}\sinh jr\cos j\sigma \\
&= \frac{\cosh kr\sinh r\cos k\sigma + \sinh kr\sin k\sigma\sin\sigma - \sinh r}{\cosh r - \cos\sigma} + \sinh kr\cos k\sigma.
\end{aligned}
$$

Thus

$$
\begin{aligned}
&2\sum_{j=0}^{k}\cosh(kt+jr)\cos j\sigma \\
&\quad = 2\cosh kt\sum_{j=0}^{k}\cosh jr\cos j\sigma + 2\sinh kt\sum_{j=0}^{k}\sinh jr\cos j\sigma \\
&\quad = \frac{\sinh ks\sinh r\cos k\sigma + \cosh ks\sin k\sigma\sin\sigma - \sinh r\sinh kt}{\cosh r - \cos\sigma} \\
&\quad\quad + \cosh ks\cos k\sigma + \cosh kt,
\end{aligned}
$$

which implies (19). We can also show (20) in a similar way.

We assume that $\langle f, \Phi_{k,\rho,\gamma,3}\rangle = 0$, and show $\Psi = O(|s|)$, $\nabla_{(s,\theta)}\Psi = O(1)$ as $|s| \to \infty$. From [8, Theorem 2.1], we see that

$$
\begin{aligned}
&\left|\int_0^{2\pi}\int_{-\infty}^{\infty}\Gamma_0(s,\theta,t,\tau)\Gamma_1(s,\theta,t,\tau)f(t,\tau)\,dt\,d\tau\right| \\
&\quad \le C\int_0^{2\pi}\int_{-\infty}^{\infty}\bigl|\Gamma_0(s,\theta,t,\tau)f(t,\tau)\bigr|\,dt\,d\tau \\
&\quad \le C\bigl(1+|s|\bigr)\int_0^{2\pi}\int_{-\infty}^{\infty}\bigl(1+|t|\bigr)\bigl|f(t,\tau)\bigr|\,dt\,d\tau
\end{aligned}
$$

for large $|s|$. Here C is a positive constant independent of (s,θ). Moreover,

$$
\begin{aligned}
&\bigl|4\pi\Gamma_2(s,\theta,t,\tau) - \cosh ks\sin k\theta\,\Phi_{k,\rho,\gamma,3}(t,\tau)\bigr| \\
&\quad \le C\bigl(1+|s|\bigr) + C\sum_{j=1}^{k-1}\frac{\cosh js\cosh jt + \rho\cosh(k-j)s\cosh jt}{(\cosh ks - \rho\cos k\theta)(\cosh kt - \rho\cos k\tau)} \\
&\quad \le C\bigl(1+|s|\bigr),
\end{aligned}
$$

and hence, from $\langle f, \Phi_{k,\rho,\gamma,3}\rangle = 0$,

$$\left|\int_0^{2\pi}\int_{-\infty}^{\infty} \Gamma_2(s,\theta,t,\tau) f(t,\tau)\,dt\,d\tau\right| \le C\big(1+|s|\big)\int_0^{2\pi}\int_{-\infty}^{\infty}\big|f(t,\tau)\big|\,dt\,d\tau.$$

Thus we have $\Psi = O(|s|)$ as $|s| \to \infty$. Boundedness of the gradient of Ψ is obtained by Lemma 1. □

Remark 1 f does not need to have a compact support since Γ_1, Γ_2 are smooth and do not grow as $|t| \to \infty$. It is sufficient to assume that f has exponential decay as $|s| \to \infty$.

Next we prove Theorem 2.

Proof of Theorem 2 First we show (i) and (ii). Let $\varphi_j(s)$, $\psi_j(s)$ denote the Fourier coefficients of Φ with respect to θ, that is, $\Phi(s,\theta) = \varphi_0(s)/2 + \sum_{j=1}^{\infty}(\varphi_j(s)\cos\theta + \psi_j(s)\sin\theta)$. Then φ_j satisfies

$$(\varphi_j)_{ss} + \left(\frac{2K^2}{\cosh^2 Ks} - j^2\right)\varphi_j = 0, \quad s \in \mathbb{R}. \tag{21}$$

For $j = 0$, the functions $\Phi_{K,1}$ and $\Phi_{K,2}$ satisfy (21), and hence they are the fundamental system. If $j \neq 0$ and $j \neq K$, we can check that the functions

$$\varphi_{j,1}(s) := \frac{(K-j)e^{(K+j)s} - (K+j)e^{-(K-j)s}}{\cosh Ks},$$
$$\varphi_{j,2}(s) := \frac{(K+j)e^{(K-j)s} - (K-j)e^{-(K+j)s}}{\cosh Ks}$$

are linearly independent solutions of (21). These functions diverge exponentially as $s \to \infty$ or $s \to -\infty$, and hence $\varphi_{j,1}, \varphi_{j,2} \notin X$. This implies that $\varphi_j \equiv 0$ if $j \neq 0$, $j \neq K$. Since ψ_j satisfies the same equation as φ_j, we also have $\psi_j \equiv 0$ for $j \neq 0$, $j \neq K$. Thus (i) holds. In the case where $K = k \in \mathbb{N}$, the fundamental solutions of (21) for $j = k$ are given by

$$\varphi_{k,1}(s) := \frac{1}{\cosh ks}, \quad \varphi_{k,2}(s) := \frac{ks}{\cosh ks} + \sinh ks.$$

Since $\varphi_{k,1} \in X$ and $\varphi_{k,2} \notin X$, (ii) follows.

Next we verify (iii). From (17), $\Delta_{(s,\theta)}\Phi = -8k^2 e^{U_{k,\rho,\gamma}}\Phi = O(e^{-\mu_0|s|})$ as $|s| \to \infty$ for arbitrary fixed $0 < \mu_0 < 2k$. Hence Lemma 1 shows that there are a positive constant μ and real numbers c_+, c_-, d_+, d_- such that $\Phi(s,\theta) - c_\pm s - d_\pm = O(e^{-\mu|s|})$, $\Phi_s(s,\theta) - c_\pm = O(e^{-\mu|s|})$ as $s \to \pm\infty$. Furthermore it is easily seen that $\Phi_{k,\rho,\gamma,1}(s,\theta) = \pm 1 + O(e^{-2k|s|})$, $\Phi_{k,\rho,\gamma,2}(s,\theta) = 1 + O(e^{-2k|s|})$, $(\Phi_{k,\rho,\gamma,j})_s(s,\theta) = O(e^{-2k|s|})$ $(j = 1, 2)$ as $s \to \pm\infty$. Therefore multiplying both sides of (17) by $\Phi_{k,\rho,\gamma,j}$ and integrating over $\mathbb{R} \times (0, 2\pi)$ yield

$$\begin{aligned} 0 &= \int_0^{2\pi}\big[\Phi_s(s,\theta)\Phi_{k,\rho,\gamma,j}(s,\theta) - \Phi(s,\theta)(\Phi_{k,\rho,\gamma,j})_s(s,\theta)\big]_{s=-\infty}^{\infty}\,d\theta \\ &= \begin{cases} 2\pi(c_+ + c_-) & \text{if } j = 1, \\ 2\pi(c_+ - c_-) & \text{if } j = 2. \end{cases} \end{aligned}$$

From this we have $c_+ = c_- = 0$. Let $c_1, c_2 \in \mathbb{R}$ satisfy $c_1 + c_2 = d_+$, $-c_1 + c_2 = d_-$ and put $c_3 = \langle \Phi - c_1\Phi_{k,\rho,\gamma,1} - c_2\Phi_{k,\rho,\gamma,2}, \Phi_{k,\rho,\gamma,3}\rangle / \|\Phi_{k,\rho,\gamma,3}\|_{L^2(\mathbb{R}\times(0,2\pi))}$, where $\langle\cdot,\cdot\rangle$ denote the inner product in $L^2(\mathbb{R}\times(0,2\pi))$. Then $\hat{\Phi} := \Phi - c_1\Phi_{k,\rho,\gamma,1} - c_2\Phi_{k,\rho,\gamma,2} - c_3\Phi_{k,\rho,\gamma,3}$ satisfies $\hat{\Phi}(s,\theta) = O(e^{-\mu|s|})$ (as $|s| \to \infty$) and $\langle\hat{\Phi}, \Phi_{k,\rho,\gamma,3}\rangle = 0$. Now we prove that $\hat{\Phi} \equiv 0$. To this end we consider the equation $\Delta_{(s,\theta)}\Psi + 8k^2(1-\rho^2)e^{U_{k,\rho,\gamma}}\Psi = \hat{\Phi}$. According to Proposition 3 and Remark 1, this equation has a solution Ψ such that $\Psi(s,\theta) = O(|s|)$, $\Psi_s(s,\theta) = O(1)$ as $|s| \to \infty$. Therefore, by multiplying the above equation by $\hat{\Phi}$ and integrating over $\mathbb{R}\times(0,2\pi)$, we have $\|\hat{\Phi}\|^2_{L^2(\mathbb{R}\times(0,2\pi))} = 0$. Thus we conclude that $\hat{\Phi} \equiv 0$, which gives (iii). □

4.2 Proof of Theorem 1

What we have to take care of in the proof of Theorem 1 is the rotational invariance of the equation, that is, the fact that $\hat{\Phi} := \partial U_{k,\rho,\gamma}/\partial\gamma$ satisfies $\Delta_{(s,\theta)}\hat{\Phi} + 8k^2(1-\rho^2)e^{U_{k,\rho,\gamma}}\hat{\Phi} = 0$. To avoid this difficulty, we decompose the function $u \in X$ which is in a neighborhood of $U_{k,\rho,\gamma}$ as $u = U_{k,\tilde{\rho},\tilde{\gamma}} + V$, where $(\tilde{\rho},\tilde{\gamma})$ is taken so that $\langle V, \hat{\Phi}\rangle = 0$, and we consider V instead of u. This approach is based on [10].

Proof of Theorem 1 We prove the theorem by contradiction. Suppose that there exist a solution $(A_0, B_0, u_0) \in \mathscr{C}$ and a sequence $\{(A_n, B_n, u_n)\}_{n=1}^{\infty} \subset \mathscr{S} \setminus \mathscr{C}$ such that $(A_n, B_n, u_n) \to (A_0, B_0, u_0)$ in $\mathbb{R}\times\mathbb{R}\times X$ as $n \to \infty$. Then the following three cases are possible:

(i) $(A_0, B_0, u_0) = (8K^2, 2K, U_K)$ for some $K \geq 1$, $K \notin \mathbb{N}$;
(ii) $(A_0, B_0, u_0) = (8k^2, 2k, U_k)$ for some $k \in \mathbb{N}$;
(iii) $(A_0, B_0, u_0) = (8k^2(1-\rho^2), 2k, U_{k,\rho,\gamma})$ for some $k \in \mathbb{N}$, $\rho \in (0,1)$, $\gamma \in S^1$.

We first consider (i). Let $K_n \geq 1$ satisfy $A_n = 8K_n^2$ and put

$$(b_n, v_n) = \frac{(B_n - 2K_n, u_n - U_{K_n})}{|B_n - 2K_n| + \|u_n - U_{K_n}\|_X}.$$

Then $K_n \to K$, $|b_n| + \|v_n\|_X = 1$ and

$$\begin{cases} \Delta_{(s,\theta)}v_n + 8K_n^2 e^{U_{K_n}}\left\{\displaystyle\int_0^1 e^{\eta(u_n - U_{K_n})}d\eta\right\}v_n = 0, & (s,\theta) \in \mathbb{R}\times S^1, \\ v_n(s,\theta) = -b_n|s| + o(1) & \text{as } |s| \to \infty. \end{cases} \tag{22}$$

From the fact that $u_n - U_{K_n} = (u_n - U_K) - (U_{K_n} - U_K) \to 0$ in X as $n \to \infty$, we have, for any fixed $0 < \mu_0 < 2K$,

$$|\Delta_{(s,\theta)}v_n| \leq 8K_n^2 e^{U_{K_n}} e^{|u_n - U_{K_n}|}|v_n| \leq Ce^{-2K_n|s|}e^{\|u_n - U_{K_n}\|_X|s|}\big(1+|s|\big) \leq Ce^{-\mu_0|s|}$$

provided that n is sufficiently large. Here C denotes a positive constant independent of s, θ and n. Therefore the L^p estimate for the Laplacian and the Sobolev embedding theorem show that a subsequence of $\{v_n\}$ converges locally uniformly on $\mathbb{R} \times S^1$ to a solution v of the equation

$$\Delta_{(s,\theta)} v + 8K^2 e^{U_K} v = 0, \quad (s,\theta) \in \mathbb{R} \times S^1.$$

Furthermore, from Lemma 1, there are constants $C > 0$, $\mu > 0$ such that $|v_n(s,\theta) + b_n|s|| \leq Ce^{-\mu|s|}$ for all $(s,\theta) \in \mathbb{R} \times S^1$ and large n. Hence, by taking a subsequence if necessary, we see that $|v(s,\theta) + b|s|| \leq Ce^{-\mu|s|}$, where $b := \lim_{n\to\infty} b_n \in [-1,1]$. On the other hand, Proposition 2 implies that $v = c_1 \Phi_{K,1} + c_2 \Phi_{K,2}$ for some $c_1, c_2 \in \mathbb{R}$. In particular, we have $v(s,\theta) = kc_2|s| \pm c_1 - c_2 + o(1)$ as $s \to \pm\infty$. This leads to $kc_2 = -b$, $c_1 - c_2 = -c_1 - c_2 = 0$, and hence $b = c_1 = c_2 = 0$, $v \equiv 0$. Moreover,

$$\begin{aligned}
\limsup_{n\to\infty} \|v_n\|_X &\leq \limsup_{n\to\infty} \left(\sup_{|s|\leq R,\, \theta\in S^1} \frac{|v_n(s,\theta)|}{1+|s|} + \sup_{|s|\geq R,\, \theta\in S^1} \frac{|v_n(s,\theta)|}{1+|s|} \right) \\
&\leq \limsup_{n\to\infty} \left(\sup_{|s|\geq R, \theta\in S^1} \frac{|v_n(s,\theta) + b_n|s||}{1+|s|} + |b_n| \right) \\
&\leq Ce^{-\mu R} \to 0 \quad (R \to \infty).
\end{aligned}$$

Thus $\lim_{n\to\infty}(|b_n| + \|v_n\|_X) = 0$, which contradicts the fact that $|b_n| + \|v_n\|_X = 1$.

Next we consider (ii), which is the most difficult case. We derive a contradiction in two steps. In Step 1, we show that for some $c_2 \neq 0$,

$$u_n = U_k + c_2(\Phi_{k,2} + 1)\delta_n + o(\delta_n) \tag{23}$$

in X as $n \to \infty$. Here $\delta_n := |A_n - 8k^2| + |B_n - 2k| + \|u_n - U_k\|_X$. On the other hand, in Step 2, we derive another formula

$$u_n = U_k + (c_3 \Phi_{k,3} + c_4 \Phi_{k,4})\delta_n + o(\delta_n), \tag{24}$$

where c_3, c_4 are some real constants which satisfy $(c_3, c_4) \neq (0,0)$. As a consequence of Steps 1 and 2, we have $c_2(\Phi_{k,2} + 1) = c_3 \Phi_{k,3} + c_4 \Phi_{k,4}$. From this a contradiction is derived, because multiplying by $\Phi_{k,j}$ $(j = 3,4)$ and integrating over $\mathbb{R} \times (0, 2\pi)$ yield $c_2 = c_3 = c_4 = 0$. Therefore what we need to show is (23) and (24).

Step 1. We first verify (23). For any large n, we can choose $\rho_n \in [0,1)$, $\gamma_n \in S^1$ such that

$$\langle u_n - U_{k,\rho_n,\gamma_n}, \Phi_{k,j} \rangle = 0 \quad \text{for } j = 3,4 \tag{25}$$

and $|\rho_n| \leq C\|u_n - U_k\|_X$ for some positive constant C independent of n. Indeed, this can be shown as follows. Let $h_n : (-1,1) \times (-1,1) \ni (\xi,\zeta) \mapsto h_n(\xi,\zeta) \in \mathbb{R}^2$ be defined by the relation

$$\begin{gathered}
h_n(\xi,\zeta) = \big(\langle U_{k,\rho,\gamma} - u_n, \Phi_{k,3} \rangle, \langle U_{k,\rho,\gamma} - u_n, \Phi_{k,4} \rangle \big), \\
(\xi,\zeta) = (\rho\cos\gamma, -\rho\sin\gamma).
\end{gathered}$$

Then h_n is of class C^1 near $(\xi,\zeta) = (0,0)$ and we have

$$\begin{aligned}|h_n(0,0)| &= \left|\left(\langle U_k - u_n, \Phi_{k,3}\rangle, \langle U_k - u_n, \Phi_{k,4}\rangle\right)\right| \\ &\le \left\{\left\|(1+|s|)\Phi_{k,3}\right\|_{L^1(\mathbb{R}\times(0,2\pi))} + \left\|(1+|s|)\Phi_{k,4}\right\|_{L^1(\mathbb{R}\times(0,2\pi))}\right\} \\ &\quad \times \|U_k - u_n\|_X,\end{aligned}$$

$$\begin{aligned}&\lim_{\substack{\xi,\zeta\to 0\\ n\to\infty}} \frac{\partial h_n}{\partial(\xi,\zeta)}(\xi,\zeta) \\ &= \left.\begin{pmatrix} \langle \frac{\partial U_{k,\rho,\gamma}}{\partial \xi}, \Phi_{k,3}\rangle & \langle \frac{\partial U_{k,\rho,\gamma}}{\partial \zeta}, \Phi_{k,3}\rangle \\ \langle \frac{\partial U_{k,\rho,\gamma}}{\partial \xi}, \Phi_{k,4}\rangle & \langle \frac{\partial U_{k,\rho,\gamma}}{\partial \zeta}, \Phi_{k,4}\rangle \end{pmatrix}\right|_{(\xi,\zeta)=(\rho\cos\gamma,-\rho\sin\gamma)=(0,0)} \\ &= \begin{pmatrix} \langle 2\Phi_{k,3}, \Phi_{k,3}\rangle & \langle 2\Phi_{k,4}, \Phi_{k,3}\rangle \\ \langle 2\Phi_{k,3}, \Phi_{k,4}\rangle & \langle 2\Phi_{k,4}, \Phi_{k,4}\rangle \end{pmatrix} = \begin{pmatrix} 4\pi/k & 0 \\ 0 & 4\pi/k \end{pmatrix}.\end{aligned}$$

Since the above matrix is invertible, a similar way to prove the implicit function theorem shows that for any large n, there exists (ξ_n, ζ_n) such that $h_n(\xi_n, \zeta_n) = 0$ and $(|\xi_n|^2 + |\zeta_n|^2)^{1/2} \le C|h_n(0,0)|$. The corresponding (ρ_n, γ_n), that is, (ρ_n, γ_n) which satisfies $(\xi_n, \zeta_n) = (\rho_n \cos\gamma_n, -\rho_n \sin\gamma_n)$, has the desired properties.

We put $\tilde{\delta}_n := |A_n - 8k^2(1-\rho_n^2)| + |B_n - 2k| + \|u_n - U_{k,\rho_n,\gamma_n}\|_X$ and

$$(a_n, b_n, v_n) := \tilde{\delta}_n^{-1}\left(A_n - 8k^2(1-\rho_n^2), B_n - 2k, u_n - U_{k,\rho_n,\gamma_n}\right).$$

Then we see at once that (a_n, b_n, v_n) satisfies $|a_n| + |b_n| + \|v_n\|_X = 1$ and

$$\begin{cases} \Delta_{(s,\theta)} v_n + A_n e^{U_{k,\rho_n,\gamma_n}} \left\{\int_0^1 e^{\eta(u_n - U_{k,\rho_n,\gamma_n})}\, d\eta\right\} v_n + a_n e^{U_{k,\rho_n,\gamma_n}} = 0, \\ (s,\theta) \in \mathbb{R}\times S^1, \\ v_n(s,\theta) = -b_n|s| + o(1) \quad \text{as } |s| \to \infty. \end{cases} \tag{26}$$

Then, as in the case of (i), it can be shown that a subsequence of $\{(a_n, b_n, v_n)\}$ (we denote it by the same notation) converges to a solution (a, b, v) of

$$\begin{cases} \Delta_{(s,\theta)} v + 8k^2 e^{U_k} v + a e^{U_k} = 0, & (s,\theta) \in \mathbb{R}\times S^1, \\ v(s,\theta) = -b|s| + o(1) & \text{as } |s| \to \infty \end{cases} \tag{27}$$

in $\mathbb{R}\times\mathbb{R}\times X$. Since (27) is rewritten as $\Delta_{(s,\theta)}(v + a/(8k^2)) + 8k^2 e^{U_k}(v + a/(8k^2)) = 0$, we see from Proposition 2 that $v + a/(8k^2) = c_1\Phi_{k,1} + c_2\Phi_{k,2} + c_3\Phi_{k,3} + c_4\Phi_{k,4}$ for some $c_1, c_2, c_3\ c_4 \in \mathbb{R}$. From (25), v also satisfies $\langle v, \Phi_{k,j}\rangle = 0$ for $j = 3, 4$, and therefore $c_3 = c_4 = 0$. Furthermore, by considering the limit as $s \to \pm\infty$, we have $kc_2 = -b$, $c_1 - c_2 - a/(8k^2) = -c_1 - c_2 - a/(8k^2) = 0$. This gives $c_1 = 0$, $c_2 = -a/(8k^2) = -b/k$ and $v = c_2(\Phi_{k,2} + 1)$. In particular, this implies that $c_2 \ne 0$ since $|a| + |b| + \|v\|_X = 1$.

Now we show that $\rho_n = (\xi_n^2 + \zeta_n^2)^{1/2} = o(\tilde{\delta}_n)$. A simple calculation yields

$$\begin{aligned} U_{k,\rho_n,\gamma_n} &= U_k + \left.\left(\xi_n \frac{\partial U_{k,\rho,\gamma}}{\partial \xi} + \zeta_n \frac{\partial U_{k,\rho,\gamma}}{\partial \zeta}\right)\right|_{(\xi,\zeta)=(\rho\cos\gamma,-\rho\sin\gamma)=(0,0)} + o(\rho_n) \\ &= U_k + 2\xi_n\Phi_{k,3} + 2\zeta_n\Phi_{k,4} + o(\rho_n) \end{aligned} \tag{28}$$

in X as $n \to \infty$. Hence by multiplying both sides of (26) by $\Phi_{k,3}$ and integrating over $\mathbb{R}\times(0,2\pi)$, we have

$$
\begin{aligned}
0 &= \left\langle \Delta_{(s,\theta)} v_n + A_n e^{U_{k,\rho_n,\gamma_n}} \left\{ \int_0^1 e^{\eta(u_n - U_{k,\rho_n,\gamma_n})} \, d\eta \right\} v_n + a_n e^{U_{k,\rho_n,\gamma_n}}, \Phi_{k,3} \right\rangle \\
&= \left\langle -8k^2 e^{U_k} v_n + \left(8k^2 + a\tilde{\delta}_n\right) e^{U_k + 2\xi_n \Phi_{k,3} + 2\zeta_n \Phi_{k,4}} \left\{ \int_0^1 e^{\eta \tilde{\delta}_n v} \, d\eta \right\} v_n \right. \\
&\quad \left. + a_n e^{U_k + 2\xi_n \Phi_{k,3} + 2\zeta_n \Phi_{k,4}}, \Phi_{k,3} \right\rangle + o(\rho_n) + o(\tilde{\delta}_n) \\
&= \left\langle e^{U_k} \left\{ (2\xi_n \Phi_{k,3} + 2\zeta_n \Phi_{k,4})\left(8k^2 v + a\right) + \tilde{\delta}_n \left(4k^2 v + a\right) v + a_n \right\}, \Phi_{k,3} \right\rangle \\
&\quad + o(\rho_n) + o(\tilde{\delta}_n) \\
&= 16k^2 c_2 \xi_n \left\langle e^{U_k} \Phi_{k,2} \Phi_{k,3}, \Phi_{k,3} \right\rangle + o(\rho_n) + o(\tilde{\delta}_n) \\
&= -16k\pi c_2 \xi_n + o(\rho_n) + o(\tilde{\delta}_n).
\end{aligned}
$$

In a similar way, multiplying by $\Phi_{k,4}$ and integrating by parts give $0 = -16k\pi c_2 \zeta_n + o(\rho_n) + o(\tilde{\delta}_n)$. Thus $\rho_n = (\xi_n^2 + \zeta_n^2)^{1/2} = o(\tilde{\delta}_n)$. Furthermore, from this and (28), we have

$$
\begin{aligned}
\left|A_n - 8k^2(1 - \rho_n^2)\right| &= \left|A_n - 8k^2\right| + O\left(\rho_n^2\right) = \left|A_n - 8k^2\right| + o(\tilde{\delta}_n), \\
\|u_n - U_{k,\rho_n,\gamma_n}\|_X &= \|u_n - U_k\|_X + O(\rho_n) = \|u_n - U_k\|_X + o(\tilde{\delta}_n).
\end{aligned}
$$

This implies that $\tilde{\delta}_n = (1 + o(1))\delta_n$. Consequently,

$$
u_n = U_{k,\rho_n,\gamma_n} + \tilde{\delta}_n v + o(\tilde{\delta}_n) = U_k + c_2(\Phi_{k,2} + 1)\delta_n + o(\delta_n).
$$

Step 2. We show (24). We take $K_n \geq 1$ so that $A_n = 8K_n^2$ and we put $\hat{\delta}_n := |B_n - 2K_n| + \|u_n - U_{K_n}\|_X$, $(b_n, v_n) = \hat{\delta}_n^{-1}(B_n - 2K_n, u_n - U_{K_n})$. Then (b_n, v_n) satisfies $|b_n| + \|v_n\|_X = 1$ and the same equation as (22). Moreover, in a similar way to the case of (i), we may assume that in $\mathbb{R} \times X$, (b_n, v_n) converges to $(b, v) \in \mathbb{R} \times X$ which satisfies

$$
\begin{cases} \Delta_{(s,\theta)} v + 8k^2 e^{U_k} v = 0, & (s,\theta) \in \mathbb{R} \times S^1, \\ v(s,\theta) = -b|s| + o(1) & \text{as } |s| \to \infty. \end{cases} \tag{29}
$$

According to Proposition 2, $v = c_1 \Phi_{k,1} + c_2 \Phi_{k,2} + c_3 \Phi_{k,3} + c_4 \Phi_{k,4}$ for some c_1, c_2, c_3, $c_4 \in \mathbb{R}$. By comparing the limit as $s \to \pm\infty$, we have $b = c_1 = c_2 = 0$, and from $|b| + \|v\|_X = 1$, we see that $(c_3, c_4) \neq 0$.

We now prove that $\kappa_n := K_n - k = o(\hat{\delta}_n)$. It is a simple manner to show that

$$
U_{K_n} = U_k + \kappa_n \left. \frac{\partial U_K}{\partial K} \right|_{K=k} + o(\kappa_n) = U_k - \frac{2}{k}(\Phi_{k,2} + 1)\kappa_n + o(\kappa_n) \tag{30}
$$

in X as $n \to \infty$. From this, multiplying both sides of (22) by $\Phi_{k,3}$ and integrating over $\mathbb{R} \times (0, 2\pi)$ yields

$$
\begin{aligned}
0 &= \left\langle \Delta_{(s,\theta)} v_n + 8K_n^2 e^{U_{K_n}} \left\{ \int_0^1 e^{\eta(u_n - U_{K_n})} \, d\eta \right\} v_n, \Phi_{k,3} \right\rangle \\
&= \left\langle -8k^2 e^{U_k} v_n + \left(8k^2 + 2k\kappa_n\right) e^{U_k - 2/k(\Phi_{k,2}+1)} \left\{ \int_0^1 e^{\eta \hat{\delta}_n v} \, d\eta \right\} v_n, \Phi_{k,3} \right\rangle
\end{aligned}
$$

$$
\begin{aligned}
&+o(\kappa_n)+o(\hat{\delta}_n)\\
&=\big\langle e^{U_k}\big(-16k\kappa_n\Phi_{k,2}v+4k^2\hat{\delta}_n v^2\big),\Phi_{k,3}\big\rangle+o(\kappa_n)+o(\hat{\delta}_n)\\
&=-16kc_3\kappa_n\big\langle e^{U_k}\Phi_{k,2}\Phi_{k,3},\Phi_{k,3}\big\rangle+o(\kappa_n)+o(\hat{\delta}_n)\\
&=\big(16\pi c_3+o(1)\big)\kappa_n+o(\hat{\delta}_n).
\end{aligned}
$$

Similarly, by replacing $\Phi_{k,3}$ with $\Phi_{k,4}$, we have $0=(16\pi c_4+o(1))\kappa_n+o(\hat{\delta}_n)$. Since either c_3 or c_4 does not vanish, we obtain $\kappa_n=o(\hat{\delta}_n)$. Furthermore, from this and (30),

$$
\begin{aligned}
\big|A_n-8k^2\big|&=8|K_n+k|\kappa_n=o(\hat{\delta}_n),\\
\|u_n-U_k\|_X=\|u_n-U_{K_n}\|_X&+O(\kappa_n)=\|u_n-U_{K_n}\|_X+o(\hat{\delta}_n).
\end{aligned}
$$

This gives $\hat{\delta}_n=(1+o(1))\delta_n$. Thus we conclude that

$$
u_n=U_{K_n}+\hat{\delta}_n v+o(\hat{\delta}_n)=U_k+(c_3\Phi_{k,3}+c_4\Phi_{k,4})\delta_n+o(\delta_n).
$$

As a consequence of Step 1 and Step 2, we see that the case of (ii) is impossible.

Finally we assume (iii) and derive a contradiction. Let $\rho_n\in(0,1)$ satisfy $A_n=8k^2(1-\rho_n^2)$. Then $\rho_n\to\rho$ as $n\to\infty$. In a similar way to (ii), we can show that for any large n, there exists $\gamma_n\in S^1$ such that $\gamma_n\to\gamma$ and $\langle U_{k,\rho_n,\gamma_n}-u_n,\Phi_{k,\rho,\gamma,3}\rangle=0$. Indeed, such γ_n is found in the following way. We define $\tilde{h}_n\in C^1(\mathbb{R})$ as $\tilde{h}_n(\tilde{\gamma}):=\langle U_{k,\rho_n,\tilde{\gamma}}-u_n,\Phi_{k,\rho,\gamma,3}\rangle$. Then we have

$$
\begin{aligned}
\big|h_n(\gamma)\big|&\le\big\|(1+|s|)\Phi_{k,\rho,\gamma,3}\big\|_{L^1(\mathbb{R}\times(0,2\pi))}\\
&\quad\times\big(\|U_{k,\rho_n,\gamma}-U_{k,\rho,\gamma}\|_X+\|U_{k,\rho,\gamma}-u_n\|_X\big)\\
&\to 0\quad(n\to\infty),\\
\lim_{\substack{\tilde{\gamma}\to\gamma\\ n\to\infty}}\frac{\partial h_n}{\partial\tilde{\gamma}}(\tilde{\gamma})&=\left\langle\frac{\partial U_{k,\rho,\tilde{\gamma}}}{\partial\tilde{\gamma}}\bigg|_{\tilde{\gamma}=\gamma},\Phi_{k,\rho,\gamma,3}\right\rangle=\langle-2\rho\Phi_{k,\rho,\gamma,3},\Phi_{k,\rho,\gamma,3}\rangle\\
&=\frac{-4\pi\rho(1+\rho^2)}{k(1-\rho^2)^2}\ne 0.
\end{aligned}
$$

Thus, in a similar manner to prove the implicit function theorem, we get γ_n which we find.

We put

$$
(b_n,v_n)=\frac{(B_n-2k,u_n-U_{k,\rho_n,\gamma_n})}{|B_n-2k|+\|u_n-U_{k,\rho_n,\gamma_n}\|_X}.
$$

As is the case in (i), by taking a subsequence if necessary, we see that $(b_n,v_n)\to(b,v)$ in $\mathbb{R}\times X$ as $n\to\infty$, and (b,v) must satisfy $|b|+\|v\|_X=1$, $\langle v,\Phi_{k,\rho,\gamma,3}\rangle=0$ and the same equation as (29). This is impossible, because Proposition 2 implies that $b=0$, $v\equiv 0$.

In all the cases a contradiction is derived, and the proof is complete. □

Acknowledgements The author would like to express his thanks to Professor Irai Shafrir and Professor Juan Dávila for calling his attention to Refs. [2, 13].

Appendix

In this appendix, we prove the following lemma which is used in Sect. 4.

Lemma 1 *Let* $\tilde{\Phi} \in X$ *and suppose that there are positive constants* C_0, μ_0 *such that the inequality* $|\Delta_{(s,\theta)}\tilde{\Phi}(s,\theta)| \le C_0 e^{-\mu_0|s|}$ *holds for all* $(s,\theta) \in \mathbb{R} \times S^1$. *Then there exist positive constants* C, μ, s_0 *and real numbers* c_+, c_-, d_+, d_- *such that*

$$\left|\tilde{\Phi}(s,\theta) - c_\pm s - d_\pm\right| \le Ce^{-\mu|s|}, \tag{31}$$

$$\left|\tilde{\Phi}_s(s,\theta) - c_\pm\right| + \left|\tilde{\Phi}_\theta(s,\theta)\right| \le Ce^{-\mu|s|} \tag{32}$$

for all $\pm s \ge s_0$ *and* $\theta \in S^1$.

Proof Let $\tilde{\varphi}_0(s)/2 + \sum_{j=1}^{\infty}(\tilde{\varphi}_j(s)\cos j\theta + \tilde{\psi}_j(s)\sin j\theta)$ be the Fourier series of $\tilde{\Phi}$ with respect to θ and put $F(s,\theta) := \Delta_{(s,\theta)}\tilde{\Phi}(s,\theta)$. Then $\tilde{\varphi}_j$ satisfies

$$(\tilde{\varphi}_j)_{ss} - j^2\tilde{\varphi}_j = f_j(s), \quad s \in \mathbb{R},$$

where $f_j(s) = \pi^{-1}\int_0^{2\pi} F(s,\tau)\cos j\tau d\tau$. In particular, we have $(\tilde{\varphi}_0)_{ss}(s) = O(e^{-\mu_0|s|})$ as $|s| \to \infty$. From this it follows easily that for some constants c_+, c_-, d_+, d_-, $\varphi_0(s)/2 - c_\pm s - d_\pm = O(e^{-\mu_0|s|})$, $(\varphi_0)_s(s)/2 - c_\pm = O(e^{-\mu_0|s|})$ as $s \to \pm\infty$. We estimate $\tilde{\varphi}_j$ for $j \ge 1$. Since $\tilde{\Phi} \in X$, $\tilde{\varphi}_j$ does not have exponential growth. Hence $\tilde{\varphi}_j$ is given by

$$\tilde{\varphi}_j(s) = -\frac{1}{2j}\int_{-\infty}^{\infty} e^{-j|s-t|} f_j(t)\,dt.$$

From the assumption, we have

$$\begin{aligned}
\left|\tilde{\varphi}_j(s)\right| &\le \frac{C_0}{j}\int_{-\infty}^{\infty} e^{-j|s-t|-\mu_0|t|}\,dt = \frac{2C_0}{j(j^2-\mu_0^2)}\left(je^{-\mu_0|s|} - \mu_0 e^{-j|s|}\right) \\
&= \frac{2C_0}{j(j+\mu_0)}\left(e^{-\mu_0|s|} + \mu_0|s|\int_0^1 e^{-\{(1-\eta)j+\eta\mu_0\}|s|}\,d\eta\right) \\
&\le \frac{2C_0}{j^2}\left(1+\mu_0|s|\right)e^{-\min\{\mu_0,1\}|s|}.
\end{aligned}$$

Therefore we see that for any fixed $0 < \mu < \min\{\mu_0, 1\}$, there exist a constant $C > 0$ independent of s and j such that $|\tilde{\varphi}_j(s)| \le Cj^{-2}e^{-\mu|s|}$. Since the same estimate holds also for $\tilde{\psi}_j$, we have $|\tilde{\Phi}(s,\theta) - \tilde{\varphi}_0(s)/2| \le C(\sum_{j=1}^{\infty} j^{-2})e^{-\mu|s|}$. Thus (31) follows.

We verify (32). Since

$$(\tilde{\varphi}_j)_s(s) = \frac{1}{2}\int_0^{2\pi}\int_{-\infty}^{\infty} \frac{s-t}{|s-t|}e^{-j|s-t|}F(t,\tau)\cos j\tau\,dt\,d\tau,$$

$$(\tilde{\psi}_j)_s(s) = \frac{1}{2}\int_0^{2\pi}\int_{-\infty}^{\infty} \frac{s-t}{|s-t|}e^{-j|s-t|}F(t,\tau)\sin j\tau\,dt\,d\tau$$

for $j \geq 1$, we have

$$\left|\tilde{\Phi}_s(s,\theta) - \frac{(\tilde{\varphi}_0)_s(s)}{2}\right| = \left|\sum_{j=1}^{\infty}\big((\tilde{\varphi}_j)_s(s)\cos j\theta + (\tilde{\psi}_j)_s(s)\sin j\theta\big)\right|$$
$$\leq \frac{1}{2}\int_0^{2\pi}\int_{-\infty}^{\infty}\left|\sum_{j=1}^{\infty} e^{-j|s-t|}\cos j(\theta-\tau)\right|\big|F(t,\tau)\big|\,dt\,d\tau$$
$$\leq \frac{C_0}{2}\int_0^{2\pi}\int_{-\infty}^{\infty}\left|\mathrm{Re}\left(\frac{1}{e^z-1}\right)\right| e^{-\mu_0|t|}\,dt\,d\tau.$$

Here $z = |s-t| + \sqrt{-1}(\theta-\tau)$. In a similar way,

$$\big|\tilde{\Phi}_\theta(s,\theta)\big| = \left|\sum_{j=1}^{\infty} j\big(-\tilde{\varphi}_j(s)\sin j\theta + \tilde{\psi}_j(s)\cos j\theta\big)\right|$$
$$\leq \frac{1}{2}\int_0^{2\pi}\int_{-\infty}^{\infty}\left|\sum_{j=1}^{\infty} e^{-j|s-t|}\sin j(\theta-\tau)\right|\big|F(t,\tau)\big|\,dt\,d\tau$$
$$\leq \frac{C_0}{2}\int_0^{2\pi}\int_{-\infty}^{\infty}\left|\mathrm{Im}\left(\frac{1}{e^z-1}\right)\right| e^{-\mu_0|t|}\,dt\,d\tau.$$

These estimates yield

$$\left|\tilde{\Phi}_s(s,\theta) - \frac{(\tilde{\varphi}_0)_s(s)}{2}\right| + \big|\tilde{\Phi}_\theta(s,\theta)\big|$$
$$\leq C_0\int_0^{2\pi}\int_{-\infty}^{\infty}\left|\frac{1}{e^z-1}\right| e^{-\mu_0|t|}\,dt\,d\tau$$
$$= C_0\int_0^{2\pi}\int_{-\infty}^{\infty}\frac{e^{-\mu_0|t|}}{\{e^{2|s-t|} - 2e^{|s-t|}\cos(\theta-\tau)+1\}^{1/2}}\,dt\,d\tau$$
$$\leq C_0\int_0^{2\pi}\int_{-\infty}^{\infty}\frac{e^{-\mu|s-t|}}{(e^{2|t|} - 2e^{|t|}\cos\tau+1)^{1/2}}\,dt\,d\tau$$
$$\leq C_0 e^{-\mu|s|}\int_0^{2\pi}\int_{-\infty}^{\infty}\frac{e^{\mu|t|}}{(e^{2|t|} - 2e^{|t|}\cos\tau+1)^{1/2}}\,dt\,d\tau,$$

where $0 < \mu < \min\{\mu_0, 1\}$. The last integral of the above inequality is finite, and consequently (32) is verified. □

References

1. Dancer, E.N.: Global breaking of symmetry of positive solutions on two-dimensional annuli. Differ. Integral Equ. **5**, 903–913 (1992)
2. del Pino, M., Esposito, P., Musso, M.: Nondegeneracy of entire solutions of a singular Liouville equation. Proc. Am. Math. Soc. **140**, 581–588 (2012)

3. del Pino, M., Kowalczyk, M., Musso, M.: Singular limits in Liouville-type equations. Calc. Var. Partial Differ. Equ. **24**, 47–81 (2005)
4. Gidas, B., Ni, W.M., Nirenberg, L.: Symmetry and related properties via the maximum principle. Commun. Math. Phys. **68**, 209–243 (1979)
5. Holzmann, M., Kielhöfer, H.: Uniqueness of global positive solution branches of nonlinear elliptic problems. Math. Ann. **300**, 221–241 (1994)
6. Kielhöfer, H.: Bifurcation Theory. An Introduction with Applications to PDEs. Applied Mathematical Sciences, vol. 156, Springer, New York (2004), viii+346 pp.
7. Lin, S.-S.: On non-radially symmetric bifurcation in the annulus. J. Differ. Equ. **80**, 251–279 (1989)
8. Masaki, S.: Local existence and WKB approximation of solutions to Schrodinger-Poisson system in the two-dimensional whole space. Commun. Partial Differ. Equ. **35**, 2253–2278 (2010)
9. Miyamoto, Y.: Global branches of non-radially symmetric solutions to a semilinear Neumann problem in a disk. J. Funct. Anal. **256**, 747–776 (2009)
10. Miyamoto, Y.: Non-existence of a secondary bifurcation point for a semilinear elliptic problem in the presence of symmetry. J. Math. Anal. Appl. **357**, 89–97 (2009)
11. Miyamoto, Y.: Nondegeneracy of the second bifurcating branches for the Chafee-Infante problem on a planar symmetric domain. Proc. Am. Math. Soc. **139**, 975–984 (2011)
12. Nagasaki, K., Suzuki, T.: Radial and nonradial solutions for the nonlinear eigenvalue problem $\Delta u + \lambda e^u = 0$ on annuli in $\mathbb{R}^2$. J. Differ. Equ. **87**, 144–168 (1990)
13. Prajapat, J., Tarantello, G.: On a class of elliptic problems in $\mathbb{R}^2$: symmetry and uniqueness results. Proc. R. Soc. Edinb., Sect. A **131**, 967–985 (2001)

A Priori Estimates and Comparison Principle for Some Nonlinear Elliptic Equations

Anna Mercaldo

Abstract We present a priori estimates and comparison principle for second order quasilinear elliptic operators in divergence form with a first order term. We deduce existence and uniqueness results for weak solutions or "solution obtained as limit of approximations" to Dirichlet problems related to these types of operators when data belong to suitable Lorentz spaces. Moreover it is also shown how the summability of these solutions increases when the summability of the datum increases.

Keywords A priori estimates · Existence · Comparison principle · Uniqueness · Nonlinear elliptic operators

1 Introduction

Let us consider the homogeneous Dirichlet problem

$$\begin{cases} -\text{div}\big(\mathbf{a}(x,u,\nabla u)\big) = H(x,\nabla u) + f & \text{in } \Omega \\ u = 0 & \text{on } \partial\Omega, \end{cases} \tag{1}$$

where Ω is a bounded open subset of $\mathbf{R}^N$, $N \geq 2$. We assume that

$$\mathbf{a} : \Omega \times \mathbf{R} \times \mathbf{R}^N \to \mathbf{R}^N$$

and

$$H : \Omega \times \mathbf{R}^N \to \mathbf{R}$$

are Carathéodory functions which satisfy the ellipticity condition

$$\mathbf{a}(x,s,\xi)\cdot\xi \geq \alpha|\xi|^p, \quad \alpha > 0, \tag{2}$$

the monotonicity condition

$$\big(\mathbf{a}(x,s,\xi) - \mathbf{a}(x,s,\eta)\big)\cdot(\xi-\eta) > 0, \quad \xi \neq \eta, \tag{3}$$

A. Mercaldo (✉)
Dipartimento di Matematica e Applicazioni "R. Caccioppoli", Università di Napoli Federico II, Complesso di Monte S. Angelo, 80126 Napoli, Italy
e-mail: mercaldo@unina.it

R. Magnanini et al. (eds.), *Geometric Properties for Parabolic and Elliptic PDE's*, Springer INdAM Series 2, DOI 10.1007/978-88-470-2841-8_14,

and the growth conditions

$$\big|\mathbf{a}(x,s,\xi)\big| \le a_0|\xi|^{p-1} + a_1|s|^{p-1} + a_2, \quad a_0, a_1, a_2 > 0, \tag{4}$$

$$\big|H(x,\xi)\big| \le h|\xi|^q, \quad h > 0, \tag{5}$$

with $1 < p < +\infty$, $p - 1 < q < p$ for almost every $x \in \mathbf{R}^N$, for every $s \in \mathbf{R}$, for every $\xi, \eta \in \mathbf{R}^N$. Finally we assume that f belongs to suitable Lorentz spaces.

The purpose of the present note is to announce some recent results dealing with existence, uniqueness and regularity for solutions to problem (1). The notion of solution to which we refer depends on the summability of the datum. We use the classical weak solution when the datum f is an element of $W^{-1,p'}(\Omega)$, the dual space of $W_0^{1,p}(\Omega)$, while, if f is not in such a space, a different notion of solution has to be adopted. Various equivalent notion of solutions are available in literature, i.e. renormalized solution [24, 27], entropy solution [6] or "solution obtained as limit of approximations" ([15], see also [16]). We will refer to "solution obtained as limit of approximations", whose definition is based on a procedure of passage to the limit which we recall in Sect. 2.

Existence results for solution to problem (1) are well-known in literature (see e.g. [18, 20, 21]). Their proofs are based on a priori estimates for solutions and its gradients which are the first step of a standard approach, whose main idea consists in a passage to the limit in a sequence of approximated problems having regular data. By using the a priori estimates, one can prove that weak solutions to such problems converge in a suitable sense to a function which is the solution to problem (1). A priori estimates for weak solutions or renormalized solutions to (1) are proved in [18] and in [21]; in these papers the solutions belong to a class of functions which satisfy further summability conditions. In Sect. 2 we present some sharp a priori estimates for weak solution or "solution obtained as limit of approximations" proved in [5] when the datum f belongs to sharp Lorentz spaces. The optimality of our results are shown in [5] by examples concerning the radial case. The proofs of a priori estimates are based on the choice of a suitable test function, built on the level sets of the solution to (1), and classical symmetrization methods introduced by Talenti and Maz'ya (see, for instance, [29] and [26]). Further sharp estimates are also proved in [25].

A few words on the bounds $p - 1 < q < p$ on the growth of the function H are in order. The existence of weak solutions to problem (1) when $q = p - 1$ is well-known; it is a consequence of the theory of monotone operators (see, for example, [23]) at least when h is small enough. When h is large this framework is not applicable, since in general the elliptic operator is not coercive. Nevertheless a priori estimates, and therefore existence results, for weak solutions have been proved in various papers [14, 17, 29] and the existence of a "solution obtained as limit of approximations" or a renormalized solution is obtained, for instance, in [2] and [9].

The study of the existence in the limit case $q = p$ has been faced by various authors, who proved a priori estimates for bounded weak solutions or unbounded weak solutions satisfying further regularity conditions. Here we just recall [19] and references therein.

For what concerns the regularity of solutions to (1), in Sect. 2 we show an example of our results proved in [11]. Starting from a pointwise estimate of the rearrangement of the gradient of a solution, we study how the summability of the gradient of a solution increases when the summability of the datum f, belonging to Lorentz spaces, increases. This type of study is faced also in [21] and in [12] when the datum f is in Lebesgue spaces. The previous pointwise estimate is obtained by adapting the classical symmetrization methods; an analogous result is proved in [3].

As far as the uniqueness concerns we consider elliptic operators which satisfy some further conditions. Firstly we assume that the function $\mathbf{a}$ does not depend on u, that is we consider Dirichlet problems of the type:

$$\begin{cases} -\mathrm{div}\big(\mathbf{a}(x,\nabla u)\big) = H(x,\nabla u) + f & \text{in } \Omega \\ u = 0 & \text{on } \partial\Omega. \end{cases} \tag{6}$$

We change the monotonicity condition (3) in the following "strong monotonicity" condition

$$\big(\mathbf{a}(x,\xi) - \mathbf{a}(x,\eta)\big)\cdot(\xi-\eta) \ge \beta\big(\varepsilon + |\xi| + |\eta|\big)^{p-2}|\xi-\eta|^2, \tag{7}$$

for some $\beta > 0$, with ε nonnegative and strictly positive if $p > 2$. Moreover we assume the following locally Lipschitz condition on H

$$\big|H(x,\xi) - H(x,\eta)\big| \le \gamma\big(\eta + |\xi| + |\eta|\big)^{q-1}|\xi-\eta|, \tag{8}$$

where $\gamma > 0$, η is nonnegative and strictly positive if $1 < p < 2$.

When the growth condition on H is satisfied with $q = p - 1$, uniqueness results for weak solutions have been proved in [4, 7, 10], while uniqueness of a renormalized solution or "solution obtained as limit of approximations" are proved, for example, in [2, 8] and [28]. In some cases these results are consequence of the continuity with respect to the data, as in the papers [2, 7], in some other cases they are obtained by proving a comparison principle, as in [4, 28]. In Sect. 3 we present comparison principles proved in [11] for a weak solution or a "solution obtained as limit of approximations" to problem (6) when $p - 1 < q < p$. Actually we assume $q \le \frac{p}{N} + p - 1$ since classical examples show that uniqueness for weak solutions does not hold when $q > \frac{p}{N} + p - 1$. These results complete the study started in [4] in the case where $q = p - 1$ and the datum f is an element of the dual space $W^{-1,p'}(\Omega)$. Our results improve the comparison principle proved in [28] for renormalized solutions.

2 A Priori Estimates

When dealing with the problem of existence of solutions to problem (1) some necessary conditions are required on the data. In [1] and [22] explicit criteria for the existence are given by involving geometric capacitary terms for the Laplacian operator, i.e. $p = 2$. When the datum f belongs to some Lebesgue space, such condition means that $f \in L^m(\Omega)$, $m \geq \frac{N(q-1)}{q}$ and that its norm is sufficiently small. For arbitrary p the sharp condition in Lebesgue spaces is $f \in L^m(\Omega)$, $m \geq \frac{N(q-p+1)}{q}$. In [5] existence results are proven when the datum f belongs to the Lorentz space $L(\frac{N(q-p+1)}{q}, 1)$. This is the optimal space when one refers to "intermediate" spaces between Lebesgue spaces.

We recall that a function f belongs to the Lorentz space $L(m,k)$ if it is a measurable functions on Ω such that the quantity

$$\|f\|_{m,k} = \left(\int_0^{|\Omega|} [f^*(s)s^{\frac{1}{m}}]^k \frac{ds}{s} \right)^{\frac{1}{k}}$$

is finite. Here f^* denotes the decreasing rearrangement of f.

The Lorentz space $L(m,\infty)$, $m > 1$, coincides with the Marcinkiewicz space, that is the set of all measurable functions on Ω such that the quantity

$$\|f\|_{m,\infty} = \sup_{s\in(0,|\Omega|)} \frac{f^*(s)}{s^{\frac{1}{m}}}$$

is finite.

These spaces are refinements of Lebesgue spaces in the sense that $L(m,m) = L^m(\Omega)$ for any $m > 1$. Moreover fixed the first index m, they become larger as the second index increases, i.e.

$$L(m,k_1) \hookrightarrow L(m,k_2) \quad \text{when } m > 0,\ k_1 \leq k_2$$

while, if Ω is bounded, we have

$$L(m_1,k_1) \hookrightarrow L(m_2,k_2) \quad \text{when } m_1 > m_2,\ k_1, k_2 > 0.$$

In order to discuss the existence of solutions, we have to distinguish three ranges for the values of q in the growth condition on the function H, i.e.

$$p - 1 + \frac{p}{N} \leq q < p,$$
$$\frac{N(p-1)}{N-1} \leq q < p - 1 + \frac{p}{N},$$
$$p - 1 \leq q < \frac{N(p-1)}{N-1}.$$

Moreover we assume $p < N$ even if the case $p \geq N$ is also studied in [5].

When $p - 1 + \frac{p}{N} \leq q < p$, we assume that f belongs to the Lorentz space $L(\frac{N(q-p+1)}{q}, 1)$. Observe that the critical value of summability of f, $\frac{N(q-p+1)}{q}$, is greater then $(p^*)' = \frac{Np}{N(p-1)+p}$. This means that data belonging to this Lorentz space are in turn elements of $W^{-1,p'}(\Omega)$, the dual space of $W_0^{1,p}(\Omega)$. This leads to consider the standard notion of weak solution. A function $u \in W_0^{1,p}(\Omega)$ is a weak solution to problem (1) if

$$\int_\Omega \mathbf{a}(x,u,\nabla u) \cdot \nabla\varphi\, dx = \int_\Omega H(x,\nabla u)\varphi\, dx + \int_\Omega f\varphi\, dx, \tag{9}$$

for every $\varphi \in W_0^{1,p}(\Omega) \cap L^\infty(\Omega)$.

Our first result gives a priori estimates for bounded weak solutions to problem (1) with regular data. This is the fundamental step of a limit procedure which allows to prove the existence of a weak solution to (1) and which we briefly recall below. The details of the proof of a priori estimates stated in Theorem 1 below are given in [5].

Theorem 1 *Assume* (2)–(5) *with*

$$p-1+\frac{p}{N}\leq q<p.$$

Let $u\in W_0^{1,p}(\Omega)\cap L^\infty(\Omega)$ *be a weak solution to problem* (1) *with* $f\in C^\infty$. *If the norm of* f *in* $L(\frac{N(q-p+1)}{q},1)$ *is small enough, then*

$$\|\nabla u\|_{L^p}\leq C,$$
$$\|u\|_{t^*,p}\leq C,$$

for every $t<N(q-p+1)$, *where* C *is a positive constant which depends only on* $p,q,N,|\Omega|$ *and on the norm of the datum* f.

As consequence of these a priori estimates, in a standard way, we obtain the following existence result for weak solutions to problem (1). We just give a sketch of its proof (see [5, 21]).

Theorem 2 *Assume* (2)–(5) *with*

$$p-1+\frac{p}{N}\leq q<p.$$

If the norm of f *in* $L(\frac{N(q-p+1)}{q},1)$ *is small enough, then at least a weak solution* u *exists such that* $u\in L(t^*,p)$, *for every* $t<N(q-p+1)$.

Proof Let $(f_n)_n$ be a sequence of C^∞-functions which converges strongly to f in $L(\frac{N(q-p+1)}{q},1)$. We can assume that the $L(\frac{N(q-p+1)}{q},1)$ norm of f_n is small enough. Let $u_n\in W_0^{1,p}(\Omega)\cap L^\infty(\Omega)$ be a weak solution to the approximated problem

$$\begin{cases}-\operatorname{div}\big(\mathbf{a}(x,u_n,\nabla u_n)\big)=T_n\big(H(x,\nabla u_n)\big)+f_n & \text{in } \Omega,\\ u_n=0 & \text{on } \partial\Omega\end{cases}\tag{10}$$

i.e.

$$\int_\Omega \mathbf{a}(x,u_n,\nabla u_n)\cdot\nabla\phi\,dx=\int_\Omega T_n\big(H(x,\nabla u_n)\big)\phi\,dx+\int_\Omega f_n\phi\,dx\tag{11}$$

for every $\phi\in W_0^{1,p}(\Omega)$.

Here T_n denotes the usual truncation at level $n>0$, i.e. defined, for a given $n>0$, as

$$T_n(s)=\begin{cases}s & \text{if } |s|\leq n\\ n\,\operatorname{sign} s & \text{if } |s|>n.\end{cases}$$

The existence of the approximated solutions u_n is assured by classical results on monotone operators (see [23]). By Theorem 1 we deduce that u_n are bounded in $W_0^{1,p}(\Omega)$. Therefore there exists a subsequence, which we denote $(u_n)_n$, such that

$$u_n \to u \qquad \text{weakly in } W_0^{1,p}(\Omega), \text{ strongly in } L^p(\Omega), \text{ a.e. in } \Omega. \tag{12}$$

Moreover $|\nabla u_n|^q$ is bounded in $L^{\frac{p}{q}}(\Omega)$ and therefore, by assumption (5), $T_n(H(x,\nabla u_n))$ is bounded in $L^1(\Omega)$. This implies that the right-hand side of approximated problem (10) is bounded in $L^1(\Omega)$, that is

$$\left\| T_n\big(H(x,\nabla u_n)\big) + f_n \right\|_{L^1} \le C. \tag{13}$$

The assumptions (2)–(3), conditions (12) and (13) allow to apply the compactness result proved in [13], which states that, up to a subsequence,

$$\nabla u_n \to \nabla u \quad \text{a.e. in } \Omega. \tag{14}$$

We deduce that $(\mathbf{a}(x,u_n,\nabla u_n))$ converges to $(\mathbf{a}(x,u,\nabla u))$ a.e. in Ω and $T_n(H(x,\nabla u_n))$ converges to $H(x,\nabla u)$ a.e. in Ω. By Vitali theorem we can pass to the limit at left-hand side of (11) and in the first integral to the right-hand side. This proves that u is a weak solution to (1). □

The definition of weak solution does not fit the case when $p-1<q<p-1+\frac{p}{N}$. In this range we assume that the datum f is in $L(\frac{N(q-p+1)}{q},1)$ and the bounds on q imply that $\frac{N(q-p+1)}{q} < (p^*)'$. Therefore f in general is not an element of $W^{-1,p'}(\Omega)$. This leads to consider the more general framework of "solution obtained as limit of approximations".

We recall that a measurable function $u:\Omega \to \mathbf{R}$, finite almost everywhere in Ω is a "*solution obtained as limit of approximations*" to (1) with $f \in L(m,k)$, $m \ge 1$, $k \ge 0$ if

(i) $T_k(u)$, its truncation to the level k, belongs to $W_0^{1,p}(\Omega)$, for every $k>0$;
(ii) a sequence of functions $f_n \in C_0^\infty(\Omega)$ exists such that $f_n \to f$ strongly in $L(m,k)$;
(iii) the sequence of the weak solutions $u_n \in W_0^{1,p}(\Omega)$ to the approximated problem (10) satisfies

$$u_n \to u \quad \text{a.e. in } \Omega.$$

As pointed out a crucial step in the proof of the existence of a "solution obtained as limit of approximations" to problem (1) consists in proving a priori estimates for the weak solutions u_n to the approximated problems (10) with regular data. Such estimates allows to prove that u_n and its gradients ∇u_n converge to a function u and its gradient ∇u respectively. Then one can prove that u is the "solution obtained as limit of approximations" to (1). We explicitly remark that the gradient ∇u in general is not the usual gradient used in Sobolev spaces since u is just a measurable function. Nevertheless according to Lemma 2.1 in [6], since the truncations of u belong to the Sobolev space $W_0^{1,p}(\Omega)$, there exists a measurable function $v:\Omega \to \mathbf{R}^N$ such that

$\nabla T_k(u) = v\chi_{|u|\le k}$ almost everywhere in Ω, for every $k > 0$. This function v is the generalized gradient of u, i.e. $v = \nabla u$, which coincides with the distributional gradient when $v \in L^1(\Omega)$.

The second main result concerns the a priori estimates in the case where q is in the range $\frac{N(p-1)}{N-1} \le q < p - 1 + \frac{p}{N}$. The details of the proofs are given in [5].

Theorem 3 *Assume* (2)–(5) *with*

$$\frac{N(p-1)}{N-1} \le q < p - 1 + \frac{p}{N}.$$

Let $u \in W_0^{1,p}(\Omega) \cap L^\infty(\Omega)$ *be a weak solution to problem* (1) *with* $f \in C^\infty$. *If* $q > \frac{N(p-1)}{N-1}$ *and the norm of* f *in* $L(\frac{N(q-p+1)}{q}, 1)$ *is small enough, then*

$$\|\nabla u\|_{t,p} \le C,$$
$$\|u\|_{t^*,p} \le C,$$

for every $t < N(q - p + 1)$, *where* C *is a positive constant which depends only on* $p, q, N, |\Omega|$ *and on the norm of the datum* f.

If $q = \frac{N(p-1)}{N-1}$ *and the norm of* f *in* $L(m, 1)$, *with* $m > 1$, *is small enough, then*

$$\|\nabla u\|_{m^*(p-1),p} \le C,$$

where C *is a positive constant which depends only on* $p, q, N, m, |\Omega|$ *and on the norm of the datum* f.

As in the previous case, Theorem 3, applied to the approximated solutions to (10), implies an existence result for "solution obtained as limit of approximations" to problem (1), whose proof is contained in [5].

Theorem 4 *Assume* (2)–(5) *with*

$$\frac{N(p-1)}{N-1} \le q < p - 1 + \frac{p}{N}.$$

If $q > \frac{N(p-1)}{N-1}$ *and the norm of* f *in* $L(\frac{N(q-p+1)}{q}, 1)$ *is small enough, or* $q = \frac{N(p-1)}{N-1}$ *and the norm of* f *in* $L(m, 1)$, *with* $m > 1$, *is small enough, then a "solution obtained as limit of approximations" to* (1) *exists.*

Our approach allows to face also the last interval of values of q, $p - 1 < q < \frac{N(p-1)}{N-1}$. The a priori estimates, which we prove in [5] and we state below, overlap with the result obtained in [21].

Theorem 5 *Assume* (2)–(5) *with* $1 < p < N$ *and*

$$p - 1 < q < \frac{N(p-1)}{N-1}.$$

Let $u \in W_0^{1,p}(\Omega) \cap L^\infty(\Omega)$ *be a weak solution to problem* (1) *with* $f \in C^\infty$. *If the norm of* f *in* $L^1(\Omega)$ *is small enough, then*

$$\|\nabla u\|_{\frac{N(p-1)}{N-1},\infty} \le C,$$
$$\|u\|_{\frac{N(p-1)}{N-p},\infty} \le C,$$

where C is a positive constant which depends only on $p, q, N, |\Omega|$ and on the norm of the datum.

In a standard way this result implies the existence of a "solution obtained as limit of approximations" to (1) stated in the following theorem (see [5, 21]).

Theorem 6 *Assume* (2)–(5) *with $1 < p < N$ and*

$$p - 1 < q < \frac{N(p-1)}{N-1}.$$

If the norm of f in $L^1(\Omega)$ is small enough, then a "solution obtained as limit of approximations" to (1) *exists such that $u \in L(\frac{N(p-1)}{N-p}, \infty)$ and $|\nabla u| \in L(\frac{N(p-1)}{N-1}, \infty)$.*

A few words about the method that we use in [5] in proving previous a priori estimates are in order. The basic ingredient is the choice of a test function which is built on the level sets of the solution u. For instance the test function used in the proof of Theorem 1 is given by

$$\varphi(x) = \int_0^{u(x)} [\mu(t)]^\gamma \, dt,$$

where $\mu(t) = |\{x \in \Omega : |u(x)| \ge t\}|$, $t \ge 0$ is the distribution function of u and γ is a suitable constant. This test function, some Sobolev-type inequalities and classical symmetrization methods are the main tools of the proofs.

As pointed out in the Introduction, in [11] we study how the summability of gradient of a solution to (1), and therefore the summability of a solution, increases when we increase the summability of the datum f, assuming that it belongs to Lorentz spaces. The starting point of this study is the following pointwise estimate of the decreasing rearrangement of the gradient of a weak solution to approximated problems (3), whose proof is contained in [11]. A similar result is proved in [3].

Proposition 1 *Let $u \in W_0^{1,p}(\Omega) \cap L^\infty(\Omega)$ be a weak solution to* (1) *with $f \in C^\infty(\Omega)$ under the assumptions* (2)–(5). *Then for every $\lambda > \frac{N-p}{N(p-1)}$, we have*

$$\begin{aligned}\big(|\nabla u|^*(s)\big)^p \le \frac{\lambda+1}{\alpha(N\omega_N^{\frac{1}{N}})^{p'}} \Bigg\{ &\frac{1}{s^{\lambda+1}} \int_0^s r^{\lambda - \frac{p'}{N'}} \left[\int_0^r \psi(r,\sigma) f^*(\sigma)\, d\sigma \right]^{\frac{p}{p-1}} dr \\ &+ \frac{1}{s} \int_s^{|\Omega|} \frac{1}{r^{\frac{p'}{N'}}} \left[\int_0^r \psi(r,\sigma) f^*(\sigma)\, d\sigma \right]^{\frac{p}{p-1}} dr \Bigg\},\end{aligned} \tag{15}$$

where

$$\psi(r,\sigma) = \exp\left(\frac{h}{(N\omega_N^{\frac{1}{N}})^{p-q}} \int_\sigma^r \frac{[(-u^*)'(z)]^{q-p+1}}{z^{\frac{N-1}{N}(p-q)}} \, dz \right). \tag{16}$$

By using Proposition 1 one can prove the following two results which concern the regularity of the gradient of a "solution obtained as limit of approximations" to (1) when $q \geq \frac{N(p-1)}{N-1}$; they are proved in [11] where all the ranges of q are considered.

Theorem 7 *Assume* (2)–(5) *with*

$$\frac{N(p-1)}{N-1} < q < p - 1 + \frac{p}{N}.$$

Let u be a "solution obtained as limit of approximations" to (1) *with $f \in L(\frac{N(q-p+1)}{q}, 1)$ having norm small enough. If $f \in L(m,k)$ with*

$$\max\left\{\frac{N(q-p+1)}{q}, \frac{N}{Np-N+1}\right\} < m < (p^*)', \quad 0 < k \leq +\infty,$$

then

$$\|\nabla u\|_{m^*(p-1),k(p-1)} \leq C, \tag{17}$$

where C is a positive constant depending on N, p, q, $|\Omega|$, and the norm of f.

Theorem 8 *Assume* (2)–(5) *with*

$$q = \frac{N(p-1)}{N-1}.$$

Let u be a "solution obtained as limit of approximations" to (1) *with $f \in L(\frac{N(q-p+1)}{q}, 1)$ having norm small enough. If $f \in L^m(\Omega)$ with*

$$\max\left\{1, \frac{N}{Np-N+1}\right\} < m < (p^*)',$$

then (17) *holds true.*

3 Comparison Principle

In this section we present comparison principles which in turn imply uniqueness of solutions to problem (6) under the assumptions (7) of "strong monotonicity" condition on $\mathbf{a}$ and (8) of locally Lipschitz condition on H. We fix our attention on the case $p \leq 2$, even if in [11] also the case $p > 2$ is studied.

We assume $q \leq p - 1 + \frac{p}{N}$ since a well-known example shows that uniqueness of a weak solution to (6) does not hold if $q > p - 1 + \frac{p}{N}$ (see also [28]). So we start by assuming $q = p - 1 + \frac{p}{N}$. For this value of q, we have to assume that the datum f belongs to the Lorentz space $L(\frac{N(q-p+1)}{q}, 1)$, i.e. $L((p^*)', 1)$. This means that f is an element of the dual space $W^{-1,p'}(\Omega)$ and we have to refer to weak solution. Here for sake of simplicity we assume that the data belong to Lebesgue spaces.

Let us denote

$$Qu \equiv -\text{div}\big(\mathbf{a}(x, \nabla u)\big) - H(x, \nabla u).$$

By comparison principle we mean that if u, v are weak solutions to the following Dirichlet problems respectively

$$u \in W_0^{1,p}(\Omega), \quad Qu = f \quad \text{in } \Omega, \tag{18}$$

$$v \in W_0^{1,p}(\Omega), \quad Qv = g \quad \text{in } \Omega, \tag{19}$$

with $f, g \in L^m(\Omega)$, $m > (p^*)'$ and

$$f \le g \quad \text{in } \mathscr{D}'(\Omega), \tag{20}$$

then

$$u \le v \quad \text{a.e. in } \Omega. \tag{21}$$

Observe that (18), (19) and (20) imply

$$\int_\Omega \big[\mathbf{a}(x, \nabla u) - \mathbf{a}(x, \nabla v)\big] \cdot \nabla\varphi \, dx + \int_\Omega \big[H(x, \nabla u) - H(x, \nabla v)\big]\varphi \, dx \le 0 \tag{22}$$

for all nonnegative $\varphi \in W_0^{1,p}(\Omega) \cap L^\infty(\Omega)$.

Our first result is the following (see [11] for the proof).

Theorem 9 *Let* $\frac{2N}{N+2} \le p \le 2$. *Assume* (4), (7) *and* (8) *with*

$$q = p - 1 + \frac{p}{N}. \tag{23}$$

If u, v *satisfy* (18), (19) *with* $f, g \in L^m(\Omega)$, $m > (p^*)'$ *which verify* (20), *then* (21) *holds.*

Under the same hypotheses of Theorem 9 we can deduce uniqueness for weak solutions to the Dirichlet problem (6) when $f \in L^m(\Omega)$, $m > (p^*)'$.

A comparison principle holds true also for the range

$$\frac{N(p-1)}{N-1} \le q < p - 1 + \frac{p}{N}.$$

As shown in the previous section, for these bounds on q, we have to refer to "solution obtained as limit of approximations" to (6); we assume again that the data belong to Lebesgue spaces.

So by comparison principle we mean that if u, v are "solution obtained as limit of approximations" to the following Dirichlet problems respectively

$$u \in W_0^{1,1}(\Omega), \quad Qu = f \quad \text{in } \Omega, \tag{24}$$

$$v \in W_0^{1,1}(\Omega), \quad Qv = g \quad \text{in } \Omega, \tag{25}$$

with $f, g \in L^m(\Omega)$, $m > \frac{N(q-p+1)}{q}$ and

$$f \le g \quad \text{a.e. in } \Omega, \tag{26}$$

then

$$u \le v \quad \text{a.e. in } \Omega. \tag{27}$$

Observe that, since u, v are "solution obtained as limit of approximations" to problem (6), with f, g respectively, then

(i) $T_k(u), T_k(v) \in W_0^{1,p}(\Omega)$, for every $k > 0$;
(ii) two sequence of functions $f_n, g_n \in C_0^\infty(\Omega)$ exist such that $f_n \to f$ strongly in $L^m(\Omega)$ and $g_n \to g$ strongly in $L^m(\Omega)$;
(iii) the sequence of the weak solutions $u_n, v_n \in W_0^{1,p}(\Omega)$ to the approximated problems

$$\begin{cases} Q_n u_n \equiv -\mathrm{div}\big(\mathbf{a}(x, \nabla u_n)\big) - T_n\big(H(x, \nabla u_n)\big) = f_n & \text{in } \Omega \\ u_n = 0 & \text{on } \partial\Omega, \end{cases}$$
$$\begin{cases} Q_n v_n \equiv -\mathrm{div}\big(\mathbf{a}(x, \nabla v_n)\big) - T_n\big(H(x, \nabla v_n)\big) = g_n & \text{in } \Omega \\ v_n = 0 & \text{on } \partial\Omega \end{cases}$$

satisfies

$$u_n \to u \quad \text{a.e. in } \Omega \quad \text{and} \quad v_n \to v \quad \text{a.e. in } \Omega.$$

Moreover, under the hypotheses of Theorems 7, 8, by assuming $m = k$, gradients of u, v satisfy a better summability condition, that is $|\nabla u|, |\nabla v| \in L^{m^*(p-1)}(\Omega)$.

Our comparison result can be stated (see [11] for the proof).

Theorem 10 *Let* $\frac{2N}{N+2} \leq p \leq 2$. *Assume* (4), (7) *and* (8) *with*

$$\frac{N(p-1)}{N-1} \leq q < p - 1 + \frac{p}{N}. \tag{28}$$

If u, v *satisfy* (24), (25) *with* $f, g \in L^m(\Omega)$ *which verify* (26) *with*

$$m > \max\left\{\frac{N(q-p+1)}{q}, \frac{N(2-p)}{p}\right\}, \quad \text{if } q > \frac{N(p-1)}{N-1},$$
$$m > \max\left\{1, \frac{N(2-p)}{p}\right\}, \quad \text{if } q = \frac{N(p-1)}{N-1},$$

then (27) *holds.*

As in the previous case under the assumptions of Theorem 10, a uniqueness result for "solution obtained as limit of approximations" to problem (6) holds.

References

1. Alaa, N., Pierre, M.: Weak solutions of some quasilinear elliptic equations with data measures. SIAM J. Math. Anal. **24**, 23–35 (1993)
2. Alvino, A., Mercaldo, A.: Nonlinear elliptic equations with lower order terms and symmetrization methods. Boll. Unione Mat. Ital. **1**, 645–662 (2008)
3. Alvino, A., Ferone, V., Trombetti, G.: Estimates for the gradient of solutions of nonlinear elliptic equations with L^1 data. Ann. Mat. Pura Appl. **178**, 129–142 (2000)

4. Alvino, A., Betta, M.F., Mercaldo, A.: Comparison principle for some class of nonlinear elliptic equations. J. Differ. Equ. **249**, 3279–3290 (2010)
5. Alvino, A., Ferone, V., Mercaldo, A.: Sharp a-priori estimates for a class of nonlinear elliptic equations with lower order terms (in preparation)
6. Bénilan, Ph., Boccardo, L., Gallouët, Th., Gariepy, R., Pierre, M., Vázquez, J.L.: An L^1 theory of existence and uniqueness of solutions of nonlinear elliptic equations. Ann. Sc. Norm. Super. Pisa. Cl. Sci. (4) **22**, 241–273 (1995)
7. Betta, M.F., Mercaldo, A.: Uniqueness results for nonlinear elliptic equations via symmetrization methods. Atti Accad. Naz. Lincei Cl. Sci. Fis. Mat. Natur. Rend. Lincei (9) Mat. Appl. **21**, 1–14 (2010)
8. Betta, M.F., Mercaldo, A., Murat, F., Porzio, M.M.: Uniqueness of renormalized solutions to nonlinear elliptic equations with lower-order term and right-hand side in $L^1(\Omega)$. A tribute to J.-L. Lions. ESAIM Control Optim. Calc. Var. **8**, 239–272 (2002) (electronic)
9. Betta, M.F., Mercaldo, A., Murat, F., Porzio, M.M.: Existence of renormalized solutions to nonlinear elliptic equations with a lower order term and right-hand side measure. J. Math. Pures Appl. **82**, 90–124 (2003)
10. Betta, M.F., Mercaldo, A., Murat, F., Porzio, M.M.: Uniqueness results for nonlinear elliptic equations with a lower order term. Nonlinear Anal. **63**, 153–170 (2005)
11. Betta, M.F., Di Nardo, R., Mercaldo, A., Perrotta, A.: Gradient estimates and comparison principle for some nonlinear elliptic equations (in preparation)
12. Bidaut-Véron, M.F., Hamid, H.A.: Correlation between two quasilinear elliptic problems with a source terms of order 0 or 1. Commun. Contemp. Math. **12**, 727–788 (2010)
13. Boccardo, L., Murat, F.: Almost everywhere convergence of the gradients of solutions to elliptic and parabolic equations. Nonlinear Anal. **19**, 581–597 (1992)
14. Bottaro, G., Marina, M.E.: Problema di Dirichlet per equazioni ellittiche di tipo variazionale su insiemi non limitati. Boll. Unione Mat. Ital. **8**, 46–56 (1973)
15. Dall'Aglio, A.: Approximated solutions of equations with L^1 data. Application to the H-convergence of quasi-linear parabolic equations. Ann. Mat. Pura Appl. **170**, 207–240 (1996)
16. Dal Maso, G., Malusa, A.: Some properties of reachable solutions of nonlinear elliptic equations with measure data. Ann. Sc. Norm. Super. Pisa. Cl. Sci. (4) **25**, 375–396 (1997)
17. Del Vecchio, T., Porzio, M.M.: Existence results for a class of non coercive Dirichlet problems. Ric. Mat. **44**, 421–438 (1995)
18. Ferone, V., Messano, B.: Comparison and existence results for classes of nonlinear elliptic equations with general growth in the gradient. Adv. Nonlinear Stud. **7**, 31–46 (2007)
19. Ferone, V., Murat, F.: Nonlinear problems having natural growth in the gradient: an existence result when the source terms are small. Nonlinear Anal. **42**, 1309–1326 (2000)
20. Grenon, N., Murat, F., Porretta, A.: Existence and a priori estimate for elliptic problems with subquadratic gradient dependent terms. C. R. Math. Acad. Sci. Paris **342**, 23–28 (2006)
21. Grenon, N., Murat, F., Porretta, A.: A priori estimates and existence for elliptic equations with gradient dependent term. Ann. Sc. Norm. Super. Pisa (in press)
22. Hansson, K., Maz'ya, V., Verbisky, I.: Criteria of solvability for multidimensional Riccati's equation. Ark. Mat. **37**, 247–276 (1999)
23. Leray, J., Lions, J.-L.: Quelques résultats de Visik sur les problémes elliptiques non linéaires par les méthodes de Minty Browder. Bull. Soc. Math. Fr. **93**, 97–107 (1965)
24. Lions, P.-L., Murat, F.: Sur les solutions renormalisées d'equations elliptiques non linéaires. Manuscript
25. Maderna, C., Salsa, S.: Dirichlet problem for elliptic equations with nonlinear first order terms: a comparison result. Ann. Mat. Pura Appl. **148**, 277–288 (1987)
26. Maz'ya, V.G.: On weak solutions of the Dirichlet and Neumann problems. Trans. Mosc. Math. Soc. **20**, 135–172 (1969)
27. Murat, F.: Soluciones renormalizadas de EDP elipticas no lineales. Preprint 93023, Laboratoire d'Analyse Numérique de l'Université Paris VI, (1993)

28. Porretta, A.: On the Comparison Principle for p-Laplace Operators with First Order Terms. Quaderni di Matematica, vol. 23. Department of Mathematics, Seconda Università di Napoli, Caserta (2008).
29. Talenti, G.: Nonlinear elliptic equations, rearrangements of functions and Orlicz spaces. Ann. Mat. Pura Appl. **120**, 160–184 (1979)

Existence and Uniqueness of the n-Dimensional Helfrich Flow

Takeyuki Nagasawa

Abstract The n-dimensional Helfrich variational problem is, roughly speaking, as follows: minimize the integral of squared mean curvature among n-dimensional hypersurfaces with the prescribed area and enclosed volume. In this article the n-dimensional Helfrich flow is considered. This is the L^2-gradient flow associated with the n-dimensional Helfrich variational problem. The equation of the flow is described as a projected gradient flow. A local existence result and partial uniqueness results are presented. In particular the latter improves previous results.

Keywords Helfrich variational problem · Gradient flow · Constraints

1 The Helfrich Variational Problem and Its Gradient Flow

Let $\Sigma \subset \mathbb{R}^{n+1}$ be a closed and oriented hypersurface. We do not assume that the immersion $\Sigma \subset \mathbb{R}^{n+1}$ is an embedding. We discuss the variational problem for the functional

$$\mathcal{W}(\Sigma) = \frac{n}{2}\int_\Sigma (H - c_0)^2\, dS$$

under the constraints

$$\mathcal{A}(\Sigma) = \mathcal{A}_0, \qquad \mathcal{V}(\Sigma) = \mathcal{V}_0.$$

This is the n-dimensional Helfrich problem. Here, H is the mean curvature, and c_0, $\mathcal{A}_0$, $\mathcal{V}_0$ are given constants. The functionals $\mathcal{A}$ and $\mathcal{V}$ are given by

$$\mathcal{A}(\Sigma) = \int_\Sigma dS,$$

The author partly supported by Grant-in-Aid for Scientific Research (A) (No. 22244010-01), Japan Society for the Promotion Science.

T. Nagasawa (✉)
Department of Mathematics, Graduate School of Science and Engineering, Saitama University, Saitama 338-8570, Japan
e-mail: tnagasaw@rimath.saitama-u.ac.jp

R. Magnanini et al. (eds.), *Geometric Properties for Parabolic and Elliptic PDE's*, Springer INdAM Series 2, DOI 10.1007/978-88-470-2841-8_15,

and

$$\mathcal{V}(\Sigma) = -\frac{1}{n+1}\int_{\Sigma} \boldsymbol{f}\cdot\boldsymbol{\nu}\, dS.$$

The vectors $\boldsymbol{f}$ and $\boldsymbol{\nu}$ are the position vector of a point on Σ and the unit normal vector at that point, respectively. That is, $\mathcal{A}$ is the area of Σ, and $\mathcal{V}$ is the "signed" volume. Indeed, if $\Sigma \subset \mathbb{R}^{n+1}$ is the boundary of a domain $\Omega \subset \mathbb{R}^{n+1}$ and if $\boldsymbol{\nu}$ is the inner normal, then $\mathcal{V}$ is the volume of the domain enclosed by Σ.

The original Helfrich variational problem is a model of shape transformation theory of a human red blood cell [5]. For this case, n is 2 and c_0 is called the *spontaneous curvature*, which is determined by the molecular structure of cell membrane. That is, Helfrich considered the cell membrane a closed and oriented surface Σ in $\mathbb{R}^3$, and he proposed that the shape of cell is a minimizer of $\mathcal{W}$ under the prescribed area and enclosed volume.

The one-dimensional case, that is, the corresponding problem for closed curves, was considered by Watanabe [15, 16] as a spectral optimization problem of a plane domain. Let Ω be a bounded plane domain, and let Σ be its boundary. The function $G(x, y, t)$ is the Green function for the heat equation on $\Omega \times (0, T)$ under the Dirichlet boundary condition. The asymptotic expansion

$$\int_{\Omega} G(x,x,t)\, dx = \frac{1}{4\pi t}\left(a_0 + a_1 t^{\frac{1}{2}} + a_2 t + a_3 t^{\frac{3}{2}} + \cdots\right) \quad \text{as } t \to +0$$

is well-known as the trace formula. Here,

$$a_0 = \mathcal{V}(\Sigma), \qquad a_1 = -\frac{\sqrt{\pi}}{2}\mathcal{A}(\Sigma), \qquad a_2 = \frac{1}{3}\int_{\Sigma} \kappa\, ds, \qquad a_3 = \frac{\sqrt{\pi}}{64}\int_{\Sigma} \kappa^2\, ds.$$

Here, $\kappa (= H)$ is the curvature of the curve Σ, and s is the arc-length parameter. a_2 is determined by the topology of Ω. Hence, the one-dimensional Helfrich problem is equivalent to the following problem: For given a_0 and a_1, find the Ω among domains with fixed topology (*i.e.*, fixed a_2) that minimizes a_3, that is, $\frac{32}{\sqrt{\pi}}\mathcal{W}$ with $c_0 = 0$.

According to [3], a shape transformation of a closed loop of plastic tape between two parallel flat plates is governed by the one-dimensional Helfrich variational problem (with $c_0 = 0$).

Remark 1 Without loss of generality, we may assume that $c_0 = 0$ for the one-dimensional case. Indeed, the functional is

$$\frac{1}{2}\int_{\Sigma} \kappa^2\, ds - c_0 \int_{\Sigma} \kappa\, ds + \frac{1}{2}c_0^2 \int_{\Sigma} ds.$$

If we consider our variational problem under the constraints of $\mathcal{A} = \int_{\Sigma} ds$ and $\mathcal{V}$ among curves with fixed rotation number $\int_{\Sigma} \kappa\, ds$, then we can replace the functional with the first integral $\frac{1}{2}\int_{\Sigma} \kappa^2\, ds$ only.

By the method of Lagrange multipliers, the Helfrich variational problem is described by

$$\delta\mathcal{W}(\Sigma) + \lambda_1 \delta\mathcal{A}(\Sigma) + \lambda_2 \delta\mathcal{V}(\Sigma) = 0.$$

Here, λ_js are Lagrange multipliers, and δ is the first variation. Performing calculations similar to [4], we obtain

$$\begin{aligned}\delta\mathcal{W}(\Sigma) &= \Delta_g H + (H - c_0)\left\{\frac{n^2}{2}H(H + c_0) - R\right\},\\ \delta\mathcal{A}(\Sigma) &= -nH, \qquad \delta\mathcal{V} = -1.\end{aligned} \tag{1}$$

Here, Δ_g is the Laplace-Beltrami operator with respect to the metric induced by $\Sigma \subset \mathbb{R}^{n+1}$, and R is the scalar curvature. Using these, the above equation becomes

$$\Delta_g H + (H - c_0)\left\{\frac{n^2}{2}H(H + c_0) - R\right\} - \lambda_1 nH - \lambda_2 = 0. \tag{2}$$

Regarding Σ as the image $\boldsymbol{f}(\Sigma_0)$ of a fixed n-dimensional manifold Σ_0, we reduce (2) to a quasilinear elliptic equation of the forth order.

The two-dimensional Helfrich problem has a long history, and there are several known facts. It is easy to see that spheres are critical points. In 1977, Jenkins [6] numerically found bifurcating solutions from spheres. Subsequently, Peterson [13] and Ou-Yang-Helfrich [12] formally investigated their stability/instability. Their arguments were justified mathematically by Takagi and the author in [9]. Au-Wan [2] considered critical points far from spheres but with rotational symmetry. Critical points without rotational symmetry were constructed by Takagi and the author in [10].

In this article, we approach our problem by the method of steepest descent, that is, the L^2-gradient flow associated with the problem (the *Helfrich flow*).

The equation of the *Helfrich flow* is

$$V(t) = -\delta\mathcal{W}(\Sigma(t)) - \lambda_1\delta\mathcal{A}(\Sigma(t)) - \lambda_2\delta\mathcal{V}(\Sigma(t)). \tag{3}$$

The function $V = \partial_t \boldsymbol{f} \cdot \boldsymbol{\nu}$ is the normal deformation velocity of the hypersurface family $\Sigma(t)$. The multipliers are unknown functions of t.

Let us recall some previous results for the low-dimensional Helfrich flow, which was studied in [7] for $n = 2$ and in [8] for $n = 1$.

In [7], the multipliers λ_j are given as "known" constants. That is, given $\{\lambda_1, \lambda_2, \Sigma_0\}$ as the data, solutions of (3) were constructed in anticipation of the convergence of solutions to critical points as $t \to \infty$. However, for this choice of λ_j, solutions do not satisfy $\frac{d}{dt}A(\Sigma(t)) \equiv 0$, $\frac{d}{dt}V(\Sigma(t)) \equiv 0$, and we cannot expect global existence. Indeed, there exist solutions blowing up in finite/infinite time. The problem is shifted to the investigation of triples $\{\lambda_1, \lambda_2, \Sigma_0\}$ so that the solution can extend globally in time. In [7], the existence of such triples was proved near spheres. Furthermore, such triples form a finite-dimensional center manifold. The class of initial surfaces is the little Hölder space $h^{2+\alpha}$ for some $\alpha \in (0, 1)$. Here, the little Hölder space is defined as follows. For an open set U, the space $h^\beta(U)$ is the closure of $BUC^\infty(U)$ in $BUC^\beta(U)$. The space $h^\beta = h^\beta(\Sigma_0)$ is defined by means of an atlas for Σ_0 as a submanifold in $\mathbb{R}^{n+1}$.

In [8], the gradient flow $\{\Sigma_\varepsilon(t)\}$ associated with the functional

$$\mathcal{W}(\Sigma_\varepsilon) + \frac{1}{2\varepsilon}\big(\mathcal{A}(\Sigma_\varepsilon) - \mathcal{A}_0\big)^2 + \frac{1}{2\varepsilon}\big(\mathcal{V}(\Sigma_\varepsilon) - \mathcal{V}_0\big)^2$$

was constructed. The solution of (3) was obtained as the limit of $\{\Sigma_\varepsilon(t)\}$ as $\varepsilon \to +0$. This is a global solution, and it satisfies

$$\frac{d}{dt}\mathcal{W}(\Sigma(t)) \equiv -\|V(t)\|^2_{L^2(\Sigma(t))}, \qquad \frac{d}{dt}\mathcal{A}(\Sigma(t)) \equiv 0, \qquad \frac{d}{dt}\mathcal{V}(\Sigma(t)) \equiv 0. \quad (4)$$

The class of initial curves is C^∞, but the uniqueness was uncertain.

We consider the general n-dimensional case. The Lagrange multipliers are determined so that $\frac{d}{dt}\mathcal{A}(\Sigma(t)) = \frac{d}{dt}\mathcal{V}(\Sigma(t)) = 0$. In the next section, we see that these relations determine the linear combination $\lambda_1(\Sigma)\delta\mathcal{A}(\Sigma) + \lambda_2(\Sigma)\delta\mathcal{V}(\Sigma)$ even if two constraints are degenerate. The meaning of *degeneracy* is given there.

Theorem 1 *Let $P(\Sigma)$ be the orthogonal projection from $L^2(\Sigma)$ to the linear subspace* $(\mathrm{span}_{L^2(\Sigma)}\{\delta\mathcal{A}(\Sigma), \delta\mathcal{V}(\Sigma)\})^\perp$. *Then, the Helfrich flow equation can be written as*

$$V(t) = -P(\Sigma(t))\delta\mathcal{W}(\Sigma(t)) \quad \textit{for } t > 0. \qquad (5)$$

Solutions, if they exist, satisfy (4).

We next obtain the existence and uniqueness of the initial value problem.

Theorem 2 *Let $G(\Sigma)$ be the Gramian of* $(\delta\mathcal{A}(\Sigma), \delta\mathcal{V}(\Sigma))$, *which is defined in Sect.* 2.

(i) *Assume that Σ_0 is in the class of $h^{3+\alpha}$ for some $\alpha \in (0,1)$, and that $G(\Sigma_0) \neq 0$. Then, there exist $T > 0$ and a unique solution $\{\Sigma(t)\}_{0 \leqq t < T}$ of* (5) *satisfying $\Sigma(0) = \Sigma_0$.*

(ii) *Assume that $G(\Sigma_0) = 0$. Then, Σ_0 has a constant mean curvature. If Σ_0 is a (single/multi-fold) sphere, or if its mean curvature is c_0 everywhere, then there exists a global solution $\{\Sigma(t)\}_{t \geqq 0}$ of* (5) *satisfying $\Sigma(0) = \Sigma_0$.*

We show Theorem 1 in Sect. 2. In Sect. 3, we provide the proof of Theorem 2. For the statement Theorem 2 (i), we have already proved it in [11], and therefore we provide only a sketch of the proof. Instead of Theorem 2 (ii), the following statement was shown in [11]:

(ii)$'$ *Assume that $G(\Sigma_0) = 0$. Let H_0 and R_0 be the mean curvature and the scalar curvature of Σ_0 respectively. If $(\bar{H}_0 - c_0)\tilde{R}_0 \equiv 0$, then there exists a global solution $\{\Sigma(t)\}_{t \geqq 0}$ of* (5) *satisfying $\Sigma(0) = \Sigma_0$.*

We show the equivalence between (ii)$'$ and (ii) in Sect. 3.2. (ii) seems easier to understand than (ii)$'$.

The uniqueness is uncertain in case (ii), except for several special cases. We discuss the uniqueness in Sect. 4. Some open problems are presented in Sect. 5.

In the following sections, for a function ϕ on Σ, we define $\bar{\phi}$ and $\tilde{\phi}$ by

$$\bar{\phi} = \frac{1}{\mathcal{A}(\Sigma)}\int_\Sigma \phi\, dS, \quad \tilde{\phi} = \phi - \bar{\phi}.$$

2 Proof of Theorem 1

Theorem 1 is a special case of the general theory of *projected gradient flows* [14]. We denote $\Sigma(t)$ simply by Σ. $\|\cdot\|$ and $\langle\cdot,\cdot\rangle$ stand for the $L^2(\Sigma)$-norm and the $L^2(\Sigma)$-inner product, respectively. We have

$$\frac{d}{dt}\mathcal{A}(\Sigma) = \big\langle \delta\mathcal{A}(\Sigma), V\big\rangle, \qquad \frac{d}{dt}\mathcal{V}(\Sigma) = \big\langle \delta\mathcal{V}(\Sigma), V\big\rangle.$$

From these and (3), it follows that

$$\begin{pmatrix} \langle\delta\mathcal{A}(\Sigma),\delta\mathcal{A}(\Sigma)\rangle & \langle\delta\mathcal{V}(\Sigma),\delta\mathcal{A}(\Sigma)\rangle \\ \langle\delta\mathcal{A}(\Sigma),\delta\mathcal{V}(\Sigma)\rangle & \langle\delta\mathcal{V}(\Sigma),\delta\mathcal{V}(\Sigma)\rangle \end{pmatrix} \begin{pmatrix} \lambda_1 \\ \lambda_2 \end{pmatrix} = -\begin{pmatrix} \langle\delta\mathcal{A}(\Sigma),\delta\mathcal{W}(\Sigma)\rangle \\ \langle\delta\mathcal{V}(\Sigma),\delta\mathcal{W}(\Sigma)\rangle \end{pmatrix}. \tag{6}$$

Denote the Gramian of the left-hand side by $G(\Sigma)$, that is,

$$G(\Sigma) = \det\begin{pmatrix} \langle\delta\mathcal{A}(\Sigma),\delta\mathcal{A}(\Sigma)\rangle & \langle\delta\mathcal{V}(\Sigma),\delta\mathcal{A}(\Sigma)\rangle \\ \langle\delta\mathcal{A}(\Sigma),\delta\mathcal{V}(\Sigma)\rangle & \langle\delta\mathcal{V}(\Sigma),\delta\mathcal{V}(\Sigma)\rangle \end{pmatrix}.$$

We say that the constraints $\mathcal{A}(\Sigma) = \mathcal{A}_0$ and $\mathcal{V}(\Sigma) = \mathcal{V}_0$ are *degenerate* when $G(\Sigma) = 0$.

For the non-degenerate case, the multipliers are uniquely determined by the relation (6) from Σ. Hence, we can write

$$\lambda_j = \lambda_j(\Sigma).$$

When the constraints are degenerate, the multipliers are not uniquely determined, but we can show that $\lambda_1\delta\mathcal{A}(\Sigma) + \lambda_2\delta\mathcal{V}(\Sigma)$ is uniquely determined.

Set

$$\tilde{H} = H - \frac{1}{A(\Sigma)}\int_\Sigma H\,dS, \qquad H_* = \begin{cases} \frac{\tilde{H}}{\|\tilde{H}\|} & \text{if } \tilde{H} \not\equiv 0, \\ 0 & \text{if } \tilde{H} \equiv 0, \end{cases} \qquad 1_* = \frac{1}{\|1\|}.$$

Note that $\langle H_*, 1_*\rangle = 0$. Using (1), we have

$$\mathrm{span}_{L^2(\Sigma)}\big\{\delta\mathcal{A}(\Sigma), \delta\mathcal{V}(\Sigma)\big\} = \mathrm{span}_{L^2(\Sigma)}\{H, 1\} = \mathrm{span}_{L^2(\Sigma)}\{H_*, 1_*\}.$$

Hence, (3) becomes

$$V = -\delta\mathcal{W}(\Sigma) - \lambda_1\delta\mathcal{A}(\Sigma) - \lambda_2\delta\mathcal{V}(\Sigma) = -\delta\mathcal{W}(\Sigma) - \mu_1 1_* - \mu_2 H_* \tag{7}$$

for some μ_j. From $\frac{d}{dt}\mathcal{A}(\Sigma) = \frac{d}{dt}\mathcal{V}(\Sigma) = 0$, it follows that

$$\langle 1_*, V\rangle = \langle H_*, V\rangle = 0.$$

Taking the $L^2(\Sigma)$-inner product (7) and 1_*, H_*, we obtain

$$\begin{aligned} 0 &= \langle 1_*, V\rangle = -\big\langle 1_*, \delta\mathcal{W}(\Sigma)\big\rangle - \mu_1, \\ 0 &= \langle H_*, V\rangle = -\big\langle H_*, \delta\mathcal{W}(\Sigma)\big\rangle - \mu_2\|H_*\|^2. \end{aligned}$$

Regardless of whether $H_* \equiv 0$, it holds that

$$-\mu_1 1_* - \mu_2 H_* = \big\langle 1_*, \delta\mathcal{W}(\Sigma)\big\rangle 1_* + \big\langle H_*, \delta\mathcal{W}(\Sigma)\big\rangle H_*.$$

From this equation, together with (7), we obtain (5).

It holds for solutions of (5) that

$$\frac{d}{dt}\mathcal{W}(\Sigma) = \langle \delta\mathcal{W}(\Sigma), V\rangle = \langle \delta\mathcal{W}(\Sigma), -P(\Sigma)\delta\mathcal{W}(\Sigma)\rangle$$
$$= -\left\| P(\Sigma)\delta\mathcal{W}(\Sigma)\right\|^2 = -\|V\|^2.$$

Since $V \in (\mathrm{span}_{L^2(\Sigma)}\{\delta\mathcal{A}(\Sigma), \delta\mathcal{V}(\Sigma)\})^\perp$, we have

$$\frac{d}{dt}\mathcal{A}(\Sigma) = \langle \delta\mathcal{A}(\Sigma), V\rangle = 0, \qquad \frac{d}{dt}\mathcal{V}(\Sigma) = \langle \delta\mathcal{V}(\Sigma), V\rangle = 0.$$

3 Proof of Theorem 2

3.1 Sketch of the Proof of Theorem 2 (i)

The local existence for the case $G(\Sigma_0) \neq 0$ is shown in a manner similar to that of [7]. If the Helfrich flow with $\Sigma(0) = \Sigma_0$ exists, and if Σ is close to Σ_0 in a C^2-sense for small $t > 0$, then $G(\Sigma) \neq 0$. From (6), it follows that

$$\begin{pmatrix} \lambda_1(\Sigma) \\ \lambda_2(\Sigma) \end{pmatrix} = -\frac{1}{G(\Sigma)} \begin{pmatrix} \langle \delta\mathcal{V}(\Sigma), \delta\mathcal{V}(\Sigma) & -\langle \delta\mathcal{V}(\Sigma), \delta\mathcal{A}(\Sigma) \\ -\langle \delta\mathcal{A}(\Sigma), \delta\mathcal{V}(\Sigma) & \langle \delta\mathcal{A}(\Sigma), \delta\mathcal{A}(\Sigma) \end{pmatrix}$$
$$\times \begin{pmatrix} \langle \delta\mathcal{A}(\Sigma), \delta\mathcal{W}(\Sigma) \\ \langle \delta\mathcal{V}(\Sigma), \delta\mathcal{W}(\Sigma) \end{pmatrix}. \tag{8}$$

From (1), we have

$$\langle \delta\mathcal{A}(\Sigma), \delta\mathcal{A}(\Sigma)\rangle = n^2 \int_\Sigma H^2\, dS, \qquad \langle \delta\mathcal{A}(\Sigma), \delta\mathcal{V}(\Sigma)\rangle = n \int_\Sigma H\, dS,$$
$$\langle \delta\mathcal{V}(\Sigma), \delta\mathcal{V}(\Sigma)\rangle = \int_\Sigma dS,$$
$$\langle \delta\mathcal{A}(\Sigma), \delta\mathcal{W}(\Sigma)\rangle = n \int_\Sigma \left(|\nabla_g H|^2 - \frac{1}{2}n^2 H^4 + H^2 R - c_0 H R + \frac{1}{2} n^2 c_0^2 H^2 \right) dS,$$
$$\langle \delta\mathcal{V}(\Sigma), \delta\mathcal{W}(\Sigma)\rangle = \int_\Sigma \left(-\frac{1}{2}n^2 H^3 + HR - c_0 R + \frac{1}{2} n^2 c_0^2 H \right) dS,$$

$$G(\Sigma) = \int_\Sigma n^2 H^2\, dS \int_\Sigma dS - \left(\int_\Sigma nH\, dS \right)^2 = n^2 \int_\Sigma dS \int_\Sigma \tilde{H}^2\, dS. \tag{9}$$

Inserting these into (8), we have the explicit expression of $\lambda_j(\Sigma)$s. Thus, we obtain the following proposition.

Proposition 1 *If $G(\Sigma) \neq 0$, the Lagrange multipliers $\lambda_1(\Sigma)$ and $\lambda_2(\Sigma)$ in (3) have explicit expressions that depend analytically on*

$$\int_\Sigma |\nabla_g H|^2\, dS, \qquad \int_\Sigma H^p\, dS \quad (p=0,1,2,3,4),$$
$$\int_\Sigma H^q R\, dS \quad (q=0,1,2).$$

To prove Theorem 2 (i), we regard Σ as the perturbation of Σ_0 in the normal direction with signed distance ϱ. This is possible for a short time-interval. Let $\bigcup_{\ell=1}^m U_\ell$ be an open covering of Σ_0. We denote the unit normal vector fields of Σ_0 by $\boldsymbol{\nu}_0$. The mapping $X_\ell : U_\ell \times (-a,a) \ni (s,r) \to s + r\boldsymbol{\nu}_0(s) \in \mathbb{R}^{n+1}$ is a C^∞-diffeomorphism from $U_\ell \times (-a,a)$ to $\mathcal{R}_\ell = \mathrm{Im}(X_\ell)$, provided that $a>0$ is sufficiently small. Denote the inverse mapping X_ℓ^{-1} by (S_ℓ, Λ_ℓ), where $S_\ell(X_\ell(s,r)) = s \in U_\ell$ and $\Lambda_\ell(X_\ell(s,r)) = r \in (-a,a)$.

When $\Sigma(t)$ is sufficiently close to Σ_0 for small $t>0$, we can represent it as a graph of a function on Σ_0 as

$$\Sigma_{\varrho(t)} = \Sigma(t) = \bigcup_{\ell=1}^m \mathrm{Im}\big(X_\ell\big(\cdot, \varrho(\cdot,t)\big) : U_\ell \to \mathbb{R}^{n+1},\ \big[s \mapsto X_\ell\big(s, \varrho(s,t)\big)\big]\big).$$

Conversely, for a given function $\varrho : \Sigma_0 \times [0,T) \to (-a,a)$ we define the mapping $\Phi_{\ell,\varrho}$ from $\mathcal{R}_\ell \times [0,T)$ to $\mathbb{R}$ by

$$\Phi_{\ell,\varrho}(x,t) = \Lambda_\ell(x) - \varrho\big(S_\ell(x),t\big).$$

Then, $(\Phi_{\ell,\varrho}(\cdot,t))^{-1}(0)$ gives the surface $\Sigma_{\varrho(t)}$.

Let $L_\varrho = L_\varrho(s,t)$ be the Euclidean norm of $\nabla_x \Phi_{\ell,\varrho}(x,t)$ at $(x,t) = (X_\ell(s, \varrho(s,t)),t)$:

$$L_\varrho = \big\|\nabla_x \Phi_{\ell,\varrho}\big(X_\ell\big(s,\varrho(s,t)\big),t\big)\big\|.$$

The velocity in the direction of normal vector field of $\Sigma = \{\Sigma_{\varrho(t)} \,|\, t \in [0,T)\}$ at $(x,t) = (X_\ell(s,\varrho(s,t)),t)$ is given by

$$V(s,t) = -\frac{\partial_t \Phi_{\ell,\varrho}(X_\ell(s,\varrho(s,t)),t)}{L_\varrho} = \frac{\partial_t \varrho(s,t)}{L_\varrho}.$$

We can express the Laplace-Beltrami operator, the mean curvature, the scalar curvature, and the Lagrange multipliers in terms of the function ϱ and its derivatives, denoted by Δ_ϱ, $H(\varrho)$, $R(\varrho)$, and $\lambda_j(\varrho)$ respectively. Then, Eq. (5) is represented as

$$\partial_t \varrho = L_\varrho\bigg(-\Delta_\varrho H(\varrho) - \frac{1}{2}n^2 H^3(\varrho) + H(\varrho)R(\varrho) - c_0 R(\varrho) + \frac{1}{2}n^2 c_0^2 H(\varrho) + \lambda_1(\varrho) n H(\varrho) + \lambda_2(\varrho)\bigg). \tag{10}$$

We can find the expression of not only Δ_ϱ, $H(\varrho)$ but also the Gaussian curvature $K(\varrho)$ in [7] for the case $n=2$. In our case, the expression of Δ_ϱ and $H(\varrho)$ is the same as in [7], and we can obtain those of $R(\varrho)$ and $\lambda_j(\varrho)$ in a similar way. In particular, $\lambda_j(\varrho)$ can be written in terms of ϱ and its derivatives up to the third order. Combining Proposition 1, we can see that the right-hand side of (10) is linear with respect to the fourth-order derivative of ϱ, but not linear with respect to lower derivatives. The principal term $-L_\varrho \Delta_\varrho H(\varrho)$ is the same as the equation dealt with in [7, (2.1)]. We fix $0 < \alpha < \beta < 1$. Then, for $\beta_0 \in (\alpha, \beta)$ and $a > 0$, we set

$$\mathcal{U} = \{\varrho \in h^{3+\beta_0}(\Sigma_0) \mid \|\varrho\|_{C^2(\Sigma_0)} < a\}.$$

For two Banach spaces E_0 and E_1 satisfying $E_1 \hookrightarrow E_0$, the set $\mathcal{H}(E_1, E_0)$ is the class of $A \in \mathcal{L}(E_1, E_0)$ such that $-A$, considered an unbounded operator in E_0, generates a strongly continuous analytic semigroup on E_0.

Proposition 2 *There exist*

$$Q \in C^\infty(\mathcal{U}, \mathcal{H}(h^{4+\alpha}(\Sigma_0), h^\alpha(\Sigma_0))), \qquad F \in C^\infty(\mathcal{U}, h^{\beta_0}(\Sigma_0))$$

such that Eq. (10) *is in the form*

$$\varrho_t + Q(\varrho)\varrho + F(\varrho) = 0.$$

Applying [1, Theorem 12.1] with $X_\beta = \mathcal{U}$, $E_1 = h^{4+\alpha}(\Sigma_0)$, $E_0 = h^\alpha(\Sigma_0)$, and $E_\gamma = h^{\beta_0}(\Sigma_0)$, we obtain assertion (i) in Theorem 2.

Remark 2 The equation dealt with in [7] is a similar fourth-order equation, but is linear with respect to the third-order derivative of ϱ. Therefore it was solvable for initial data in the class $h^{2+\alpha}$. Since our equation contains the third-order derivative of ϱ nonlinearly, we need more regularity of Σ_0 than does [7].

3.2 Proof of Theorem 2 (ii)

In this subsection we prove (ii) in Theorem 2. The aim is two-fold. The first is to show the equivalence between (ii) and (ii)$'$. Because (ii)$'$ has already been shown in [7], we obtain the proof of (ii). The second is to give the direct proof of (ii).

Assume $G(\Sigma_0) = 0$. From (9) it follows that Σ_0 has a constant mean curvature.

Proposition 3 *Assume that a closed and oriented hypersurface has a constant mean curvature. Then, it has a constant scalar curvature, if and only if it is a* (*single/multi-fold*) *sphere.*

To prove Proposition 3, we state and prove the following lemma. Let κ_i ($i = 1, \ldots, n$) be the principal curvatures.

Lemma 1 *For every closed and oriented hypersurface Σ, it holds that*

$$\sum_i (\kappa_i - H)^2 = n(n-1)(\tilde{H} + 2\bar{H})\tilde{H} - \tilde{R} \\ - \frac{\bar{H}}{A(\Sigma)} \int_\Sigma \tilde{R} f \cdot \nu \, dS + \frac{\bar{R}}{A(\Sigma)} \int_\Sigma \tilde{H} f \cdot \nu \, dS.$$

Proof First, we show that

$$\int_\Sigma R\boldsymbol{f}\cdot\boldsymbol{\nu}\,dS = -n(n-1)\int_\Sigma H\,dS, \qquad \int_\Sigma H\boldsymbol{f}\cdot\boldsymbol{\nu}\,dS = -\mathcal{A}(\Sigma). \tag{11}$$

Though these can be shown by direct calculation, it is clever to use the following argument. Let Σ_ε be the hypersurface defined by $(1+\varepsilon)\boldsymbol{f}$, and let H_ε and dS_ε be its mean curvature and the volume element, respectively. From the scaling argument it follows that

$$\int_{\Sigma_\varepsilon} H_\varepsilon\,dS_\varepsilon = (1+\varepsilon)^{n-1}\int_\Sigma H\,dS, \qquad \mathcal{A}(\Sigma_\varepsilon) = (1+\varepsilon)^n\mathcal{A}(\Sigma).$$

Since $\boldsymbol{f}\cdot\boldsymbol{\nu}$ is the normal variation generated by $\boldsymbol{f}\mapsto(1+\varepsilon)\boldsymbol{f}$, and since

$$\delta\left(\int_\Sigma H\,dS\right) = -\frac{R}{n}, \qquad \delta\mathcal{A}(\Sigma) = -nH$$

(see the argument in [4]), we obtain

$$\begin{aligned}
\int_\Sigma R\boldsymbol{f}\cdot\boldsymbol{\nu}\,dS &= -n\delta\left(\int_\Sigma H\,dS\right)[\boldsymbol{f}\cdot\boldsymbol{\nu}] \\
&= -n\frac{d}{d\varepsilon}\int_{\Sigma_\varepsilon} H_\varepsilon\,dS_\varepsilon\bigg|_{\varepsilon=0} = -n(n-1)\int_\Sigma H\,dS, \\
\int_\Sigma H\boldsymbol{f}\cdot\boldsymbol{\nu}\,dS &= -\frac{1}{n}\delta\mathcal{A}(\Sigma)[\boldsymbol{f}\cdot\boldsymbol{\nu}] = -\frac{1}{n}\frac{d}{d\varepsilon}\mathcal{A}(\Sigma_\varepsilon)\bigg|_{\varepsilon=0} = -\mathcal{A}(\Sigma).
\end{aligned}$$

By (11) and the definition $\int_\Sigma \boldsymbol{f}\cdot\boldsymbol{\nu}\,dS = -(n+1)\mathcal{V}(\Sigma)$, we have

$$\begin{aligned}
\int_\Sigma \tilde{R}\boldsymbol{f}\cdot\boldsymbol{\nu}\,dS &= \int_\Sigma (R-\bar{R})\boldsymbol{f}\cdot\boldsymbol{\nu}\,dS \\
&= -n(n-1)\int_\Sigma H\,dS + (n+1)\bar{R}\mathcal{V}(\Sigma) \\
&= -n(n-1)\bar{H}\mathcal{A}(\Sigma) + (n+1)\bar{R}\mathcal{V}(\Sigma),
\end{aligned}$$

and

$$\mathcal{A}(\Sigma) - (n+1)\bar{H}\mathcal{V}(\Sigma) = -\int_\Sigma (H-\bar{H})\boldsymbol{f}\cdot\boldsymbol{\nu}\,dS = -\int_\Sigma \tilde{H}\boldsymbol{f}\cdot\boldsymbol{\nu}\,dS.$$

Therefore, it holds that

$$\begin{aligned}
\bar{H}\int_\Sigma \tilde{R}\boldsymbol{f}\cdot\boldsymbol{\nu}\,dS &= -n(n-1)\bar{H}^2\mathcal{A}(\Sigma) + (n+1)\bar{H}\bar{R}\mathcal{V}(\Sigma) \\
&= -n(n-1)\bar{H}^2\mathcal{A}(\Sigma) + \bar{R}\left(\mathcal{A}(\Sigma) + \int_\Sigma \tilde{H}\boldsymbol{f}\cdot\boldsymbol{\nu}\,dS\right) \\
&= \left\{-n(n-1)\bar{H}^2 + \bar{R}\right\}\mathcal{A}(\Sigma) + \bar{R}\int_\Sigma \tilde{H}\boldsymbol{f}\cdot\boldsymbol{\nu}\,dS \\
&= \left\{-n(n-1)H^2 + R + n(n-1)\left(H^2-\bar{H}^2\right) - \tilde{R}\right\}\mathcal{A}(\Sigma) \\
&\quad + \bar{R}\int_\Sigma \tilde{H}\boldsymbol{f}\cdot\boldsymbol{\nu}\,dS
\end{aligned}$$

$$= \left\{ -n(n-1)H^2 + R + n(n-1)(\tilde{H} + 2\bar{H})\tilde{H} - \tilde{R} \right\} \mathcal{A}(\Sigma) + \bar{R} \int_\Sigma \tilde{H} f \cdot \nu \, dS.$$

Since

$$n(n-1)H^2 - R = (nH)^2 - nH^2 - R = \left(\sum_i \kappa_i \right)^2 - nH^2 - 2\sum_{i<j} \kappa_i \kappa_j = \sum_i \kappa_i^2 - nH^2 = \sum_i (\kappa_i - H)^2,$$

we obtain the assertion. □

Proof of Proposition 3 If $\tilde{H} = 0$ and if $\tilde{R} = 0$, then the hypersurface is totally umbilic by Lemma 1. Such a closed hypersurface is a (single/multi-fold) sphere. □

The equivalence between the assertion of Theorem 2 (ii) with (ii)′ is clear by Proposition 3.

Next we give the direct proof of Theorem 2 (ii).

Proof of Theorem 2 (ii) Assume that Σ_0 is a (single/multi-fold) sphere or that $H \equiv c_0$. Because the mean curvature is constant, we have $P(\Sigma_0)(\delta\mathcal{W}(\Sigma_0)) = \widetilde{\delta\mathcal{W}(\Sigma_0)}$ and

$$\delta\mathcal{W}(\Sigma_0) = (\bar{H}_0 - c_0)\left\{ \frac{n^2}{2} \bar{H}_0(\bar{H}_0 + c_0) - \bar{R}_0 - \tilde{R}_0 \right\}.$$

Hence, we obtain

$$P(\Sigma_0)\big(\delta\mathcal{W}(\Sigma_0)\big) = -(\bar{H}_0 - c_0)\tilde{R}_0.$$

When Σ_0 is a sphere, it holds that $\tilde{R}_0 \equiv 0$. Thus, $P(\Sigma_0)(\delta\mathcal{W}(\Sigma_0)) \equiv 0$ when Σ_0 is a sphere or $H_0 \equiv c_0$. Hence, Σ_0 is a stationary solution of the Helfrich flow, and therefore $\Sigma(t) \equiv \Sigma_0$ is a global Helfrich flow with $\Sigma(0) = \Sigma_0$. □

As a consequence of Proposition 3, we obtain the following fact concerning the stationary Helfrich problem.

Theorem 3 *Assume that Σ is a solution of the stationary Helfrich problem $P(\Sigma)(\delta\mathcal{W}(\Sigma)) = 0$ with constant mean curvature. Then either Σ is a (single/ multi)-fold sphere or $H \equiv c_0$.*

Proof By the assumption, Σ satisfies

$$0 = P(\Sigma)\big(\delta\mathcal{W}(\Sigma)\big) = (\bar{H} - c_0)\tilde{R}.$$

If $H = \bar{H} \neq c_0$, then $\tilde{R} \equiv 0$. Consequently, Σ is a sphere by Proposition 3. □

4 On the Uniqueness for the Degenerate Case

We discuss the uniqueness for case (ii) in Theorem 2. Assume that Σ_0 is a (single/multi)-fold sphere or that the mean curvature is c_0 everywhere. Then, $\Sigma(t) \equiv \Sigma_0$ is a solution of the Helfrich flow, that is, a stationary solution. We look for cases in which this solution is unique with $\Sigma(0) = \Sigma_0$. The uniqueness is derived from $G(t) \equiv 0$ and $(\bar{H}(t) - c_0)\tilde{R}(t) = 0$. Indeed, $G(t) = 0$ implies that the mean curvature is constant, and that

$$\begin{aligned} V(t) &= -P\big(\Sigma(t)\big)\big(\delta\mathcal{W}(\Sigma(t))\big) \\ &= -\delta\mathcal{W}\big(\Sigma(t)\big) + \frac{1}{\mathcal{A}(\Sigma(t))}\int_\Sigma \delta\mathcal{W}\big(\Sigma(t)\big)\,dS \\ &= \big(\bar{H}(t) - c_0\big)\tilde{R}(t) = 0. \end{aligned}$$

We have three cases in which the uniqueness holds.

Theorem 4 *Assume that $G(\Sigma_0) = 0$. Then, $\{\Sigma(t) \equiv \Sigma_0\}$ is the unique global solution of the Helfrich flow with $\Sigma(0) = \Sigma_0$ provided one of the following conditions holds.*

(1) $n = 1$.
(2) Σ_0 *is a single-fold sphere.*
(3) $H_0 \equiv c_0$.

Proof (1) When $n = 1$, the integral $\int_\Sigma H\,dS$ is a constant multiple of the rotation number. Therefore, it does not depend on t. From (9), it holds that

$$G\big(\Sigma(t)\big) = \mathcal{A}_0 \int_\Sigma \tilde{H}^2\,dS = \mathcal{A}_0 \int_\Sigma (H - \bar{H})^2\,dS = \mathcal{A}_0\Big(\int_\Sigma H^2\,dS - \mathcal{A}_0\bar{H}^2\Big).$$

We have

$$2\mathcal{W}(\Sigma) = \int_\Sigma H^2 dS - 2c_0\mathcal{A}_0\bar{H} + c_0^2\mathcal{A}_0$$

by the definition of $\mathcal{W}$. Consequently, we obtain

$$\frac{d}{dt}G\big(\Sigma(t)\big) = \mathcal{A}_0\frac{d}{dt}\int_\Sigma H^2\,dS = 2\mathcal{A}_0\frac{d}{dt}\mathcal{W}\big(\Sigma(t)\big) = -2\mathcal{A}_0\big\|V(t)\big\|^2 \leqq 0.$$

Combining this with $G(\Sigma(t)) \geqq 0$ (see (9)), it holds that $G(\Sigma(t)) \equiv 0$ provided that $G(\Sigma_0) = 0$. Using the above relation again, we have $V(t) \equiv 0$, that is, $\Sigma(t)$ is stationary.

(2) Let Σ_0 be a single-fold sphere. For a short time-interval, the immersion $\Sigma(t) \subset \mathbb{R}^{n+1}$ is an embedding. If follows from the isoperimetric inequality and constraints $\mathcal{A}(\Sigma(t)) \equiv \mathcal{A}_0$, $\mathcal{V}(\Sigma(t)) \equiv \mathcal{V}$ that $\Sigma(t)$ is congruent with Σ_0. Consequently, $G(t) = 0$ and $\tilde{R}(t) = 0$ on the interval. Repeating the argument, we have $\Sigma(t) \equiv \Sigma_0$ on $[0, \infty)$.

(3) Let $H_0 \equiv c_0$. Since $\mathcal{W}(\Sigma_0) = 0$, Σ_0 is a global minimizer. Because $\mathcal{W}(\Sigma(t))$ is not increasing, the mean curvature of $\Sigma(t)$ is c_0 everywhere. Consequently, $G(t) = 0$ and $(\bar{H}(t) - c_0)\tilde{R}(t) = 0$ hold. □

5 Open Problems

The Helfrich flow can be constructed, at least locally, provided that the Gramian is positive. When the Gramian vanishes, we face difficulties. Indeed, the following points are unclear.

Problem 1 The solvability and uniqueness when $G(\Sigma_0) = 0$ but $(\bar{H}_0 - c_0)\tilde{R}_0 \neq 0$. The Wente torus is an example of such a Σ_0.

Problem 2 The uniqueness in case (ii) of Theorem 2: we have only a partial answer (Theorem 4).

Problem 3 Assume $G(\Sigma_0) > 0$. Then we have constructed the Helfrich flow with $G(\Sigma(t)) > 0$ locally. We want to extend this flow. It is not clear that the Gramian $G(\Sigma(t))$ maintains the positivity as long as flow exists.

For Problem 1, we have the following fact.

Lemma 2 *Let the initial hypersurface Σ_0 satisfy $G(\Sigma_0) = 0$ but $(\bar{H}_0 - c_0)\tilde{R}_0 \neq 0$. Suppose that the (local) Helfrich flow exists with $\Sigma(0) = \Sigma_0$, and that it is smooth up to $t = 0$. Then $G(\Sigma(t)) > 0$ for small $t > 0$.*

Proof Because of (9), it suffices to show that $\tilde{H}(t) \not\equiv 0$. If $\tilde{H}(t) \equiv 0$, then $\mathcal{A}(\Sigma(t)) = (n+1)\bar{H}(t)\mathcal{V}(\Sigma(t))$ (see the proof of Lemma 1). Since $\tilde{H}_0 \equiv 0$, we have $\mathcal{A}(\Sigma_0) = (n+1)\bar{H}_0\mathcal{V}(\Sigma_0)$. Since both $\mathcal{A}(\Sigma(t))$ and $\mathcal{V}(\Sigma(t))$ are independent of t, we have $\bar{H}(t) = \bar{H}_0$. On the other hand, because of $V(0) = (\bar{H}_0 - c_0)\tilde{R}_0 \neq 0$, $\delta \int_\Sigma H\, dS = -\frac{1}{n} R$, and $\int_\Sigma \tilde{R}_0 dS = 0$, we have

$$\begin{aligned} \frac{d}{dt}\bar{H}(t)\, dt\bigg|_{t=0} &= \mathcal{A}(\Sigma_0)\frac{d}{dt}\int_{\Sigma(t)} H(t)\, dS\bigg|_{t=0} = -\frac{\mathcal{A}(\Sigma_0)}{n}\int_{\Sigma_0} R_0 V(0)\, dS \\ &= -\frac{\mathcal{A}(\Sigma_0)}{n}\int_{\Sigma_0} (\bar{R}_0 + \tilde{R}_0)(\bar{H}_0 - c_0)\tilde{R}_0\, dS \neq 0 \\ &= -\frac{\mathcal{A}(\Sigma_0)}{n}\int_{\Sigma_0} (\bar{H}_0 - c_0)\tilde{R}_0^2\, dS \neq 0. \end{aligned}$$

Hence, $\bar{H}(t) \neq \bar{H}_0$ for small $t > 0$. □

Consequently, assuming smoothness, we may consider

$$V = -\delta\mathcal{W}(\Sigma) - \lambda_1\delta\mathcal{A}(\Sigma) - \lambda_2\mathcal{V}(\Sigma),$$

where

$$\lambda_1 = \frac{\langle \delta\mathcal{W}(\Sigma), \tilde{H}\rangle}{n\|\tilde{H}\|^2_{L^2(\Sigma)}}, \qquad \lambda_2 = \frac{\langle \delta\mathcal{W}(\Sigma), 1\rangle}{\mathcal{A}(\Sigma)} - \lambda_1 n\bar{H}.$$

This is the same equation as found in case (i) in Theorem 1. $G(\Sigma_0) = 0$ implies that $\tilde{H} \to 0$ as $t \to +0$. Hence, both the numerator and the denominator of λ_1 also tend to 0. Therefore, the boundedness of the Lagrange multiplier λ_j is unclear. Hence, to construct the flow, we must investigate the precise behavior of λ_j.

For Problem 3, we assume that there exists $T_* \in (0, \infty)$ such that $G(\Sigma(t)) > 0$ for $t < T_*$ and that $G(\Sigma(t))$ tends to 0 as $t \uparrow T_*$. Then the boundedness of λ_j near $t = T_*$ is unclear. Therefore, the situation is similar to Problem 1: we must study the behavior of λ_j as $t \uparrow T_*$. Furthermore, even if such T_* exists, the flow is extendable beyond T_* provided that Problem 1 is solved affirmatively.

We have assumed the smoothness up to $t = 0$ or $t = T_*$ in the above argument. Hence we must investigate the regularity or the singular behavior of Helfrich flow.

Acknowledgements The author thanks referees for various useful comments.

References

1. Amann, H.: Nonhomogeneous linear and quasilinear elliptic and parabolic boundary value problems. In: Schmeisser, H.-J., Triebel, H. (eds.) Function Spaces, Differential Operators and Nonlinear Analysis, Friedrichroda, 1992. Teubner-Text Math., vol. 133, pp. 9–126. Trubner, Stuttgart (1993)
2. Au, T.K.-K., Wan, T.Y.-H.: Analysis on an ODE arisen from studying the shape of a red blood cell. J. Math. Anal. Appl. **282**(1), 279–295 (2003)
3. Brailsfold, J.D.: Mechnoelastic properties of biological membranes. In: Membrane Fluidity in Biology, vol. 1. Concepts of Membrane Structure, pp. 291–319. Academic Press, San Diego (1983)
4. Chen, B.-y.: On a variational problem on hypersurfaces. J. Lond. Math. Soc. (2) **6**, 321–325 (1973)
5. Helfrich, W.: Elastic properties of lipid bilayers: Theory and possible experiments. Z. Naturforsch. **28c**, 693–703 (1973)
6. Jenkins, J.T.: Static equilibrium configurations of a model red blood cell. J. Math. Biol. **4**, 149–169 (1977)
7. Kohsaka, Y., Nagasawa, T.: On the existence for the Helfrich flow and its center manifold near spheres. Differ. Integral Equ. **19**(2), 121–142 (2006)
8. Kurihara, T., Nagasawa, T.: On the gradient flow for a shape optimization problem of plane curves as a singular limit. Saitama Math. J. **24**, 43–75 (2006/2007)
9. Nagasawa, T., Takagi, I.: Bifurcating critical points of bending energy with constraints related to the shape of red blood cells. Calc. Var. Partial Differ. Equ. **16**(1), 63–111 (2003)
10. Nagasawa, T., Takagi, I.: in preparation
11. Nagasawa, T., Yi, T.: Local existence and uniqueness for the n-dimensional Helfrich flow as a projected gradient flow. Hokkaido Math. J. **41**(2), 209–226 (2012)
12. Ou-Yang, Z.-c., Helfrich, W.: Bending energy of vesicle membranes: general expressions for the first, second, and third variation of the shape energy and applications to spheres and cylinders. Phys. Rev. A **39**(10), 5280–5288 (1989)
13. Peterson, M.A.: An instability of the red blood cell shape. J. Appl. Phys. **57**(5), 1739–1742 (1985)
14. Shikhman, V., Stein, O.: Constrained optimization: projected gradient flows. J. Optim. Theory Appl. **140**(1), 117–130 (2009)

15. Watanabe, K.: Plane domains which are spectrally determined. Ann. Glob. Anal. Geom. **18**(5), 447–475 (2000)
16. Watanabe, K.: Plane domains which are spectrally determined. II. J. Inequal. Appl. **7**(1), 25–47 (2002)

Ground States for Elliptic Equations in $\mathbb{R}^2$ with Exponential Critical Growth

Bernhard Ruf and Federica Sani

Abstract In this paper, we obtain a mountain pass characterization of ground state solutions for some class of elliptic equations in $\mathbb{R}^2$ with nonlinearities in the critical (exponential) growth range.

Keywords Ground state solutions · Elliptic equations in $\mathbb{R}^2$ · Exponential critical growth · Variational methods

1 Introduction

This paper is concerned with the existence of solutions of a nonlinear scalar field equation of the form

$$-\Delta u = g(u) \quad \text{in } \mathbb{R}^2, \ u \in H^1(\mathbb{R}^2), \tag{1}$$

and in particular we will study the following problem

$$-\Delta u + u = f(u) \quad \text{in } \mathbb{R}^2, \ u \in H^1(\mathbb{R}^2), \tag{2}$$

that is, problem (1) with $g(s) := f(s) - s$.

The study of these kind of problems is motivated by applications in many areas of mathematical physics. In particular, solutions of (2) provide stationary states for the nonlinear Klein-Gordon equation and for the nonlinear Schrödinger equation.

Problem (1) has been extensively studied starting from the fundamental papers due to Berestycki and Lions [4] and to Berestycki, Gallouët and Kavian [5]. We recall that these papers are both concerned with *subcritical* nonlinearities, in particular in [4] the authors treated nonlinearities with subcritical *polynomial* growth, while in [5] the authors treated nonlinearities with subcritical *exponential* growth. From now on, we will focus our attention on the case when the nonlinear term is of exponential

B. Ruf (✉) · F. Sani
Dipartimento di Matematica Federigo Enriques, Università degli Studi di Milano, via Saldini 50, Milano, Italy
e-mail: bernhard.ruf@unimi.it

F. Sani
e-mail: federica.sani@unimi.it

R. Magnanini et al. (eds.), *Geometric Properties for Parabolic and Elliptic PDE's*,
Springer INdAM Series 2, DOI 10.1007/978-88-470-2841-8_16,

type, since our aim is to study problem (2) with a nonlinearity f exhibiting a critical exponential growth.

The maximal growth which can be treated variationally in the Sobolev space $H^1(\mathbb{R}^2)$ is given by the *Trudinger-Moser inequality*:

Theorem 1 [10, Theorem 1.1] *There exists a constant* $C > 0$ *such that*

$$\sup_{u\in H^1(\mathbb{R}^2), \|u\|_{H^1}\le 1} \int_{\mathbb{R}^2} \left(e^{4\pi u^2} - 1\right) dx \le C \tag{3}$$

where $\|u\|_{H^1}^2 := \|\nabla u\|_2^2 + \|u\|_2^2$ *is the standard Sobolev norm. This inequality is sharp*: *if we replace the exponent* 4π *with any* $\alpha > 4\pi$ *the supremum is infinite.*

In view of this inequality we say that a nonlinearity f has *critical growth* if there exists $\alpha_0 > 0$ such that

$$\lim_{|s|\to+\infty} \frac{|f(s)|}{e^{\alpha s^2}} = \begin{cases} 0 & \text{for } \alpha > \alpha_0, \\ +\infty & \text{for } \alpha < \alpha_0. \end{cases}$$

Our aim is to obtain a mountain pass characterization of ground state solutions of problem (2). The natural functional corresponding to a variational approach to problem (2) is

$$\begin{aligned} I(u) &:= \frac{1}{2}\int_{\mathbb{R}^2} \left(|\nabla u|^2 + u^2\right) dx - \int_{\mathbb{R}^2} F(u)\, dx \\ &= \frac{1}{2}\int_{\mathbb{R}^2} |\nabla u|^2\, dx - \int_{\mathbb{R}^2} G(u)\, dx, \quad u \in H^1\left(\mathbb{R}^2\right), \end{aligned}$$

where $F(s) := \int_0^s f(t)\, dt$ and $G(s) := \int_0^s g(t)\, dt$. We will say that I has a *mountain pass geometry*, if the following conditions hold:

(I_0) $I(0) = 0$;

(I_1) there exist $\rho, a > 0$ such that $I(u) \ge a > 0$ for any $u \in H^1(\mathbb{R}^2)$ with $\|u\|_{H^1} = \rho$;

(I_2) there exists $u_0 \in H^1(\mathbb{R}^2)$ such that $\|u_0\|_{H^1} > \rho$ and $I(u_0) < 0$.

We will always denote by $c \in \mathbb{R}$ the *mountain pass value*

$$c := \inf_{\gamma\in\Gamma} \sup_{t\in[0,1]} I\left(\gamma(t)\right), \quad \Gamma := \left\{\gamma \in \mathscr{C}\left([0, 1], H^1\left(\mathbb{R}^2\right)\right) \mid \gamma(0) = 0, I\left(\gamma(1)\right) < 0\right\}.$$

We recall also that a solution u of problem (2) is a *ground state* if $I(u) = m$ with

$$m := \inf\left\{I(u) \mid u \in H^1\left(\mathbb{R}^2\right) \setminus \{0\} \text{ is a solution of (2)}\right\}.$$

In [8], Jeanjean and Tanaka obtain a mountain pass characterization of ground state solutions for the more general nonlinear scalar field equation (1) in the case when the nonlinearity g (not necessarily of the form $f(s) - s$) has a *subcritical* exponential growth.

Theorem 2 [8] *Assume*

(g_0) $g : \mathbb{R} \to \mathbb{R}$ *is continuous and odd*;

(g_1) $\lim_{s\to 0} \frac{g(s)}{s} = -\nu < 0$;
(g_2) *for any* $\alpha > 0$ *there exists* $C_\alpha > 0$ *such that* $|g(s)| \le C_\alpha e^{\alpha s^2}$ *for all* $s \ge 0$;
(g_3) *there exists* $s_0 > 0$ *such that* $G(s_0) > 0$.

Then the functional $I(u) := \frac{1}{2}\int_{\mathbb{R}^2} |\nabla u|^2\,dx - \int_{\mathbb{R}^2} G(u)\,dx$ *belongs to* $\mathscr{C}^1(H^1(\mathbb{R}^2),\mathbb{R})$ *and has a mountain pass geometry. Moreover the mountain pass value* c *is a critical value and* $0 < c = m$.

Recently Alves, Montenegro and Souto [3] improved the arguments in [8], assuming $g(s) = f(s) - s$ and considering nonlinearities with *critical* exponential growth.

Theorem 3 [3] *Assume that*

(f_0) $f:\mathbb{R}\to\mathbb{R}$ *is continuous and has critical exponential growth with* $\alpha_0 = 4\pi$;
(f_1) $\lim_{s\to 0} \frac{f(s)}{s} = 0$;
(f_2) $0 < 2F(s) \le f(s)s$ *for any* $s \in \mathbb{R}\setminus\{0\}$;
(f_η) *there exists* $\eta > 0$ *and* $q \in (2,+\infty)$ *such that* $f(s) \ge \eta s^{q-1}$ *for all* $s \ge 0$.

If (f_η) *holds with*

$$\eta > \left(\frac{q-2}{q}\right)^{\frac{q-2}{q}} C_q^{\frac{q}{2}} \tag{4}$$

where $C_q > 0$ *is the best constant of the Sobolev embedding* $H^1(\mathbb{R}^2) \hookrightarrow L^q(\mathbb{R}^2)$, *namely*

$$C_q \|u\|_q^2 \le \|u\|_{H_1}^2 \quad \forall u \in H^1(\mathbb{R}^2).$$

Then the mountain pass value c *is a critical value and gives the ground state level, namely* $0 < c = m$.

To obtain our results, we will follow some ideas introduced in [3].

2 Main Results

Our main result is concerned with the particular case when $f(s) = \lambda s e^{4\pi s^2}$ where $0 < \lambda < 1$.

Theorem 4 *Let* $0 < \lambda < 1$ *and let*

$$f(s) := \lambda s e^{4\pi s^2} \quad \forall s \in \mathbb{R}. \tag{5}$$

Then $I \in \mathscr{C}^1(H^1(\mathbb{R}^2),\mathbb{R})$ *has a mountain pass geometry, the mountain pass value* c *is a critical value and gives the ground state level, namely* $0 < c = m$.

Moreover, replacing assumption (f_η) of Theorem 3 with the following more natural assumption

(f_3) $\lim_{|s|\to+\infty} \frac{sf(s)}{e^{4\pi s^2}} \geq \beta_0 > 0$,

we obtain the same result as in [3] (see Theorem 3 above).

Theorem 5 *Assume* $(f_0), (f_1), (f_2)$ *and* (f_3). *Then* $I \in \mathscr{C}^1(H^1(\mathbb{R}^2), \mathbb{R})$ *has a mountain pass geometry, the mountain pass value* c *is a critical value and* $0 < c = m$.

Comparing Theorem 5 with the result obtained by Alves, Montenegro and Souto (see Theorem 3 above), we see that their hypothesis (f_η) about the behavior of f near zero is replaced by an assumption, i.e. (f_3), at infinity.

We recall that assumption (f_3) for bounded domains was introduced in [1] (see also [6]) to obtain an existence result for elliptic equations with nonlinearities in the critical exponential growth range in bounded domains of $\mathbb{R}^2$. In a subsequent paper, [7], (f_3) was taken into account to prove an existence result for analogous equations in the whole space $\mathbb{R}^2$.

To prove Theorems 4 and 5 we will follow the methods of [3] which improve the ideas introduced in [8]. In the proof of Theorem 2 a key argument is the existence of a solution of problem (1) given in [5]. In [5] it was shown that under the assumptions (g_0), (g_1), (g_2) and (g_3) the nonlinear scalar field equation (1) possesses a nontrivial ground state solution by means of the constrained minimization method

$$\inf\left\{ \frac{1}{2}\int_{\mathbb{R}^2} |\nabla u|^2\,dx \;\middle|\; u \in H^1(\mathbb{R}^2)\setminus\{0\},\ \int_{\mathbb{R}^2} G(u)\,dx = 0 \right\}.$$

The main difficulty, as highlighted in [3], for the proof of Theorem 3 is indeed to show that the infimum

$$A := \inf\left\{ \frac{1}{2}\int_{\mathbb{R}^2} |\nabla u|^2\,dx \;\middle|\; u \in H^1(\mathbb{R}^2)\setminus\{0\},\ \int_{\mathbb{R}^2} G(u)\,dx = 0 \right\}$$

is achieved, provided that (f_0), (f_1), (f_2) and (f_η) with $\eta > 0$ as in (4) hold. Therefore we point out that, following [3], as a by-product of the proofs of Theorems 4 and 5 we have

Proposition 1 *Assume either* f *is of the form* (5) *with* $0 < \lambda < 1$ *or assume* (f_0), (f_1), (f_2) *and* (f_3). *Then* A *is attained and the minimizer is, under a suitable change of scale, a solution of problem* (2). *In particular* $m \leq A$.

This paper is organized as follows. In Sect. 3 we show that the functional I has a mountain pass geometry and in Sect. 4 we introduce some preliminary results. In Sect. 5 we obtain a precise estimate for the mountain pass level c that will enable us to prove, in Sect. 6, Proposition 1. Finally in Sect. 7 we prove the main theorems, Theorems 4 and 5, and the following

Proposition 2 *Assume either* f *is of the form* (5) *with* $0 < \lambda < 1$ *or assume* (f_0), (f_1), (f_2) *and* (f_3). *Then the minimizer* $u \in H^1(\mathbb{R}^2)$ *of* A *is a ground state solution of problem* (2), *that is* $m = A$.

3 Mountain Pass Geometry

If f is as in (5) with $0 < \lambda < 1$ then, for fixed $q > 2$ we have the existence of two constants $c_1, c_2 > 0$ such that

$$|f(s)| \leq c_1|s| + c_2|s|^{q-1}(e^{4\pi s^2} - 1) \quad \forall s \in \mathbb{R}, \tag{6}$$

moreover, fixed $q > 2$ we have that for any $\varepsilon > 0$ there exists a constant $C(q,\varepsilon) > 0$ such that

$$F(s) \leq \left(\frac{\lambda}{2} + \varepsilon\right)s^2 + C(q,\varepsilon)|s|^q(e^{4\pi s^2} - 1) \quad \forall s \in \mathbb{R}. \tag{7}$$

Note that (7) implies that $F(u) \in L^1(\mathbb{R}^2)$ for any $u \in H^1(\mathbb{R}^2)$ and thus the functional $I : H^1(\mathbb{R}^2) \to \mathbb{R}$ is well defined. Furthermore, from (6) and using standard arguments (see [4, Theorem A.VI]), it follows that $I \in \mathscr{C}^1(H^1(\mathbb{R}^2), \mathbb{R})$.

Similarly in the case when (f_0) and (f_1) hold, fixed $q > 2$, for any $\alpha > 4\pi$ and any $\varepsilon > 0$ we have the existence of a constant $C(q,\alpha,\varepsilon) > 0$ such that

$$|f(s)| \leq \varepsilon|s| + C(q,\alpha,\varepsilon)|s|^{q-1}(e^{\alpha s^2} - 1) \quad \forall s \in \mathbb{R},$$

and if in addition (f_2) holds then

$$F(s) \leq \frac{\varepsilon}{2}s^2 + C(q,\alpha,\varepsilon)|s|^q(e^{\alpha s^2} - 1) \quad \forall s \in \mathbb{R}. \tag{8}$$

Therefore also in the case when (f_0), (f_1) and (f_2) hold we have that the functional I is well defined and of class $\mathscr{C}^1$ on $H^1(\mathbb{R}^2)$.

Obviously $I(0) = 0$, namely (I_0) holds. Now we prove that I satisfies also (I_1).

Lemma 1 *Assume either f is of the form* (5) *with $0 < \lambda < 1$ or assume (f_0), (f_1) and (f_2). Then there exist $\rho, a > 0$ such that $I(u) \geq a > 0$ for any $u \in H^1(\mathbb{R}^2)$ with $\|u\|_{H^1} = \rho$.*

Proof We begin considering the case when f is of the form (5) with $0 < \lambda < 1$. Fixed $q > 2$, for any $u \in H^1(\mathbb{R}^2)$ we have

$$\begin{aligned}\int_{\mathbb{R}^2} |u|^q (e^{4\pi u^2} - 1)\,dx &\leq \|u\|_{2q}^q \left(\int_{\mathbb{R}^2} (e^{4\pi u^2} - 1)^2\,dx\right)^{\frac{1}{2}} \\ &\leq \overline{C}_1 \|u\|_{H^1}^q \left(\int_{\mathbb{R}^2} (e^{8\pi u^2} - 1)\,dx\right)^{\frac{1}{2}}\end{aligned}$$

where $\overline{C}_1 > 0$ is a constant independent of u and we used the fact that the embedding $H^1(\mathbb{R}^2) \hookrightarrow L^{2q}(\mathbb{R}^2)$ is continuous. Moreover, recalling the *Trudinger-Moser inequality* (3), we have the existence of a constant $\overline{C}_2 > 0$ such that

$$\int_{\mathbb{R}^2} (e^{8\pi u^2} - 1)\,dx = \int_{\mathbb{R}^2} \left(e^{8\pi\|u\|_{H^1}^2 (\frac{u}{\|u\|_{H^1}})^2} - 1\right)dx \leq \overline{C}_2$$

for any $u \in H^1(\mathbb{R}^2)$ with $8\pi \|u\|_{H^1}^2 \leq 4\pi$. Therefore applying (7), we get for any $\varepsilon > 0$

$$\int_{\mathbb{R}^2} F(u)\,dx \leq \left(\frac{\lambda}{2} + \varepsilon\right)\|u\|_{H^1}^2 + \overline{C}(q,\varepsilon)\|u\|_{H^1}^q \quad \forall u \in H^1(\mathbb{R}^2),\ \|u\|_{H^1} \leq \frac{1}{\sqrt{2}}.$$

Let $0 < \rho < \frac{1}{\sqrt{2}}$. Fixed $q > 2$, for any $\varepsilon > 0$

$$I(u) \geq \frac{1}{2}(1 - \lambda - 2\varepsilon)\rho^2 - \overline{C}(q,\varepsilon)\rho^q \quad \forall u \in H^1(\mathbb{R}^2),\ \|u\|_{H^1} = \rho,$$

and choosing $\varepsilon > 0$ so that $1 - \lambda - 2\varepsilon > 0$ and ρ sufficiently small we have that

$$I(u) \geq a := \frac{1}{2}(1 - \lambda - 2\varepsilon)\rho^2 - \overline{C}(q,\varepsilon)\rho^q > 0.$$

Using (8) and arguing as before, it easy to prove that I satisfies (I_1) also in the case when (f_0), (f_1) and (f_2) hold. □

We end this section with the proof of (I_2).

Lemma 2 *Assume either f is of the form* (5) *with $0 < \lambda < 1$ or assume (f_0) and (f_2). Then there exists $u_0 \in H^1(\mathbb{R}^2)$ such that $\|u_0\|_{H^1} > \rho$ and $I(u_0) < 0$.*

Proof We begin with the case when f is of the form (5). We fix $u \in H^1(\mathbb{R}^2)$. Using the definition of F and the power series expansion of the exponential function, we get

$$I(tu) \leq \frac{1}{2}t^2\|u\|_{H^1} - \frac{\lambda}{2}t^2\|u\|_2^2 - \lambda\pi t^4\|u\|_4^4 \quad \forall t \geq 0,$$

from which we deduce that $I(tu) \to -\infty$ as $t \to +\infty$. In the case when (f_0) and (f_2) hold, in particular for any $M > 0$ there exists $C_M > 0$ such that

$$F(s) \geq M|s|^2 - C_M \quad \forall s \in \mathbb{R}.$$

Therefore, fixed $u \in \mathscr{C}_0^\infty(\mathbb{R}^2)$, for any $t \geq 0$ we can estimate

$$I(tu) \leq \frac{1}{2}t^2\|u\|_{H^1} - Mt^2\|u\|_2^2 + C_M|\operatorname{supp} u|$$

and choosing M sufficiently large we can conclude that $I(tu) \to -\infty$ as $t \to +\infty$. □

4 Preliminary Results

Let $H^1_{\mathrm{rad}}(\mathbb{R}^2)$ be the space of spherically symmetric functions belonging to $H^1(\mathbb{R}^2)$,

$$H^1_{\mathrm{rad}}(\mathbb{R}^2) := \{u \in H^1(\mathbb{R}^2) \mid u(x) = u(|x|) \text{ a.e. in } \mathbb{R}^2\}.$$

Lemma 3 *Assume that f is of the form* (5) *with $0<\lambda<1$. Let $\{u_n\}_n \subset H^1_{\mathrm{rad}}(\mathbb{R}^2)$ be a sequence satisfying*

$$\sup_n \|\nabla u_n\|_2^2 = \rho < 1 \quad \text{and} \quad \sup_n \|u_n\|_2^2 = M < +\infty. \tag{9}$$

Then $u_n \rightharpoonup u \in H^1_{\mathrm{rad}}(\mathbb{R}^2)$ in $H^1(\mathbb{R}^2)$ and

$$\int_{\mathbb{R}^2} F(u_n)\,dx - \frac{\lambda}{2}\|u_n\|_2^2 \overset{n\to+\infty}{\longrightarrow} \int_{\mathbb{R}^2} F(u)\,dx - \frac{\lambda}{2}\|u\|_2^2.$$

Before proceeding with the proof of this lemma, we point out that the *Trudinger-Moser inequality* (3) holds also if we replace the standard Sobolev norm with the modified norm

$$\|u\|^2_{H^1,\tau} := \|\nabla u\|_2^2 + \tau\|u\|_2^2 \quad \forall u \in H^1(\mathbb{R}^2)$$

where $\tau>0$. In fact in the proof of (3) given in [10] (see also [2]) the value $\tau=1$, appearing in $\|\cdot\|_{H^1} = \|\cdot\|_{H^1,1}$ as a multiplicative constant for the L^2-norm, does not play any role and can be replaced by any $\tau>0$. Therefore in [10] the author proved indeed that for any fixed $\tau>0$

$$\sup_{u\in H^1(\mathbb{R}^2),\ \|u\|_{H^1,\tau}\le 1} \int_{\mathbb{R}^2} \left(e^{4\pi u^2}-1\right) dx < +\infty. \tag{10}$$

This will enable us to prove Lemma 3.

Proof of Lemma 3 Let $\{u_n\}_n \subset H^1_{\mathrm{rad}}(\mathbb{R}^2)$ be a sequence satisfying (9) and $u_n \rightharpoonup u \in H^1_{\mathrm{rad}}(\mathbb{R}^2)$ in $H^1(\mathbb{R}^2)$. We have to show that

$$\int_{\mathbb{R}^2} P(u_n)\,dx \overset{n\to+\infty}{\longrightarrow} \int_{\mathbb{R}^2} P(u)\,dx,$$

where

$$P(s) := F(s) - \frac{\lambda}{2}s^2 = \lambda\left[\frac{1}{8\pi}\left(e^{4\pi s^2}-1\right) - \frac{1}{2}s^2\right].$$

To this aim, the idea is to apply the *compactness lemma of Strauss* (see Theorem A.I in [4]).

First, we notice that there exists $\alpha_0 > 4\pi$ such that

$$\sup_n \int_{\mathbb{R}^2} \left(e^{\alpha_0 u_n^2}-1\right) dx < +\infty. \tag{11}$$

In fact, since $\rho<1$, there exists $\sigma>0$ such that $\rho < 1-\sigma < 1$. Choosing $0<\tau<\frac{1-(\sigma+\rho)}{M}$, we have that $\|u_n\|^2_{H^1,\tau} < 1-\sigma$ for any $n\ge 1$. Therefore applying (10), we can conclude that

$$\sup_n \int_{\mathbb{R}^2} \left(e^{\alpha u_n^2}-1\right) dx < +\infty \quad \text{for any } 0<\alpha\le \frac{4\pi}{1-\sigma}$$

and, in particular, this last inequality holds for $4\pi < \alpha \le \frac{4\pi}{1-\sigma}$.

It is easy to see that

$$\lim_{s\to 0}\frac{P(s)}{e^{\alpha_0 s^2}-1}=0$$

and, since $\alpha_0>4\pi$, we have also that

$$\lim_{|s|\to+\infty}\frac{P(s)}{e^{\alpha_0 s^2}-1}=0.$$

Moreover, recalling that the embedding $H^1_{\mathrm{rad}}(\mathbb{R}^2)\hookrightarrow L^p(\mathbb{R}^2)$ is compact for any $p\in(2,+\infty)$, we have that $u_n\to u$ a.e. in $\mathbb{R}^2$ and this together with the continuity assumption on f leads us to deduce that $P(u_n)\to P(u)$ a.e. in $\mathbb{R}^2$. Finally, we can notice that $u_n(x)\to 0$ as $|x|\to+\infty$ uniformly with respect to n, as a consequence of the following *radial lemma*:

$$\big|v(x)\big|\le\frac{1}{\sqrt{2\pi}}\frac{1}{\sqrt{|x|}}\|v\|_{H^1}\quad\text{a.e. in }\mathbb{R}^2,\tag{12}$$

which holds for any $v\in H^1_{\mathrm{rad}}(\mathbb{R}^2)$.

Then, applying the *compactness lemma of Strauss*, we can conclude that $P(u_n)$ converges to $P(u)$ in $L^1(\mathbb{R}^2)$ as $n\to+\infty$. □

We recall that in [3] the authors proved the following result.

Lemma 4 *Assume* (f_0) *and* (f_1). *Let* $\{u_n\}_n\subset H^1_{\mathrm{rad}}(\mathbb{R}^2)$ *be a sequence satisfying conditions* (9) *of Lemma* 3. *Then* $u_n\rightharpoonup u\in H^1_{\mathrm{rad}}(\mathbb{R}^2)$ *in* $H^1(\mathbb{R}^2)$ *and*

$$\int_{\mathbb{R}^2}F(u_n)\,dx\to\int_{\mathbb{R}^2}F(u)\,dx.$$

We can notice that the proof of this lemma can be achieved arguing as in the proof of Lemma 3 but letting $P(s):=F(s)$.

We now prove that the infimum A is strictly positive, but before we point out that whenever we deal with a minimizing sequence for A, that is a sequence $\{u_n\}_n\subset H^1(\mathbb{R}^2)\setminus\{0\}$ such that

$$\int_{\mathbb{R}^2}G(u_n)\,dx=0\quad\forall n\ge 1\quad\text{and}\quad\frac{1}{2}\int_{\mathbb{R}^2}|\nabla u_n|^2\,dx\overset{n\to+\infty}{\longrightarrow}A;$$

without loss of generality we may assume that $\{u_n\}_n\subset H^1_{\mathrm{rad}}(\mathbb{R}^2)\setminus\{0\}$ and that $\|u_n\|_2=1$. In fact if $\{u_n\}_n\subset H^1(\mathbb{R}^2)\setminus\{0\}$ is a minimizing sequence for A then the sequence $\{u_n^*\}_n\subset H^1(\mathbb{R}^2)\setminus\{0\}$, where u_n^* is the spherically symmetric decreasing rearrangement of u_n, is a minimizing sequence too. Furthermore letting

$$v_n(x):=u_n\big(x\|u\|_2\big)\quad\text{for a.e. }x\in\mathbb{R}^2$$

for any $n\ge 1$, we have that

$$\frac{1}{2}\int_{\mathbb{R}^2}|\nabla v_n|^2=\frac{1}{2}\int_{\mathbb{R}^2}|\nabla u_n|^2,\qquad\int_{\mathbb{R}^2}G(v_n)\,dx=\frac{1}{\|u_n\|_2^2}\int_{\mathbb{R}^2}G(u_n)\,dx=0$$

and $\|v_n\|_2=1$.

Lemma 5 *Assume either f is of the form* (5) *with $0 < \lambda < 1$, or assume (f_0) and (f_1). Then $A > 0$.*

Proof In the case that we assume (f_0) and (f_1), since Lemma 4 holds, we can argue as in the proof of [3], Lemma 5.3 to conclude that $A > 0$. Therefore we only consider the case when $f(s) := \lambda s e^{4\pi s^2}$ with $0 < \lambda < 1$. Obviously $A \geq 0$ and we argue by contradiction assuming that $A = 0$. Then there exists $\{u_n\}_n \subset H^1_{\text{rad}}(\mathbb{R}^2) \setminus \{0\}$ with $\|u_n\|_2 = 1$ $\forall n \geq 1$ and

$$\int_{\mathbb{R}^2} G(u_n)\,dx = 0 \quad \forall n \geq 1, \qquad \frac{1}{2}\int_{\mathbb{R}^2} |\nabla u_n|^2\,dx \overset{n\to+\infty}{\longrightarrow} 0.$$

Let $u \in H^1_{\text{rad}}(\mathbb{R}^2)$ be the weak limit of $\{u_n\}_n$ in $H^1(\mathbb{R}^2)$, then from Lemma 3 it follows that

$$\int_{\mathbb{R}^2} F(u_n)\,dx - \frac{\lambda}{2} = \int_{\mathbb{R}^2} F(u_n)\,dx - \frac{\lambda}{2}\|u_n\|_2^2 \overset{n\to+\infty}{\longrightarrow} \int_{\mathbb{R}^2} F(u)\,dx - \frac{\lambda}{2}\|u\|_2^2.$$

Since

$$\begin{aligned} 0 = \int_{\mathbb{R}^2} G(u_n)\,dx &= \int_{\mathbb{R}^2} F(u_n)\,dx - \frac{1}{2}\|u_n\|_2^2 = \int_{\mathbb{R}^2} F(u_n)\,dx - \frac{1}{2}, \\ \text{i.e.} \int_{\mathbb{R}^2} F(u_n)\,dx &= \frac{1}{2}, \end{aligned} \tag{13}$$

we have that

$$\int_{\mathbb{R}^2} F(u)\,dx - \frac{\lambda}{2}\|u\|_2^2 = \frac{1}{2}(1-\lambda) > 0$$

from which it follows that $u \neq 0$. On the other hand, the weak convergence $u_n \rightharpoonup u$ in $H^1(\mathbb{R}^2)$ implies that

$$0 = \liminf_{n\to+\infty} \frac{1}{2}\int_{\mathbb{R}^2} |\nabla u_n|^2\,dx \geq \frac{1}{2}\int_{\mathbb{R}^2} |\nabla u|^2\,dx \geq 0,$$

that is $u \equiv 0$, which leads to a contradiction. □

We introduce the set $\mathscr{P}$ of non-trivial functions satisfying the Pohozaev identity

$$\mathscr{P} := \left\{ u \in H^1(\mathbb{R}^2) \setminus \{0\} \;\middle|\; \int_{\mathbb{R}^2} G(u)\,dx = 0 \right\}$$

and we can notice that

$$A = \inf_{u\in\mathscr{P}} \frac{1}{2}\int_{\mathbb{R}^2} |\nabla u|^2\,dx.$$

Since $A > 0$, arguing as in the proof of [8], Lemma 4.1 we have the following result.

Lemma 6 *Assume either f is of the form* (5) *with $0 < \lambda < 1$, or assume (f_0) and (f_1). Then*

$$\gamma\big([0,1]\big) \cap \mathscr{P} \neq 0, \quad \gamma \in \Gamma.$$

This lemma leads to the following relation between the infimum A and the mountain pass level c.

Lemma 7 *Assume either f is of the form (5) with $0 < \lambda < 1$, or assume (f_0) and (f_1). Then the infimum A satisfies the inequality $A \leq c$.*

Proof Let $\gamma \in \Gamma$ and let $t_0 \in (0, 1]$ be such that $\gamma(t_0) \in \mathscr{P}$, the existence of such a t_0 is guaranteed by Lemma 6. Since $\gamma(t_0) \in \mathscr{P}$, we have

$$I\big(\gamma(t_0)\big) = \frac{1}{2}\int_{\mathbb{R}^2} |\nabla u|^2\, dx$$

and thus

$$\max_{t\in[0,1]} I\big(\gamma(t)\big) \geq I\big(\gamma(t_0)\big) = \frac{1}{2}\int_{\mathbb{R}^2} |\nabla u|^2\, dx \geq A. \tag{14}$$

From the arbitrary choice of $\gamma \in \Gamma$, inequality (14) holds for any $\gamma \in \Gamma$ and hence $c \geq A$. □

5 Estimate of the Mountain Pass Level c

In order to get an upper bound for the mountain pass level c we will show the existence of $u \in H^1(\mathbb{R}^2)$ such that

$$\max_{t\geq 0} I(tu) < \frac{1}{2}. \tag{15}$$

This gives indeed more precise information about the mountain pass level c, in fact from (15) it is easy to deduce the existence of $\gamma \in \Gamma$ such that

$$c \leq \max_{t\in[0,1]} I\big(\gamma(t)\big) < \frac{1}{2}. \tag{16}$$

First, we consider the case when f is as in (5) with $0 < \lambda < 1$. To obtain the existence of $u \in H^1(\mathbb{R}^2)$ which satisfies the inequality (15), the fact that

$$\lim_{|s|\to+\infty} \frac{sf(s)}{e^{4\pi s^2}} = +\infty \tag{17}$$

plays an important role. In particular we can notice from (17) it follows that for fixed

$$\beta_0 > \frac{1}{\pi} \tag{18}$$

there exists $\overline{s} = \overline{s}(\beta_0) > 0$ such that

$$sf(s) \geq \beta_0 e^{4\pi s^2} \quad \forall |s| \geq \overline{s}. \tag{19}$$

We consider the modified *Moser sequence* introduced in [6]:

$$\overline{\omega}_n(x) := \frac{1}{\sqrt{2\pi}} \begin{cases} (\log n)^{\frac{1}{2}} & 0 \le |x| \le \frac{1}{n}, \\ \frac{\log \frac{1}{|x|}}{(\log n)^{\frac{1}{2}}} & \frac{1}{n} \le |x| \le 1, \\ 0 & |x| \ge 1. \end{cases}$$

We have $\overline{\omega}_n \in H_0^1(B_1) \subset H^1(\mathbb{R}^2)$, $\|\nabla \overline{\omega}_n\|_2 = 1$ and $\|\overline{\omega}_n\|_2^2 = \mathscr{O}(1/\log n)$. Then we define

$$\omega_n := \frac{\overline{\omega}_n}{\|\overline{\omega}_n\|_{H^1}}.$$

Lemma 8 *Assume f is of the form* (5) *with* $0 < \lambda < 1$. *Then there exists* $n \ge 1$ *such that*

$$\max_{t \ge 0} I(t\omega_n) < \frac{1}{2}.$$

Proof We argue by contradiction assuming that for any $n \ge 1$ we have $\max_{t \ge 0} I(t\omega_n) \ge \frac{1}{2}$. For any $n \ge 1$, let $t_n > 0$ be such that $I(t_n\omega_n) = \max_{t \ge 0} I(t\omega_n) \ge 1/2$, then we can estimate

$$\frac{1}{2} \le I(t_n\omega_n) = \frac{1}{2} t_n^2 \|\omega_n\|_{H^1}^2 - \int_{\mathbb{R}^2} F(t_n\omega_n)\, dx \le \frac{1}{2} t_n^2$$

and thus $t_n^2 \ge 1$, $\forall n \ge 1$. At $t = t_n$ we have

$$0 = \frac{d}{dt} I(t\omega_n)\bigg|_{t=t_n} = t_n - \int_{\mathbb{R}^2} f(t_n\omega_n)\omega_n\, dx,$$

which implies that

$$t_n^2 = \int_{\mathbb{R}^2} f(t_n\omega_n) t_n\omega_n\, dx. \tag{20}$$

We claim that $\{t_n\}_n \subset \mathbb{R}$ is bounded. In fact, since

$$t_n\omega_n = \frac{t_n}{\|\overline{\omega}_n\|_{H^1}} \frac{1}{\sqrt{2\pi}} \sqrt{\log n} \to +\infty \quad \text{in } B_{\frac{1}{n}},$$

it follows from (19) that at least for $n \ge 1$ sufficiently large

$$t_n^2 \ge \int_{B_{\frac{1}{n}}} f(t_n\omega_n) t_n\omega_n\, dx \ge \beta_0 \int_{B_{\frac{1}{n}}} e^{4\pi (t_n\omega_n)^2}\, dx = \frac{\pi}{n^2} \beta_0 e^{2\frac{t_n^2}{\|\overline{\omega}_n\|_{H^1}^2} \log n}. \tag{21}$$

Consequently

$$1 \ge \pi\beta_0 e^{2\frac{t_n^2}{\|\overline{\omega}_n\|_{H^1}^2} \log n - 2\log t_n - 2\log n}$$

for $n \ge 1$ sufficiently large, and thus $\{t_n\}_n$ must be bounded.

We claim that $t_n^2 \to 1$ as $n \to +\infty$. Since $t_n^2 \ge 1\ \forall n \ge 1$, we argue by contradiction assuming that $\lim_{n\to+\infty} t_n^2 > 1$. Recalling (21), for $n \ge 1$ sufficiently large we have

$$t_n^2 \geq \pi \beta_0 e^{2\log n(\frac{t_n^2}{\|\overline{\omega}_n\|^2_{H^1}}-1)}$$

and letting $n \to +\infty$ we get a contradiction with the boundedness of the sequence $\{t_n\}_n$.

In order to estimate (20) more precisely, we define the sets $A_n := \{x \in B_1 \mid t_n\omega_n(x) \geq \overline{s}\}$ and $C_n := B_1 \setminus A_n$ where $\overline{s} > 0$ is given in (19). With (20) and (19) we can estimate for any $n \geq 1$

$$\begin{aligned} t_n^2 &\geq \int_{B_1} f(t_n\omega_n)t_n\omega_n\,dx \\ &\geq \beta_0 \int_{B_1} e^{4\pi t_n^2\omega_n^2}\,dx + \int_{C_n} f(t_n\omega_n)t_n\omega_n\,dx - \beta_0 \int_{C_n} e^{4\pi t_n^2\omega_n^2}\,dx. \end{aligned} \tag{22}$$

Since $\omega_n \to 0$ a.e. in B_1, from the definition of C_n we obtain that the characteristic functions $\chi_{C_n} \to 1$ a.e. in B_1, and the Lebesgue dominated convergence theorem implies that

$$\int_{C_n} f(t_n\omega_n)t_n\omega_n\,dx \to 0, \qquad \int_{C_n} e^{4\pi t_n^2\omega_n^2}\,dx \to \pi \quad \text{as } n \to +\infty.$$

If we prove that

$$\lim_{n\to+\infty} \int_{B_1} e^{4\pi t_n^2\omega_n^2}\,dx \geq 2\pi \tag{23}$$

then by (22) $1 = \lim_{n\to+\infty} t_n^2 \geq \pi\beta_0$ which is in contradiction with (18). To end the proof it remains only to show that inequality (23) holds. Since $t_n^2 \geq 1$, we have

$$\int_{B_1} e^{4\pi t_n^2\omega_n^2}\,dx \geq \int_{B_1\setminus B_{\frac{1}{n}}} e^{4\pi \omega_n^2}\,dx = 2\pi \int_{\frac{1}{n}}^{1} e^{\frac{2}{\|\overline{\omega}_n\|^2_{H^1}}\frac{1}{\log n}\log^2(\frac{1}{s})} s\,ds$$

and if we make the change of variable

$$\tau = \frac{\log\frac{1}{s}}{\|\overline{\omega}_n\|_{H^1}\log n}$$

then we obtain the following estimate

$$\int_{B_1\setminus B_{\frac{1}{n}}} e^{4\pi t_n^2\omega_n^2}\,dx \geq 2\pi\,\|\overline{\omega}_n\|_{H^1}\log n \int_0^{\frac{1}{\|\overline{\omega}_n\|_{H^1}}} e^{2\log n(\tau^2 - \tau\|\overline{\omega}_n\|_{H^1})}\,d\tau.$$

Now it suffices to notice that

$$\tau^2 - \tau\|\overline{\omega}_n\|_{H^1} \geq \begin{cases} -\|\tau\overline{\omega}_n\|_{H^1}, & 0 \leq \tau \leq \frac{1}{2\|\overline{\omega}_n\|_{H^1}}, \\ (\frac{2}{\|\overline{\omega}_n\|_{H^1}} - \|\overline{\omega}_n\|_{H^1})(\tau - \frac{1}{\|\overline{\omega}_n\|_{H^1}}) + \frac{1}{\|\overline{\omega}_n\|^2_{H^1}} - 1, & \frac{1}{2\|\overline{\omega}_n\|_{H^1}} \leq \tau \leq \frac{1}{\|\overline{\omega}_n\|_{H^1}} \end{cases}$$

to conclude that (23) holds. □

Next, we consider the case when (f_2) and (f_3) hold. In this case, as a consequence of (f_3), we have that for any $\varepsilon > 0$ there exists $s_\varepsilon > 0$ such that $sf(s) \geq (\beta_0 - \varepsilon)e^{4\pi s^2}$ $\forall |s| \geq s_\varepsilon$. Let $r > 0$ be such that $\beta_0 > 1/(r^2\pi)$, we consider the *Moser sequence* introduced in [9]:

$$\overline{M}_n(x) := \frac{1}{\sqrt{2\pi}} \begin{cases} (\log n)^{\frac{1}{2}} & 0 \leq |x| \leq \frac{r}{n}, \\ \dfrac{\log \frac{r}{|x|}}{(\log n)^{\frac{1}{2}}} & \frac{r}{n} \leq |x| \leq r, \\ 0 & |x| \geq r. \end{cases}$$

Arguing as before (see also [7, Lemma 4.4]) we have the following result.

Lemma 9 *Assume (f_2) and (f_3). Then there exists $n \in \mathbb{N}$ such that*

$$\max_{t \geq 0} I(tM_n) < \frac{1}{2} \quad \textit{where } M_n := \frac{\overline{M}_n}{\|\overline{M}_n\|_{H^1}}.$$

6 The Infimum A Is Attained

In this section we will prove Proposition 1. We can notice that in either case, when f is of the form (5) with $0 < \lambda < 1$, or when $(f_0), (f_1), (f_2)$ and (f_3) hold, if the infimum A is attained then the minimizer $u \in H^1_{\mathrm{rad}}(\mathbb{R}^2) \setminus \{0\}$ is a solution of problem (2), under a suitable change of scale. In fact, if $u \in H^1_{\mathrm{rad}}(\mathbb{R}^2) \setminus \{0\}$ is such that

$$\frac{1}{2}\int_{\mathbb{R}^2} |\nabla u|^2\,dx = A \quad \text{and} \quad \int_{\mathbb{R}^2} G(u)\,dx = 0$$

then there exists a Lagrange multiplier $\theta \in \mathbb{R}$, namely

$$\int_{\mathbb{R}^2} \nabla u \cdot \nabla v\,dx = \theta \int_{\mathbb{R}^2} g(u)v\,dx \quad \forall v \in H^1(\mathbb{R}^2).$$

Claim *The Lagrange multiplier θ is positive.*

Proof of the claim First, we can notice that the case $\theta = 0$ does not occur, since by assumption $u \neq 0$. We infer that $\theta > 0$. In fact, suppose by contradiction that $\theta < 0$. Then arguing as in [4], we can find $w \in H^1(\mathbb{R}^2)$ satisfying

$$\int_{\mathbb{R}^2} G(u + \varepsilon w)\,dx > 0 \quad \text{and} \quad \|\nabla(u + \varepsilon w)\|_2^2 < \|\nabla u\|_2^2$$

for some $\varepsilon > 0$ sufficiently small. Moreover, we may assume that $u + \varepsilon w \neq 0$. Now we define $h \in \mathscr{C}([0, 1], \mathbb{R})$ as

$$h(t) := \int_{\mathbb{R}^2} G\big(t[u + \varepsilon w]\big)\,dx.$$

By construction $h(0) = 0$ and $h(1) > 0$.

Assume that f is of the form (5) with $0<\lambda<1$. Then from (7) it follows that

$$\begin{aligned}\left|\int_{\mathbb{R}^2} F\big(t[u+\varepsilon w]\big)\,dx\right| &\le \left(\frac{\lambda}{2}+\varepsilon\right)t^2\|u+\varepsilon w\|_2^2\\ &\quad + C(q,\varepsilon)t^q\int_{\mathbb{R}^2}|u+\varepsilon w|^q\big(e^{4\pi t^2(u+\varepsilon w)^2}-1\big)\,dx\\ &\le \left(\frac{\lambda}{2}+\varepsilon\right)t^2\|u+\varepsilon w\|_2^2\\ &\quad + C(q,\varepsilon)t^q\,\|u+\varepsilon w\|_{2q}^q\left(\int_{\mathbb{R}^2}\big(e^{8\pi t^2(u+\varepsilon w)^2}-1\big)\,dx\right)^{\frac12}.\end{aligned}$$

Since $0\le t^2\|u+\varepsilon w\|_{H^1}^2\to 0$ as $t\to 0$, there exists $t_0\in(0,1)$ such that for $0<t<t_0$ we have

$$\left|\int_{\mathbb{R}^2} F\big(t[u+\varepsilon w]\big)\,dx\right| \le \left(\frac{\lambda}{2}+\varepsilon\right)t^2\|u+\varepsilon w\|_2^2+\tilde{C}(q,\varepsilon)t^q\,\|u+\varepsilon w\|_{2q}^q,$$

as a consequence of the *Trudinger-Moser inequality* (3). Thus, for $0<t<t_0$

$$h(t)\le\left(\frac{\lambda}{2}+\varepsilon-\frac{1}{2}\right)t^2\|u+\varepsilon w\|_2^2+\tilde{C}(q,\varepsilon)t^q\,\|u+\varepsilon w\|_{2q}^q$$

and choosing $\varepsilon>0$ so small that $\frac{\lambda}{2}+\varepsilon-\frac12<0$, we deduce that $h(t)<0$ for sufficiently small $t>0$. In the case when (f_0), (f_1), (f_2) and (f_3) hold, we can achieve the same conclusion applying (8).

Hence, in either case, when f is of the form (5) or when (f_0), (f_1), (f_2) and (f_3) hold, we have $h(t)<0$ for sufficiently small $t>0$. Consequently, there exists $t_1\in(0,1)$ such that $h(t_1)=0$, which means that $t_1(u+\varepsilon w)\in H^1(\mathbb{R}^2)$ satisfies the constraint condition

$$\int_{\mathbb{R}^2} G\big(t_1[u+\varepsilon w]\big)\,dx=0$$

and, since u is a minimizer for A,

$$\frac12\|\nabla u\|_2^2\le\frac12\big\|t_1\nabla(u+\varepsilon w)\big\|_2^2<\frac12\big\|\nabla(u+\varepsilon w)\big\|_2^2<\frac12\|\nabla u\|_2^2.$$

This is a contradiction and θ must be positive; hence the claim is proved. □

Since $\theta>0$, we can set

$$u_\theta(x):=u\left(\frac{x}{\sqrt{\theta}}\right)\quad\text{for a.e. } x\in\mathbb{R}^2. \tag{24}$$

Then u_θ is a non-trivial solution of problem (2) and hence $m\le I(u_\theta)$. Moreover

$$\int_{\mathbb{R}^2}|\nabla u_\theta|^2\,dx=\int_{\mathbb{R}^2}|\nabla u|^2\,dx=A,\qquad \int_{\mathbb{R}^2}G(u_\theta)\,dx=\theta\int_{\mathbb{R}^2}G(u)\,dx=0,$$

from which we get $I(u_\theta)=\frac12\int_{\mathbb{R}^2}|\nabla u_\theta|^2\,dx=A$ and thus $m\le A$.

Therefore to prove Proposition 1, it remains to show that the infimum A is achieved. The proof in the case in which we assume (f_0), (f_1), (f_2) and (f_3) can be easily reduced to the proof of [3, Theorem 1.4]. It suffices to notice that from Lemma 7 and from inequality (16) it follows that $A < 1/2$, and thus we are in the same framework of the proof of [3, Theorem 1.4].

Proof of Proposition 1 in the case $f(s) := \lambda s e^{4\pi s^2}$ $\forall s \in \mathbb{R}$ *with* $0 < \lambda < 1$ From Lemma 7 and from inequality (16), it follows that $A < 1/2$.

Let $\{u_n\}_n \in H^1_{\text{rad}}(\mathbb{R}^2) \setminus \{0\}$, $\|u_n\|_2 = 1$, $\forall n \geq 1$, be a minimizing sequence for A:

$$\int_{\mathbb{R}^2} G(u_n)\,dx = 0 \quad \forall n \geq 1 \quad \text{and} \quad \frac{1}{2}\int_{\mathbb{R}^2} |\nabla u_n|^2\,dx \overset{n\to+\infty}{\longrightarrow} A. \tag{25}$$

We will prove that the weak limit $u \in H^1_{\text{rad}}(\mathbb{R}^2)$ of $\{u_n\}_n$ in $H^1(\mathbb{R}^2)$ is a minimizer for A.

Since

$$\limsup_{n\to+\infty} \int_{\mathbb{R}^2} |\nabla u_n|^2\,dx = 2A < 1,$$

the assumptions of Lemma 3 are satisfied. Arguing as in (13), we deduce that

$$\int_{\mathbb{R}^2} F(u)\,dx - \frac{\lambda}{2}\|u\|_2^2 = \frac{1}{2}(1-\lambda) > 0 \tag{26}$$

which in particular implies that $u \neq 0$.

From the weak convergence $u_n \rightharpoonup u$ in $H^1(\mathbb{R}^2)$, we get

$$A = \liminf_{n\to+\infty} \frac{1}{2}\int_{\mathbb{R}^2} |\nabla u_n|^2 \geq \frac{1}{2}\int_{\mathbb{R}^2} |\nabla u|^2.$$

Let

$$h(t) := \int_{\mathbb{R}^2} G(tu)\,dx = \int_{\mathbb{R}^2} F(tu)\,dx - \frac{t^2}{2}\|u\|_2^2 \quad \forall t > 0;$$

to conclude the proof it suffices to prove that $h(1) = 0$. Since $u_n \rightharpoonup u$ in $H^1(\mathbb{R}^2)$, we have $\|u\|_2^2 \leq \liminf_{n\to+\infty} \|u_n\|_2^2 = 1$ and this together with (26) gives

$$\begin{aligned} h(1) &= \int_{\mathbb{R}^2} F(u)\,dx - \frac{1}{2}\|u\|_2^2 = \int_{\mathbb{R}^2} F(u)\,dx - \frac{\lambda}{2}\|u\|_2^2 + \frac{1}{2}(\lambda - 1)\|u\|_2^2 \\ &= \frac{1}{2}(1-\lambda) + \frac{1}{2}(\lambda-1)\|u\|_2^2 = \frac{1}{2}(1-\lambda)\big(1 - \|u\|_2^2\big) \geq 0. \end{aligned}$$

We argue by contradiction assuming that $h(1) \neq 0$, that is $h(1) > 0$. Using the definition of F and the power series expansion of the exponential function, for any $t \in (0,1)$ we have

$$\begin{aligned} \int_{\mathbb{R}^2} F(tu)\,dx &\leq \frac{\lambda}{2}t^2\|u\|_2^2 + t^4\frac{\lambda}{8\pi}\sum_{j=2}^{+\infty}\frac{(4\pi)^j}{j!}\int_{\mathbb{R}^2} u^{2j}\,dx \\ &\leq \frac{\lambda}{2}t^2\|u\|_2^2 + t^4\frac{\lambda}{8\pi}\int_{\mathbb{R}^2}\big(e^{4\pi u^2} - 1\big)\,dx. \end{aligned}$$

Hence for any $t \in (0,1)$

$$h(t) \le \frac{1}{2}(\lambda - 1)t^2\|u\|_2^2 + t^4 \frac{\lambda}{8\pi}\int_{\mathbb{R}^2}\left(e^{4\pi u^2} - 1\right)dx,$$

from which we deduce that $h(t) < 0$ for $t > 0$ sufficiently small. But, by assumption, $h(1) > 0$ and thus there exists $t_0 \in (0,1)$ such that $h(t_0) = 0$. Consequently, recalling the definition of h, we have

$$A \le \frac{1}{2}\int_{\mathbb{R}^2}\left|\nabla(t_0 u)\right|^2 dx = \frac{1}{2}t_0^2\int_{\mathbb{R}^2}|\nabla u|^2 dx \le t_0^2 A < A$$

which is a contradiction. □

7 Proofs of Theorems 4 and 5

In order to prove Theorems 4 and 5 we can notice that, both in the case when f is of the form (5) with $0 < \lambda < 1$ and in the case when (f_0), (f_1), (f_2) and (f_3) hold, from Proposition 1 we have $m \le A$. Moreover, Lemma 7 tells us that $A \le c$ and hence $m \le c$. It remains only to show that

$$m \ge c \tag{27}$$

to conclude that the mountain pass level c gives the ground state level.

In [8] the authors proved the following result.

Theorem 6 [8, Lemma 2.1] *Assume (g_0), (g_1), (g_2) and (g_3) as in Theorem 2. Then for any solution u of (1) there exists a path $\gamma \in \Gamma$ such that $u \in \gamma([0,1])$ and*

$$\max_{t\in[0,1]} I\big(\gamma(t)\big) = I(u).$$

It is easy to see that the proof of this theorem works also under our assumptions and this leads to (27).

Indeed, we can notice that in this way we proved that $m = A = c$. Hence if $u \in H^1(\mathbb{R}^2)$ is a minimizer for A and we define u_θ as in (24) then u_θ is a ground state solution of problem (2). This gives the proof of Proposition 2.

Acknowledgements The authors wish to thank the anonymous referees for their useful remarks and suggestions.

References

1. Adimurthi: Existence results for the semilinear Dirichlet problem with critical growth for the n-Laplacian. Ann. Sc. Norm. Super. Pisa, Cl. Sci. **4**, 393–413 (1990)
2. Adimurthi, Yang, Y.: An interpolation of Hardy inequality and Trudinger-Moser inequality in $\mathbb{R}^N$ and its applications. Int. Math. Res. Not. **2010**(13), 2394–2426 (2010)

3. Alves, C.O., Souto, M.A.S., Montenegro, M.: Existence of a ground state solution for a nonlinear scalar field equation with critical growth. Calc. Var. Partial Differ. Equ. **43**(3–4), 537–554 (2012)
4. Berestycki, H., Lions, P.L.: Nonlinear scalar field equations. I. Existence of a ground state. Arch. Ration. Mech. Anal. **82**(4), 313–345 (1983)
5. Berestycki, H., Gallouët, T., Kavian, O.: Équations de champs scalaires euclidiens non linéaires dans le plan. C. R. Acad. Sci. Paris Sér. I Math. **297**(5), 307–310 (1983)
6. De Figueiredo, D.G., Miyagaki, O.H., Ruf, B.: Elliptic equations in $\mathbb{R}^2$ with nonlinearities in the critical growth range. Calc. Var. Partial Differ. Equ. **3**(2), 139–153 (1995)
7. Do Ó, J.M., Medeiros, E., Severo, U.: A nonhomogeneous elliptic problem involving critical growth in dimension two. J. Math. Anal. Appl. **345**(1), 286–304 (2008)
8. Jeanjean, L., Tanaka, K.: A remark on least energy solutions in $\mathbb{R}^N$. Proc. Am. Math. Soc. **131**(8), 2399–2408 (2003)
9. Moser, J.: A sharp form of an inequality by N. Trudinger. Indiana Univ. Math. J. **20**, 1077–1092 (1970/1971)
10. Ruf, B.: A sharp Trudinger-Moser type inequality for unbounded domains in $\mathbb{R}^2$. J. Funct. Anal. **219**(2), 340–367 (2005)

Stationary Level Surfaces and Liouville-Type Theorems Characterizing Hyperplanes

Shigeru Sakaguchi

Abstract We consider an entire graph $S : x_{N+1} = f(x)$, $x \in \mathbb{R}^N$ in $\mathbb{R}^{N+1}$ of a continuous real function f over $\mathbb{R}^N$ with $N \geq 1$. Let Ω be an unbounded domain in $\mathbb{R}^{N+1}$ with boundary $\partial\Omega = S$. Consider nonlinear diffusion equations of the form $\partial_t U = \Delta\phi(U)$ containing the heat equation $\partial_t U = \Delta U$. Let $U = U(X,t) = U(x, x_{N+1}, t)$ be the solution of either the initial-boundary value problem over Ω where the initial value equals zero and the boundary value equals 1, or the Cauchy problem where the initial datum is the characteristic function of the set $\mathbb{R}^{N+1} \setminus \Omega$. The problem we consider is to characterize S in such a way that there exists a stationary level surface of U in Ω.

We introduce a new class $\mathscr{A}$ of entire graphs S and, by using the sliding method due to Berestycki, Caffarelli, and Nirenberg, we show that $S \in \mathscr{A}$ must be a hyperplane if there exists a stationary level surface of U in Ω. This is an improvement of the previous result (Magnanini and Sakaguchi in J. Differ. Equ. 252:236–257, 2012, Theorem 2.3 and Remark 2.4). Next, we consider the heat equation in particular and we introduce the class $\mathscr{B}$ of entire graphs S of functions f such that $\{|f(x) - f(y)| : |x - y| \leq 1\}$ is bounded. With the help of the theory of viscosity solutions, we show that $S \in \mathscr{B}$ must be a hyperplane if there exists a stationary isothermic surface of U in Ω. This is a considerable improvement of the previous result (Magnanini and Sakaguchi in J. Differ. Equ. 248:1112–1119, 2010, Theorem 1.1, case (ii)).

Related to the problem, we consider a class $\mathscr{W}$ of Weingarten hypersurfaces in $\mathbb{R}^{N+1}$ with $N \geq 1$. Then we show that, if S belongs to $\mathscr{W}$ in the viscosity sense and S satisfies some natural geometric condition, then $S \in \mathscr{B}$ must be a hyperplane. This is also a considerable improvement of the previous result (Sakaguchi in Discrete Contin. Dyn. Syst., Ser. S 4:887–895, 2011, Theorem 1.1).

Keywords Nonlinear diffusion · Heat equation · Initial-boundary value problem · Cauchy problem · Liouville-type theorems · Hyperplanes · Stationary level surfaces · Stationary isothermic surfaces · Sliding method

S. Sakaguchi (✉)
Division of Mathematics, Graduate School of Information Sciences, Tohoku University, Sendai 980-8579, Japan
e-mail: sigersak@m.tohoku.ac.jp

R. Magnanini et al. (eds.), *Geometric Properties for Parabolic and Elliptic PDE's*, Springer INdAM Series 2, DOI 10.1007/978-88-470-2841-8_17,

1 Introduction

For $f \in C(\mathbb{R}^N)$ where $N \geq 1$, let Ω be the domain in $\mathbb{R}^{N+1}$ given by

$$\Omega = \{X = (x, x_{N+1}) \in \mathbb{R}^{N+1} : x_{N+1} > f(x)\}. \tag{1}$$

Throughout this paper we write $X = (x, x_{N+1}) \in \mathbb{R}^{N+1}$ for $x = (x_1, \ldots, x_N) \in \mathbb{R}^N$. Then we notice that $\partial\Omega = \partial(\mathbb{R}^{N+1} \setminus \overline{\Omega})$. Let $\phi : \mathbb{R} \to \mathbb{R}$ satisfy

$$\phi \in C^2(\mathbb{R}), \qquad \phi(0) = 0, \quad \text{and} \quad 0 < \delta_1 \leq \phi'(s) \leq \delta_2 \quad \text{for } s \in \mathbb{R}, \tag{2}$$

where δ_1, δ_2 are positive constants. Consider the unique bounded solution $U = U(X,t)$ of either the initial-boundary value problem:

$$\partial_t U = \Delta\phi(U) \quad \text{in } \Omega \times (0, +\infty), \tag{3}$$

$$U = 1 \quad \text{on } \partial\Omega \times (0, +\infty), \tag{4}$$

$$U = 0 \quad \text{on } \Omega \times \{0\}, \tag{5}$$

where $\Delta = \sum_{j=1}^{N+1} \frac{\partial^2}{\partial x_j^2}$, or the Cauchy problem:

$$\partial_t U = \Delta\phi(U) \quad \text{in } \mathbb{R}^{N+1} \times (0, +\infty) \quad \text{and} \quad U = \chi_{\Omega^c} \quad \text{on } \mathbb{R}^{N+1} \times \{0\}; \tag{6}$$

here χ_{Ω^c} denotes the characteristic function of the set $\Omega^c = \mathbb{R}^{N+1} \setminus \Omega$. Note that the uniqueness of the solution of either problem (3)–(5) or problem (6) follows from the comparison principle (see [8, Theorem A.1, p. 253]). We consider the solution $U \in C^{2,1}(\Omega \times (0, +\infty)) \cap L^\infty(\Omega \times (0, +\infty)) \cap C(\overline{\Omega} \times (0, +\infty))$ such that $U(\cdot, t) \to 0$ in $L^1_{loc}(\Omega)$ as $t \to 0^+$ for problem (3)–(5). For problem (6), we consider the solution $U \in C^{2,1}(\mathbb{R}^{N+1} \times (0, +\infty)) \cap L^\infty(\mathbb{R}^{N+1} \times (0, +\infty))$ such that $U(\cdot, t) \to \chi_{\Omega^c}(\cdot)$ in $L^1_{loc}(\mathbb{R}^{N+1})$ as $t \to 0^+$.

By the strong comparison principle, we know that

$$0 < U < 1 \quad \text{and} \quad \frac{\partial U}{\partial x_{N+1}} < 0 \quad \text{either in } \Omega \times (0, +\infty) \text{ or in } \mathbb{R}^{N+1} \times (0, +\infty). \tag{7}$$

The profile of U as $t \to 0^+$ is controlled by the function Φ defined by

$$\Phi(s) = \int_1^s \frac{\phi'(\xi)}{\xi} d\xi \quad \text{for } s > 0. \tag{8}$$

In fact, in [8, Theorem 2.1 and Remark 2.2, p. 239] (see also [6, Theorem 1.1 and Theorem 4.1, p. 940 and p. 947]) it is shown that, if U is the solution of either problem (3)–(5) or problem (6), then

$$-4t\Phi\big(U(X,t)\big) \to d(X)^2 \quad \text{as } t \to 0^+ \text{ uniformly on every compact subset of } \Omega. \tag{9}$$

Here, $d = d(X)$ is the distance function:

$$d(X) = \text{dist}(X, \partial\Omega) \quad \text{for } X = (x, x_{N+1}) \in \Omega. \tag{10}$$

Formula (9) is regarded as a nonlinear version of one obtained by Varadhan [12].

A hypersurface Γ in Ω is said to be a *stationary level surface* of U (*stationary isothermic surface* of U when $\phi(s) \equiv s$) if at each time t the solution U remains constant on Γ (a constant depending on t). Hence it follows from (9) that there exists $R > 0$ such that

$$d(X) = R \quad \text{for every } X \in \Gamma, \tag{11}$$

provided Γ is a stationary level surface of U. The following theorem characterizes the boundary $\partial\Omega$ in such a way that U has a stationary level surface Γ in Ω.

Theorem 1 *Let U be the solution of either problem* (3)–(5) *or problem* (6). *Assume that there exists a basis $\{y^1, y^2, \dots, y^N\} \subset \mathbb{R}^N$ such that for every $j = 1, \dots, N$ the function $f(x + y^j) - f(x)$ has either a maximum or a minimum in $\mathbb{R}^N$. Suppose that U has a stationary level surface Γ in Ω. Then f is affine and $\partial\Omega$ must be a hyperplane.*

Remark 1 In order to prove Theorem 1, we shall also use the sliding method due to Berestycki, Caffarelli, and Nirenberg [2]. In [8, Theorem 2.3 and Remark 2.4, p. 240], instead of the assumption on f, it is assumed that for each $y \in \mathbb{R}^N$ there exists $h(y) \in \mathbb{R}$ such that

$$\lim_{|x| \to \infty} \left[f(x + y) - f(x) \right] = h(y), \tag{12}$$

which implies the assumption on f in Theorem 1. The condition (12) is a modified version of [2, (7.2), p. 1108], in which $h(y)$ is supposed identically zero. When $N = 1$, $f(x) = ax + b + \sin x$ $(a, b \in \mathbb{R})$ satisfies the assumption on f in Theorem 1, but it does not satisfy (12) provided $\frac{y}{2\pi}$ is not an integer. Another $f(x) = ax + b + \sin x \tan^{-1} x$ $(a, b \in \mathbb{R})$ does not satisfy the assumption, but it is Lipschitz continuous on $\mathbb{R}$.

Let us consider the case where $\phi(s) \equiv s$, that is, that of the heat equation, in particular. The following theorem characterizes the boundary $\partial\Omega$ in such a way that the caloric function U has a stationary *isothermic* surface in Ω.

Theorem 2 *Let $\phi(s) \equiv s$ and let U be the solution of either problem* (3)–(5) *or problem* (6). *Assume that U has a stationary isothermic surface Γ in Ω. Then f is affine and $\partial\Omega$ must be a hyperplane, if either $N \le 2$ or $\{|f(x) - f(y)| : |x - y| \le 1\}$ is bounded.*

Remark 2 When f is Lipschitz continuous in $\mathbb{R}^N$ and Ω satisfies the uniform exterior sphere condition, this theorem was proved in [7, Theorem 1.1(ii), p. 1113]. By combining [9, Lemma 3.1] with [11, Theorem 1.1, p. 887], we see that the assumption that Ω satisfies the uniform exterior sphere condition is not needed. Also, the Lipschitz continuity of f can be replaced by the uniform continuity of f, because of Professor Hitoshi Ishii's suggestion. Namely, by essentially the same proof as in [11], it can be shown that [11, Theorem 1.1, p. 887] holds even if the Lipschitz continuity is replaced by the uniform continuity. *Here, the advantage of* Theorem 2 *is that we do not need to assume any uniform continuity of f.*

Let $F = F(s)$ be a C^1 symmetric and concave function on the positive cone Λ given by

$$\Lambda = \left\{ s = (s_1, \ldots, s_N) \in \mathbb{R}^N : \min_{1 \le j \le N} s_j > 0 \right\},$$

where $N \ge 1$. Assume that F satisfies

$$F_{s_j} \left(= \frac{\partial F}{\partial s_j} \right) > 0 \quad \text{for all } j = 1, \ldots, N \text{ in } \Lambda. \tag{13}$$

Define $G = G(s)$ by

$$G(s) = F(1/s_1, \ldots, 1/s_N) \quad \text{for } s \in \Lambda. \tag{14}$$

Assume that G is convex in Λ. Such a class of functions F is dealt with in [1, 11].

Related to Theorems 1 and 2, for $f \in C(\mathbb{R}^N)$ we consider the domain Ω given by (1). Consider the entire graph $\partial\Omega = \{(x, f(x)) \in \mathbb{R}^{N+1} : x \in \mathbb{R}^N\}$ in $\mathbb{R}^{N+1}$ of f. Let $\kappa_1(x), \ldots, \kappa_N(x)$ be the principal curvatures of $\partial\Omega$ with respect to the upward unit normal vector to $\partial\Omega$ at $(x, f(x))$ for $x \in \mathbb{R}^N$. As mentioned in [11, pp. 888–889], for two real constants $R > 0$ and c, $f \in C(\mathbb{R}^N)$ is said to satisfy

$$F(1 - R\kappa_1, \ldots, 1 - R\kappa_N) = c \quad \text{in } \mathbb{R}^N \tag{15}$$

in the viscosity sense (or to be a viscosity solution of (15)), if it is both a viscosity subsolution and supersolution of (15). Here, $f \in C(\mathbb{R}^N)$ is said to be a viscosity subsolution (respectively supersolution) of (15), if for any open set $W \subset \mathbb{R}^N$, any $\varphi \in C^2(W)$ and any point $x_0 \in W$ satisfying

$$f(x_0) = \varphi(x_0), \qquad 1 - R \max_{1 \le j \le N} \kappa_j^{\varphi}(x_0) > 0, \quad \text{and}$$
$$f \le \varphi \quad (\text{respectively } f \ge \varphi) \quad \text{in } W,$$

we have

$$F\big(1 - R\kappa_1^{\varphi}(x_0), \ldots, 1 - R\kappa_N^{\varphi}(x_0)\big) \le c \quad (\text{respectively } \ge c),$$

where $\kappa_1^{\varphi}(x_0), \ldots, \kappa_N^{\varphi}(x_0)$ denote the principal curvatures of the graph of φ with respect to the upward normal vector to the graph at $(x_0, \varphi(x_0))$. We introduce a function $g \in C(\mathbb{R}^N)$ defined by

$$g(x) = \sup_{|x-y| \le R} \left\{ f(y) + \sqrt{R^2 - |x-y|^2} \right\} \quad \text{for every } x \in \mathbb{R}^N. \tag{16}$$

Then we have

$$\big\{ (x, g(x)) \in \mathbb{R}^{N+1} : x \in \mathbb{R}^N \big\} = \big\{ X \in \mathbb{R}^{N+1} : d(X) = R \big\} \ (= \Gamma). \tag{17}$$

Moreover, let us introduce a function $f^* \in C(\mathbb{R}^N)$ defined by

$$f^*(x) = \inf_{|x-y| \le R} \left\{ g(y) - \sqrt{R^2 - |x-y|^2} \right\} \quad \text{for every } x \in \mathbb{R}^N. \tag{18}$$

Then, by setting

$$D = \big\{ X = (x, x_{N+1}) \in \mathbb{R}^{N+1} : x_{N+1} > g(x) \big\}, \tag{19}$$

we notice the following:

$$\{(x, f^*(x)) \in \mathbb{R}^{N+1} : x \in \mathbb{R}^N\} = \{X \in \mathbb{R}^{N+1} : \mathrm{dist}(X, \overline{D}) = R\}, \tag{20}$$

$$f(x) \le f^*(x) \quad \text{for every } x \in \mathbb{R}^N. \tag{21}$$

The third theorem gives a Liouville-type theorem for some Weingarten hypersurfaces in the viscosity sense.

Theorem 3 *Suppose that there exist two real constants $R > 0$ and c such that $f \in C(\mathbb{R}^N)$ satisfies* (15) *in the viscosity sense, and moreover suppose that the equality holds in* (21), *that is,*

$$f(x) = f^*(x) \quad \textit{for every } x \in \mathbb{R}^N, \tag{22}$$

where $f^* = f^*(x)$ is defined by (18). *Then, $c = F(1, \ldots, 1)$ and f is an affine function, that is, $\partial\Omega$ must be a hyperplane, provided $\{|f(x) - f(y)| : |x - y| \le 1\}$ is bounded.*

Remark 3 The case where $F(s) = (\prod_{j=1}^N s_j)^{1/N}$ or $F(s) = \sum_{j=1}^N \log s_j$ is related to Theorem 2. The assumption (22), that is,

$$f(x) = \inf_{|x-y|\le R} \{g(y) - \sqrt{R^2 - |x-y|^2}\} \quad \text{for every } x \in \mathbb{R}^N, \tag{23}$$

implies that

$$\max_{1\le j\le N} \kappa_j \le \frac{1}{R} \quad \text{in } \mathbb{R}^N \tag{24}$$

holds in the viscosity sense, because (22) yields that for every point $X \in \partial\Omega$ there exists an open ball $B_R(Y)$ with radius R and centered at $Y \in \Gamma$ satisfying

$$X \in \partial B_R(Y) \quad \text{and} \quad B_R(Y) \subset \Omega. \tag{25}$$

Inequality (24) is one of main assumptions of [11, Theorem 1.1, p. 887]. Namely, boundedness of $\{|f(x) - f(y)| : |x - y| \le 1\}$ is much weaker than Lipschitz continuity of f, but (22) is stronger than (24). Also, (22) is satisfied by every classical C^2 solution f of (15) having the strict inequality in (24), because of the implicit function theorem.

The present paper is organized as follows. In Sect. 2, we prove Theorem 1 by using the sliding method due to Berestycki, Caffarelli, and Nirenberg [2]. In Sect. 3, we prove Theorem 2 with the aid of the theory of viscosity solutions. We follow the proof of [11, Theorem 1.1, p. 887] basically, but we here need a key lemma (see Lemma 4) which gives new gradient estimates for f and g, because we do not assume any uniform continuity of f. Section 4 is devoted to a proof of Theorem 3, where gradient estimates for f and g are replaced by Lipschitz constant estimates for f and g (see Lemma 7). In Sect. 5, we give a Bernstein-type theorem for some C^2 Weingarten hypersurfaces (see Theorem 4) as a remark on Theorem 3.

2 Proof of Theorem 1

Since Γ is a stationary level surface of U, it follows from (9), (7) and the implicit function theorem that there exist a number $R > 0$ and a function $g \in C^2(\mathbb{R}^N)$ such that both (16) and (17) hold.

Conversely, let $\nu(y)$ denote the upward unit normal vector to Γ at $(y, g(y)) \in \Gamma$. The facts that g is smooth, $\partial\Omega$ is a graph, and $(y, g(y)) - R\nu(y) \in \partial\Omega$ for every $y \in \mathbb{R}^N$, imply that (22), (18), and (20) hold, namely, both (23) and (20) where f^* is replaced by f hold. Hence, we have in particular

$$\partial\Omega = \left\{\left(x, f(x)\right) \in \mathbb{R}^{N+1} : x \in \mathbb{R}^N\right\} = \left\{X \in \mathbb{R}^{N+1} : \operatorname{dist}(X, \overline{D}) = R\right\}, \tag{26}$$

where D is given by (19). Thus, it follows from (26) that for every $X \in \partial\Omega$ there exists $Y \in \Gamma$ satisfying

$$X \in \partial B_R(Y) \quad \text{and} \quad B_R(Y) \subset \Omega. \tag{27}$$

Choose j arbitrarily. By the assumption of Theorem 1, the function $f(x + y^j) - f(x)$ has either a maximum or a minimum in $\mathbb{R}^N$. Since the proof below is similar, say $f(x + y^j) - f(x)$ has a maximum M in $\mathbb{R}^N$. Then there exists $x_0 \in \mathbb{R}^N$ such that

$$f\left(x + y^j\right) - f(x) \le M = f\left(x_0 + y^j\right) - f(x_0) \quad \text{for every } x \in \mathbb{R}^N. \tag{28}$$

Let us use the sliding method due to Berestycki, Caffarelli, and Nirenberg [2]. We set

$$\Omega_{y^j,M} = \left\{(x, x_{N+1}) \in \mathbb{R}^{N+1} : \left(x + y^j, x_{N+1} + M\right) \in \Omega\right\}.$$

Then we have

$$\begin{gathered} f\left(x + y^j\right) - M \le f(x) \quad \text{for every } x \in \mathbb{R}^N, \\ \Omega_{y^j,M} \supset \Omega \text{ and } \left(x_0, f(x_0)\right) \in \partial\Omega \cap \partial\Omega_{y^j,M}. \end{gathered}$$

Suppose that $\Omega_{y^j,M} \supsetneqq \Omega$. Then, by the strong comparison principle we have

$$\begin{aligned} &U\left(x + y^j, x_{N+1} + M, t\right) < U(X, t) \\ &\quad \text{for every } (X, t) = (x, x_{N+1}, t) \in \Omega \times (0, +\infty). \end{aligned} \tag{29}$$

On the other hand, since $(x_0, f(x_0)) \in \partial\Omega \cap \partial\Omega_{y^j,M}$ and $\Omega_{y^j,M} \supset \Omega$, it follows from (27) that there exists $Y_0 = (y_0, g(y_0)) \in \Gamma$ satisfying

$$\left(x_0, f(x_0)\right) \in \partial B_R(Y_0) \quad \text{and} \quad B_R(Y_0) \subset \Omega \subset \Omega_{y^j,M}.$$

Hence, since $\Gamma = \{X \in \mathbb{R}^{N+1} : d(X) = R\}$ and Γ is a stationary level surface of U, we have

$$U\left(y_0 + y^j, g(y_0) + M, t\right) = U(Y_0, t) \quad \text{for every } t > 0,$$

which contradicts (29). Thus, we get $\Omega_{y^j,M} = \Omega$, that is,

$$f\left(x + y^j\right) - M = f(x) \quad \text{for every } x \in \mathbb{R}^N.$$

Therefore we conclude that there exist $a_1, \dots, a_N \in \mathbb{R}$ satisfying

$$f\big(x + y^j\big) = f(x) + a_j \quad \text{for every } x \in \mathbb{R}^N \text{ and for } j = 1, \dots, N, \tag{30}$$

since j is chosen arbitrarily. Since f is continuous on $\mathbb{R}^N$ and $\{y^1, y^2, \dots, y^N\}$ is a basis of $\mathbb{R}^N$, we can solve (30) as a system of functional equations and conclude that $f(x)$ is determined by its values on $E = \{\sum_{j=1}^N \beta_j y^j \in \mathbb{R}^N : 0 \le \beta_j < 1,\ j = 1, \dots, N\}$. Indeed, if $x = \sum_{j=1}^N (r_j + \beta_j) y^j$ for $r = (r_1, \dots, r_N) \in \mathbb{Z}^N$ and $\beta = (\beta_1, \dots, \beta_N) \in [0, 1)^N$, then $f(x) = f(\sum_{j=1}^N \beta_j y^j) + \sum_{j=1}^N r_j a_j$. Moreover, this property of f implies that for every $y \in \mathbb{R}^N$ the function $f(x + y) - f(x)$ has either a maximum or a minimum on $\mathbb{R}^N$. Thus, by employing the sliding method again, we get

$$f(x + y) - f(x) = f(z + y) - f(z) \quad \text{for every } x, y, z \in \mathbb{R}^N. \tag{31}$$

Since f is continuous on $\mathbb{R}^N$, we solve (31) as a system of functional equations and conclude that f is affine. This completes the proof of Theorem 1.

3 Proof of Theorem 2

Note that U is real analytic in x, since U satisfies the heat equation. Since Γ is a stationary isothermic surface of U, it follows from (7) and the implicit function theorem that Γ is the graph of a real analytic function $g = g(x)$ for $x \in \mathbb{R}^N$. Let us first quote an important lemma from [9, Lemma 3.1]. We can use this lemma, since $\partial\Omega = \partial(\mathbb{R}^{N+1} \setminus \overline{\Omega})$, Γ is already real analytic and $\Gamma = \partial D$ where D is given by (19). The interior cone condition of D in the lemma with respect to Γ is of course satisfied, but in [9] it is used only to show that Γ is smooth.

Lemma 1 [9] *The following assertions hold*:

(1) *there exists a number $R > 0$ such that $d(X) = R$ for every $X \in \Gamma$*;
(2) *Γ is a real analytic hypersurface*;
(3) *$\partial\Omega$ is also a real analytic hypersurface, such that the mapping $\partial\Omega \ni (x, f(x)) \mapsto Y(x, f(x)) \equiv (x, f(x)) + R\nu(x) \in \Gamma$, where $\nu(x)$ is the upward unit normal vector to $\partial\Omega$ at $(x, f(x)) \in \partial\Omega$, is a diffeomorphism*; *in particular, $\partial\Omega$ and Γ are parallel hypersurfaces at distance R*;
(4) *it holds that*

$$\max_{1 \le j \le N} \kappa_j(x) < \frac{1}{R} \quad \textit{for every } x \in \mathbb{R}^N, \tag{32}$$

where $\kappa_1(x), \dots, \kappa_N(x)$ are the principal curvatures of $\partial\Omega$ at $(x, f(x)) \in \partial\Omega$ with respect to the upward unit normal vector to $\partial\Omega$;
(5) *there exists a number $c > 0$ such that*

$$\prod_{j=1}^N \big(1 - R\kappa_j(x)\big) = c \quad \textit{for every } x \in \mathbb{R}^N. \tag{33}$$

Note that in Lemma 1 (1) follows from (9) and (2) follows simply from the implicit function theorem. When $N = 1$, (5) of Lemma 1 implies the conclusion of Theorem 2, since the curvature of the curve $\partial\Omega$ is constant. Let $N \geq 2$. With the aid of Lemma 1, applying [11, Lemmas 4.2 and 4.3, p. 891 and p. 892] to $F(s) = (\prod_{j=1}^{N} s_j)^{1/N}$ yields the following lemma.

Lemma 2 *$c = 1$ and $H_{\partial\Omega} \leq 0 \leq H_\Gamma$ in $\mathbb{R}^N$, where $H_{\partial\Omega}$ (resp. H_Γ) denotes the mean curvature of $\partial\Omega$ (resp. Γ) with respect to the upward unit normal vector to $\partial\Omega$ (resp. Γ).*

When $N = 2$, by setting

$$\Gamma^* = \left\{ X \in \Omega : d(X) = \frac{R}{2} \right\}, \tag{34}$$

the fact that $c = 1$ implies that Γ^* is an entire minimal graph over $\mathbb{R}^2$. Therefore, by Bernstein's theorem for the minimal surface equation, Γ^* must be a hyperplane as in [7]. (See [4, 5] for Bernstein's theorem, and for more general setting see also Theorem 4 in Sect. 5 in the present paper.) Thus it remains to consider the case where $N \geq 3$ and $\{|f(x) - f(y)| : |x - y| \leq 1\}$ is bounded.

On the other hand, (3) of Lemma 1 gives us the following geometric property.

Lemma 3 *The following two assertions hold:*

(i) *for every $Y \in \Gamma$ there exists $X \in \partial\Omega$ such that $Y \in \partial B_R(X)$ and $B_R(X) \subset \mathbb{R}^{N+1} \setminus \overline{D}$;*
(ii) *for every $X \in \partial\Omega$ there exists $Y \in \Gamma$ such that $X \in \partial B_R(Y)$ and $B_R(Y) \subset \Omega$.*

Moreover f and g have the relationship, (16) and (23). Indeed, (16) follows from (1) of Lemma 1, and (23) follows from (1) of Lemma 1 and (ii) of Lemma 3. Since $\{|f(x) - f(y)| : |x - y| \leq 1\}$ is bounded, we see that $\{|g(x) - g(y)| : |x - y| \leq 1\}$ is also bounded. By Lemma 2 we have

$$\mathscr{M}(f) \leq 0 \leq \mathscr{M}(g) \equiv \operatorname{div}\left(\frac{\nabla g}{\sqrt{1 + |\nabla g|^2}} \right) \quad \text{in } \mathbb{R}^N. \tag{35}$$

Let $B_n = \{x \in \mathbb{R}^N : |x| < n\}$ for $n \in \mathbb{N}$. Then, by [4, Theorem 16.9, pp. 407–408], for each $n \in \mathbb{N}$, there exist two functions $f_n, g_n \in C^2(B_n) \cap C(\overline{B_n})$ solving

$$\mathscr{M}(f_n) = \mathscr{M}(g_n) = 0 \quad \text{in } B_n,$$
$$f_n = f \quad \text{and} \quad g_n = g \quad \text{on } \partial B_n.$$

Hence it follows from the comparison principle that for each $n \in \mathbb{N}$ there exists $z_n \in \partial B_n$ such that

$$f_{n+1} \leq f_n \leq f < g \leq g_n \leq g_{n+1} \quad \text{and} \quad g_n - f_n \leq g(z_n) - f(z_n) \quad \text{in } B_n. \tag{36}$$

Since $\{|f(x) - f(y)| : |x - y| \leq 1\}$ is bounded, it follows from (16) that $g - f$ is bounded in $\mathbb{R}^N$ and hence with the aid of (36) there exists a constant $C_* > 0$ satisfying

$$g - C_* \leq f_n \leq f \quad \text{and} \quad g \leq g_n \leq f + C_* \quad \text{in } B_n \text{ for every } n \in \mathbb{N}. \tag{37}$$

Thus, since both $\{|f(x) - f(y)| : |x - y| \leq 1\}$ and $\{|g(x) - g(y)| : |x - y| \leq 1\}$ are bounded, by using the interior estimates for the minimal surface equation (see [4, Corollary 16.7, p. 407]) with the aid of (37) and the monotonicity with n in (36), we proceed as in [11, pp. 893–894] to see that there exist two functions $f_\infty, g_\infty \in C^\infty(\mathbb{R}^N)$ satisfying

$$\mathcal{M}(f_\infty) = \mathcal{M}(g_\infty) = 0 \quad \text{in } \mathbb{R}^N,$$

$$|\nabla f_\infty| \quad \text{and} \quad |\nabla g_\infty| \text{ are bounded on } \mathbb{R}^N,$$

$$f_n \to f_\infty \quad \text{and} \quad g_n \to g_\infty \quad \text{as } n \to \infty \text{ uniformly on every compact set in } \mathbb{R}^N.$$

Then it follows from Moser's theorem [10, Corollary, p. 591] that both f_∞ and g_∞ are affine and hence the graph of f_∞ is parallel to that of g_∞ because $f_\infty \leq g_\infty$ in $\mathbb{R}^N$. Thus there exists $\eta \in \mathbb{R}^N$ satisfying

$$f_\infty(x) = \eta \cdot x + f_\infty(0) \quad \text{and} \quad g_\infty(x) = \eta \cdot x + g_\infty(0) \quad \text{for every } x \in \mathbb{R}^N. \tag{38}$$

Moreover we have

$$f_\infty \leq f < g \leq g_\infty \quad \text{in } \mathbb{R}^N, \tag{39}$$

$$f(z_n) - f_\infty(z_n) \quad \text{and} \quad g_\infty(z_n) - g(z_n) \to 0 \quad \text{as } n \to \infty. \tag{40}$$

Indeed, (39) follows from (36). Observe that for each $n \in \mathbb{N}$

$$\begin{aligned} g_n(0) - f_n(0) &\leq g(z_n) - f(z_n) \leq g_{n+1}(z_n) - f_{n+1}(z_n) \\ &\leq g(z_{n+1}) - f(z_{n+1}) \leq g_\infty(z_{n+1}) - f_\infty(z_{n+1}) = g_\infty(0) - f_\infty(0). \end{aligned}$$

Hence letting $n \to \infty$ yields that $g(z_n) - f(z_n) \to g_\infty(0) - f_\infty(0)$ as $n \to \infty$. Thus as $n \to \infty$

$$\begin{aligned} &\big(f(z_n) - f_\infty(z_n)\big) + \big(g_\infty(z_n) - g(z_n)\big) \\ &= \big(g_\infty(0) - f_\infty(0)\big) - \big(g(z_n) - f(z_n)\big) \to 0, \end{aligned}$$

which gives (40).

It suffices to show that $f \equiv f_\infty$ and $g \equiv g_\infty$. Lemma 3 yields the following key lemma.

Lemma 4 (Gradient estimates) *There exist three constants $\varepsilon_0 > 0$, $\delta_0 > 0$, and $C_0 > 0$ such that*

(1) *if $z \in \mathbb{R}^N$ and $(0 \leq) g_\infty(z) - g(z) \leq \varepsilon_0$, then $\sup_{|y-z| \leq \delta_0} |\nabla g(y)| \leq C_0$;*
(2) *if $z \in \mathbb{R}^N$ and $(0 \leq) f(z) - f_\infty(z) \leq \varepsilon_0$, then $\sup_{|x-z| \leq \delta_0} |\nabla f(x)| \leq C_0$.*

Proof (i) of Lemma 3 yields (1) and (ii) of Lemma 3 yields (2), respectively. Let us show (1). Recall that g_∞ is affine and $\nabla g_\infty \equiv \eta$. Denote by $\mathscr{H}$ the hyperplane given by the graph of g_∞. Then $\frac{(-\eta,1)}{\sqrt{1+|\eta|^2}}$ is the upward unit normal vector to $\mathscr{H}$. By (i) of Lemma 3, for every $Y = (y, g(y)) \in \Gamma$ there exists $X = (x, f(x)) \in \partial\Omega$ such that the ball $B_R(X)$ touching Γ from below at $Y \in \Gamma$ must be below $\mathscr{H}$. Hence,

$$\text{if } Y \text{ is sufficiently close to } \mathscr{H}, \text{ then } \frac{Y-X}{R} \text{ is sufficiently close to } \frac{(-\eta,1)}{\sqrt{1+|\eta|^2}}. \tag{41}$$

Namely, for every $\mu > 0$ there exists $\lambda > 0$ such that, if $(0 \le) g_\infty(y) - g(y) \le \lambda$, then

$$\left|\frac{y-x}{R} - \frac{-\eta}{\sqrt{1+|\eta|^2}}\right|^2 + \left(\frac{g(y)-f(x)}{R} - \frac{1}{\sqrt{1+|\eta|^2}}\right)^2 < \mu^2. \tag{42}$$

Of course, at the touching point Y, $\nabla g(y)$ equals the gradient of $f(x) + \sqrt{R^2 - |y-x|^2}$ with respect to y, that is,

$$\nabla g(y) = -\frac{y-x}{\sqrt{R^2-|y-x|^2}}. \tag{43}$$

On the other hand, if a point $(z, g(z)) \in \Gamma$ is sufficiently close to $\mathscr{H}$, then by (41) there exists a uniform neighborhood $\mathscr{N}_z$ of z in $\mathbb{R}^N$ such that every point $Y = (y, g(y)) \in \Gamma$ with $y \in \mathscr{N}_z$ is sufficiently close to $\mathscr{H}$. Namely, for every $\lambda > 0$ there exist $\varepsilon > 0$ and $\delta > 0$ such that, if $(0 \le) g_\infty(z) - g(z) \le \varepsilon$ and $|y - z| < \delta$, then $(0 \le) g_\infty(y) - g(y) \le \lambda$. Thus, combining this fact with (42) and (43) yields (1). (2) is similar. □

The last lemma is

Lemma 5 *The following two assertions hold:*

(i) $g_\infty(x+z_n) - g(x+z_n) \to 0$ *as* $n \to \infty$ *uniformly on every compact set in* $\mathbb{R}^N$;
(ii) $f(x+z_n) - f_\infty(x+z_n) \to 0$ *as* $n \to \infty$ *uniformly on every compact set in* $\mathbb{R}^N$.

This lemma implies the conclusion of Theorem 2. Indeed, in view of (39) and Lemma 3, Lemma 5 yields that the graphs of g_∞ and f_∞ are parallel hyperplanes at distance R. This means that $f \equiv f_\infty$ and $g \equiv g_\infty$. Thus it remains to prove Lemma 5.

Proof of Lemma 5 Since (ii) is similar to (i), let us show (i). Set

$$G_n(x) = g(x+z_n) - g(z_n) \quad \text{for } x \in \mathbb{R}^N \text{ and } n \in \mathbb{N}.$$

Then $G_n(0) = 0$ for every $n \in \mathbb{N}$. Since by (40) $g_\infty(z_n) - g(z_n) \to 0$ as $n \to \infty$, it follows from (1) of Lemma 4 that there exists $N_0 \in \mathbb{N}$ such that $\{G_n : n \ge N_0\}$ is equicontinuous and bounded on $\overline{B_{\delta_0}(0)}$ ($\subset \mathbb{R}^N$). The Arzelà-Ascoli theorem gives us that there exist a subsequence $\{G_{n'}\}$ and a function $G_\infty \in C(\overline{B_{\delta_0}(0)})$ such that

$$G_{n'} \to G_\infty \quad \text{as } n \to \infty \text{ uniformly on } \overline{B_{\delta_0}(0)}. \tag{44}$$

Notice that $G_\infty(0) = 0$. Since $\mathscr{M}(G_n) \geq 0$ in $\mathbb{R}^N$ by (35), we have that $\mathscr{M}(G_\infty) \geq 0$ in $B_{\delta_0}(0)$ in the viscosity sense. Observe that

$$\begin{aligned} G_{n'}(x) &= g(x+z_{n'}) - g(z_{n'}) \leq g_\infty(x+z_{n'}) - g(z_{n'}) \\ &= \left\{g_\infty(x+z_{n'}) - g_\infty(z_{n'})\right\} + \left\{g_\infty(z_{n'}) - g(z_{n'})\right\} \\ &= \eta \cdot x + \left\{g_\infty(z_{n'}) - g(z_{n'})\right\}. \end{aligned}$$

Hence, by (40) and (44), letting $n' \to \infty$ yields

$$G_\infty(x) \leq \eta \cdot x \quad \text{in } \overline{B_{\delta_0}(0)}. \tag{45}$$

Therefore, since $\mathscr{M}(\eta \cdot x) = 0 \leq \mathscr{M}(G_\infty)$ in $B_{\delta_0}(0)$ in the viscosity sense and $\eta \cdot 0 = 0 = G_\infty(0)$, by the strong comparison principle of Giga and Ohnuma [3, Theorem 3.1, p. 173] we see that

$$G_\infty(x) \equiv \eta \cdot x \quad \text{in } \overline{B_{\delta_0}(0)}.$$

Thus G_∞ is uniquely determined independently of the choice of the subsequence and therefore from (44) we conclude that

$$G_n(x) \to \eta \cdot x \quad \text{as } n \to \infty \text{ uniformly on } \overline{B_{\delta_0}(0)}. \tag{46}$$

Then, since

$$\begin{aligned} g_\infty(x+z_n) - g(x+z_n) &= \left\{g_\infty(x+z_n) - g_\infty(z_n)\right\} - G_n(x) + \left\{g_\infty(z_n) - g(z_n)\right\} \\ &= \eta \cdot x - G_n(x) + \left\{g_\infty(z_n) - g(z_n)\right\}, \end{aligned}$$

we get from (40) and (46)

$$g_\infty(x+z_n) - g(x+z_n) \to 0 \quad \text{as } n \to \infty \text{ uniformly on } \overline{B_{\delta_0}(0)}. \tag{47}$$

Moreover, by using (1) of Lemma 4 again for any point $z \in \partial B_{\delta_0}(0)$ and repeating the same argument as above, we see that (47) holds even if $B_{\delta_0}(0)$ is replaced by $B_{\frac{3}{2}\delta_0}(0)$. Thus, repeating this argument as many times as one wants yields conclusion (i). □

Remark 4 For the proof of Theorem 3, we give a remark for the case where $N = 1$. Even when $N = 1$, all Lemmas 2–5 hold true. Indeed, when $N = 1$, $\mathscr{M}(g) = g''(1+(g')^2)^{-\frac{3}{2}}$ in (35). Hence the graphs of f_n and g_n are line segments and without using Moser's theorem we can get two affine functions f_∞ and g_∞ in (38).

4 Proof of Theorem 3

We follow the proof of Theorem 2. By [11, Lemmas 4.2 and 4.3, p. 891 and p. 892], we have instead of Lemma 2.

Lemma 6 $c = F(1, \dots, 1)$ *and* $H_{\partial\Omega} \le 0 \le H_\Gamma$ *in* $\mathbb{R}^N$ *in the viscosity sense, where* $H_{\partial\Omega}$ (resp. H_Γ) *denotes the mean curvature of* $\partial\Omega$ (resp. Γ) *with respect to the upward unit normal vector to* $\partial\Omega$ (resp. Γ).

Also, in view of (16) and (23) coming from (22), we see that Lemma 3 also holds. Then proceeding as in the proof of Theorem 2 yields two affine functions f_∞ and g_∞ satisfying (38), (39), and (40). Hence, it suffices to show that $f \equiv f_\infty$ and $g \equiv g_\infty$. Lemma 3 yields the following key lemma instead of Lemma 4.

Lemma 7 (Lipschitz constant estimates) *There exist three constants* $\varepsilon_0 > 0$, $\delta_0 > 0$, *and* $C_0 > 0$ *such that*

(1) *if* $z \in \mathbb{R}^N$ *and* $(0 \le) g_\infty(z) - g(z) \le \varepsilon_0$, *then* $\sup_{x,y \in B_{\delta_0}(z),\, x \ne y} \frac{|g(x)-g(y)|}{|x-y|} \le C_0$;
(2) *if* $z \in \mathbb{R}^N$ *and* $(0 \le) f(z) - f_\infty(z) \le \varepsilon_0$, *then* $\sup_{x,y \in B_{\delta_0}(z),\, x \ne y} \frac{|f(x)-f(y)|}{|x-y|} \le C_0$.

Proof We adjust the proof of Lemma 4 to this situation. (i) of Lemma 3 yields (1) and (ii) of Lemma 3 yields (2), respectively. Let us show (1). Recall that g_∞ is affine and $\nabla g_\infty \equiv \eta$. Denote by $\mathscr{H}$ the hyperplane given by the graph of g_∞. Then $\frac{(-\eta,1)}{\sqrt{1+|\eta|^2}}$ is the upward unit normal vector to $\mathscr{H}$. By (i) of Lemma 3, for every $Y = (y, g(y)) \in \Gamma$ there exists $X = (x, f(x)) \in \partial\Omega$ such that the ball $B_R(X)$ touching Γ from below at $Y \in \Gamma$ must be below $\mathscr{H}$. Hence,

$$\text{if } Y \text{ is sufficiently close to } \mathscr{H}, \text{ then } \frac{Y-X}{R} \text{ is sufficiently close to } \frac{(-\eta,1)}{\sqrt{1+|\eta|^2}}. \tag{48}$$

Namely, for every $\mu > 0$ there exists $\lambda > 0$ such that, if $(0 \le) g_\infty(y) - g(y) \le \lambda$, then

$$\left| \frac{y-x}{R} - \frac{-\eta}{\sqrt{1+|\eta|^2}} \right|^2 + \left(\frac{g(y)-f(x)}{R} - \frac{1}{\sqrt{1+|\eta|^2}} \right)^2 < \mu^2. \tag{49}$$

On the other hand, if a point $(z, g(z)) \in \Gamma$ is sufficiently close to $\mathscr{H}$, then by (48) there exists a uniform neighborhood $\mathscr{N}_z$ of z in $\mathbb{R}^N$ such that every point $Y = (y, g(y)) \in \Gamma$ with $y \in \mathscr{N}_z$ is sufficiently close to $\mathscr{H}$. Namely, for every $\lambda > 0$ there exist $\varepsilon > 0$ and $\delta > 0$ such that, if $(0 \le) g_\infty(z) - g(z) \le \varepsilon$ and $|y - z| < \delta$, then $(0 \le) g_\infty(y) - g(y) \le \lambda$.

Moreover, in view of (48), by choosing $\frac{\pi}{2} > \theta > 0$ sufficiently small and introducing a cone $\mathscr{V}$ defined by

$$\mathscr{V} = \left\{ \Xi = (\xi, \xi_{N+1}) \in \mathbb{R}^{N+1} : \xi_{N+1} > |\Xi| \cos\theta \right\},$$

we see that, if $Y \in \Gamma$ is sufficiently close to $\mathscr{H}$, then $\mathscr{V} + Y = \{\Xi + Y : \Xi \in \mathscr{V}\} \subset D$, where $\mathscr{V} + Y$ is a cone with vertex Y. Here D is given by (19). Indeed, if $\mathscr{V} + Y \not\subset D$, then there exists another point $\tilde{Y} (\ne Y) \in \Gamma \cap (\mathscr{V} + Y)$. However, in view of (48), a ball $B_R(\tilde{X})$ touching Γ from below at $\tilde{Y}$ might contain Y since $\theta > 0$

is small. This is a contradiction. Namely, if $Y = (y, g(y)) \in \Gamma$ with $y \in \mathcal{N}_z$, then, with the aid of (i) of Lemma 3, we must have

$$\mathcal{V} + Y \subset D, \qquad B_R(X) \subset \mathbb{R}^{N+1} \setminus \overline{D}, \quad \text{and} \quad Y \in \partial(\mathcal{V} + Y) \cap \partial B_R(X).$$

This gives (1). (2) is similar. □

Hence, by using Lemma 7 instead of Lemma 4, we can proceed as in the proof of Theorem 2 to see that Lemma 5 also holds. Therefore, (39), Lemma 3 and Lemma 5 yield the conclusion of Theorem 3.

5 Concluding Remarks

When $N = 2$, we have a Bernstein-type theorem for some C^2 Weingarten hypersurfaces related to Theorem 3.

Theorem 4 *Suppose that there exist two real constants $R > 0$ and c such that $f \in C^2(\mathbb{R}^2)$ satisfies*

$$F(1 - R\kappa_1, 1 - R\kappa_2) = c \quad \text{and} \quad \max_{1 \le j \le 2} \kappa_j(x) < \frac{1}{R} \quad \text{in } \mathbb{R}^2. \tag{50}$$

Then, $c = F(1, 1)$ and f is an affine function, that is, $\partial\Omega$ must be a hyperplane.

Proof Here we have Lemma 6. We consider Γ^* defined by (34) as in Sect. 3. Then $\partial\Omega$, Γ^*, and Γ are parallel hypersurfaces. Denote by $\kappa_1^*(Z), \kappa_2^*(Z)$ the principal curvatures of Γ^* with respect to the upward unit normal vector $\nu^*(Z)$ to Γ^* at $Z \in \Gamma^*$, and denote by $\hat\kappa_1(Y)$, $\hat\kappa_2(Y)$ the principal curvatures of Γ with respect to the upward unit normal vector at $Y = Z + \frac{R}{2}\nu^*(Z) \in \Gamma$. Also, here for the principal curvatures of $\partial\Omega$ we use the notation $\kappa_1(X)$, $\kappa_2(X)$ instead of $\kappa_1(x)$, $\kappa_2(x)$ with $(x, f(x)) = X = Z - \frac{R}{2}\nu^*(Z) \in \partial\Omega$. These principal curvatures have the following relationship:

$$\kappa_j(X) = \frac{\kappa_j^*(Z)}{1 + \frac{R}{2}\kappa_j^*(Z)} \quad \text{and} \quad \hat\kappa_j(Y) = \frac{\kappa_j^*(Z)}{1 - \frac{R}{2}\kappa_j^*(Z)} \quad \text{for each } j = 1, 2.$$

Since $\max_{1 \le j \le 2} \kappa_j(X) < \frac{1}{R}$ and $1 - R\kappa_j(X) = \frac{1}{1 + R\hat\kappa_j(Y)}$, we see that

$$-\frac{2}{R} < \kappa_j^*(Z) < \frac{2}{R} \quad \text{for each } j = 1, 2. \tag{51}$$

On the other hand, by Lemma 6, we have

$$\sum_{j=1}^{2} \frac{\kappa_j^*(Z)}{1 + \frac{R}{2}\kappa_j^*(Z)} \le 0 \le \sum_{j=1}^{2} \frac{\kappa_j^*(Z)}{1 - \frac{R}{2}\kappa_j^*(Z)}.$$

This gives

$$\kappa_1^* + \kappa_2^* + R\kappa_1^*\kappa_2^* \le 0 \le \kappa_1^* + \kappa_2^* - R\kappa_1^*\kappa_2^*,$$

and hence

$$\kappa_1^*\kappa_2^* \le 0 \quad \text{and} \quad R\kappa_1^*\kappa_2^* \le \kappa_1^* + \kappa_2^* \le -R\kappa_1^*\kappa_2^*.$$

Then, with the aid of (51), we conclude that

$$\left(\kappa_1^*\right)^2 + \left(\kappa_2^*\right)^2 \le 2 \cdot (-3)\kappa_1^*\kappa_2^*.$$

Hence the Gauss map of Γ^* is $(-3, 0)$-quasiconformal on $\mathbb{R}^2$ (see [4, (16.88), p. 424]) and hence by [4, Corollary 16.19, p. 429] Γ^* must be a hyperplane. □

Acknowledgements This research was partially supported by a Grant-in-Aid for Scientific Research (B) (♯ 20340031) of Japan Society for the Promotion of Science.

References

1. Andrews, B.: Pinching estimates and motion of hypersurfaces by curvature functions. J. Reine Angew. Math. **608**, 17–33 (2007)
2. Berestycki, H., Caffarelli, L.A., Nirenberg, L.: Monotonicity for elliptic equations in unbounded Lipschitz domains. Commun. Pure Appl. Math. **50**, 1089–1111 (1997)
3. Giga, Y., Ohnuma, M.: On strong comparison principle for semicontinuous viscosity solutions of some nonlinear elliptic equations. Int. J. Pure Appl. Math. **22**, 165–184 (2005)
4. Gilbarg, D., Trudinger, N.S.: Elliptic Partial Differential Equations of Second Order, 2nd edn. Springer, Berlin (1983)
5. Giusti, E.: Minimal Surfaces and Functions of Bounded Variations. Birkhäuser, Boston (1984)
6. Magnanini, R., Sakaguchi, S.: Nonlinear diffusion with a bounded stationary level surface. Ann. Inst. Henri Poincaré, Anal. Non Linéaire **27**, 937–952 (2010)
7. Magnanini, R., Sakaguchi, S.: Stationary isothermic surfaces and some characterizations of the hyperplane in the N-dimensional Euclidean space. J. Differ. Equ. **248**, 1112–1119 (2010)
8. Magnanini, R., Sakaguchi, S.: Interaction between nonlinear diffusion and geometry of domain. J. Differ. Equ. **252**, 236–257 (2012)
9. Magnanini, R., Sakaguchi, S.: Matzoh ball soup revisited: the boundary regularity issue. Math. Methods Appl. Sci. doi:10.1002/mma.1551
10. Moser, J.: On Harnack's theorem for elliptic differential equations. Commun. Pure Appl. Math. **14**, 577–591 (1961)
11. Sakaguchi, S.: A Liouville-type theorem for some Weingarten hypersurfaces. Discrete Contin. Dyn. Syst., Ser. S **4**, 887–895 (2011)
12. Varadhan, S.R.S.: On the behavior of the fundamental solution of the heat equation with variable coefficients. Commun. Pure Appl. Math. **20**, 431–455 (1967)

Nonexistence of Multi-bubble Solutions for a Higher Order Mean Field Equation on Convex Domains

Futoshi Takahashi

Abstract In this note, we are concerned with the blowing-up behavior of solutions to the $2p$-th order mean field equation under the Navier boundary condition:

$$\begin{cases} (-\Delta)^p u = \rho \frac{V(x)e^u}{\int_\Omega V(x)e^u\,dx} & \text{in } \Omega, \\ (-\Delta)^j u = 0 & \text{on } \partial\Omega,\ (j = 0, 1, \dots, p-1), \end{cases}$$

where Ω is a smooth bounded domain in $\mathbb{R}^{2p}$ for $p \in \mathbb{N}$. By using a new Pohozaev type identity for the Green function of $(-\Delta)^p$ under the Navier boundary condition, we show that the set of blow up points for any blowing-up solution sequence must be a singleton on *convex* domains, under some assumptions on the weight function V.

Keywords Blowing-up solution · Higher order mean field equation · Green's function

1 Introduction

Recently, many authors have been interested in the study of nonlinear elliptic partial differential equations involving higher order differential operators, because of its connection to the conformal geometry. One of the most important conformally invariant differential operators on a four-dimensional Riemannian manifold (M, g) is a Paneitz operator, defined as

$$P_g = \Delta_g^2 - \delta_g \left(\frac{2}{3} S_g - 2Ric_g \right) d$$

where Δ_g denotes the Laplace-Beltrami operator with respect to g, δ_g the co-differential, d the exterior differential, S_g and Ric_g denote the scalar and Ricci

F. Takahashi (✉)
Department of Mathematics & OCAMI, Osaka City University, Sumiyoshi-ku, Osaka 558-8585, Japan
e-mail: futoshi@sci.osaka-cu.ac.jp

R. Magnanini et al. (eds.), *Geometric Properties for Parabolic and Elliptic PDE's*, Springer INdAM Series 2, DOI 10.1007/978-88-470-2841-8_18,

curvature of the metric g. By this symbol, the equation of prescribing Q-curvature on (M, g) is described as

$$P_g u + 2Q_g = 2\bar{Q}_{g_u} e^{4u}$$

where Q_g is the Q-curvature of the original metric g, $\bar{Q}_{g_u}$ is the Q-curvature of the new metric $g_u = e^{4u} g$. For the recent development of finding conformal metrics with prescribed Q-curvature on compact 4-manifolds and the bubbling behavior of non-compact solution sequences, see, for instance, [7, 9, 15]. If (M, g) is $\mathbb{R}^4$ with its standard Euclidean metric, the Paneitz operator P_g is nothing but $\Delta^2 = \Delta\Delta$ where $\Delta = \sum_{i=1}^4 \frac{\partial^2}{\partial x_i^2}$ is the Laplacian in $\mathbb{R}^4$, and the equation of prescribing Q-curvature becomes of the form

$$\Delta^2 u = \rho \frac{V(x)e^{4u}}{\int_\Omega V(x)e^{4u}\,dx}.$$

See [8, 10, 12] and the references therein.

In this paper, we consider a generalization of it, namely, we are concerned with the following $2p$-th order mean field equation ($p \in \mathbb{N}$)

$$\begin{cases} (-\Delta)^p u = \rho \frac{V(x)e^u}{\int_\Omega V(x)e^u\,dx} & \text{in } \Omega, \\ (-\Delta)^j u = 0 & \text{on } \partial\Omega,\ (j = 0, 1, \ldots, p-1), \end{cases} \tag{1}$$

where Ω is a smooth bounded domain in $\mathbb{R}^{2p}$, ρ is a positive parameter and $V \in C^{2,\beta}(\Omega)$, $(0 < \beta < 1)$ is a positive function. Let us define the variational functional $I_\rho : X \to \mathbb{R}$,

$$I_\rho(u) = \frac{1}{2}\int_\Omega \left|(-\Delta)^{\frac{p}{2}} u\right|^2 dx - \rho \log \int_\Omega V(x)e^u\,dx$$

where

$$X = H^p(\Omega) \cap \left\{ u \,\middle|\, (-\Delta)^j u \in H_0^1(\Omega),\ j = 0, 1, \ldots, \left[\frac{p-1}{2}\right] \right\},$$

and

$$(-\Delta)^{\frac{p}{2}} u = \begin{cases} \nabla(-\Delta)^{k-1} u, & (p = 2k-1), \\ (-\Delta)^k u, & (p = 2k), \end{cases}$$

for $k \in \mathbb{N}$. Then (1) is the Euler-Lagrange equation of I_ρ.

In the following, let $\alpha_0(p)$ denote the best constant for the Adams version Trudinger-Moser inequality [1]: there exists $C(\Omega) < +\infty$ such that for any $\alpha \le \alpha_0(p)$ and $u \in C_0^\infty(\Omega)$ with

$$\|(-\Delta)^{\frac{p}{2}} u\|_{L^2(\Omega)} \le 1,$$

there holds

$$\int_\Omega e^{\alpha u^2}\,dx \le C(\Omega).$$

The same holds for $u \in X$ by standard density argument. It is known that $\alpha_0(p) = (4\pi)^p p!$. Note also that $2\alpha_0(p) = 2p\Lambda$, where

$$\Lambda = (2p-1)!|S^{2p}| = \int_{S^{2p}} Q_{g_{S^{2p}}} \, dV_{g_{S^{2p}}}$$

is the total Q-curvature of the standard sphere $(S^{2p}, g_{S^{2p}})$. In the sequel, $G = G(x,y)$ will denote the Green function of $(-\Delta)^p$ under the Navier boundary condition:

$$\begin{cases} (-\Delta)^p G(\cdot, y) = \delta_y & \text{in } \Omega \subset \mathbb{R}^{2p}, \\ G(\cdot, y) = (-\Delta)^j G(\cdot, y) = 0 & \text{on } \partial\Omega, \ (j = 1, \dots, p-1). \end{cases}$$

We decompose G as $G(x,y) = \Gamma(x,y) - H(x,y)$, where $\Gamma(x,y)$ is the fundamental solution of $(-\Delta)^p$ on $\mathbb{R}^{2p}$, defined as

$$\Gamma(x,y) = C_p \log \frac{1}{|x-y|}, \qquad C_p = \frac{1}{\{2^{p-1}(p-1)!\}^2 |S^{2p-1}|},$$

and $H = H(x,y) \in C^\infty(\Omega \times \Omega)$ is called the regular part of the Green function. Finally, let $R(y) = H(y,y)$ denote the Robin function associated with the Green function of $(-\Delta)^p$ under the Navier boundary condition.

On the asymptotic behavior of blowing-up solutions to (1), C.-S. Lin and J.-C. Wei proved, among others, the following result; see [13, 14, 18].

Proposition 1 *Assume $V \in C^{2,\beta}(\Omega)$, $\inf_\Omega V > 0$. Let u_{ρ_n} be a solution sequence to* (1) *with $\rho = \rho_n > 0$ such that $\|u_{\rho_n}\|_{L^\infty(\Omega)} \to \infty$ while $\rho_n = O(1)$ as $n \to \infty$. Then there exists a subsequence (again denoted by ρ_n) and a set of m-points $\mathscr{S} = \{a_1, \dots, a_m\} \subset \Omega$* (blow up set) *such that*

$$\rho_n \to 2\alpha_0(p) m, \quad \textit{(mass quantization)}$$

$$u_{\rho_n} \to 2\alpha_0(p) \sum_{j=1}^m G(\cdot, a_j) \quad \textit{in } C^{2p}_{\text{loc}}(\overline{\Omega} \setminus \mathscr{S}),$$

$$\rho_n \frac{V(x) e^{u_{\rho_n}}}{\int_\Omega V(x) e^{u_{\rho_n}} dx} \rightharpoonup 2\alpha_0(p) \sum_{i=1}^m \delta_{a_i} \quad \textit{in the sense of measures}$$

as $n \to \infty$. Finally, each blow up point $a_i \in \mathscr{S}$ must satisfy

$$\frac{1}{2} \nabla R(a_i) - \sum_{j=1, j\neq i}^m \nabla_x G(a_i, a_j) - \frac{1}{2\alpha_0(p)} \nabla \log V(a_i) = \mathbf{0}, \tag{2}$$

for $i = 1, \dots, m$. (Characterization of blow up points.)

The main difficulty in the proof is to show that the blow up set $\mathscr{S}$ consists of only interior points of Ω. In [13, 14], the authors used the local version of the method of moving planes to overcome this difficulty. After showing that $\mathscr{S} \subset \Omega$, the rest of claims can be proved by the argument in [18]. As for the asymptotic study for the higher order mean field equation under the Dirichlet boundary condition, we refer the reader to [17] and [16].

Concerning the actual existence of multi-bubble solutions to (1) ($m \geq 2$) which exhibit the asymptotic behavior described in Proposition 1, there are some affirmative results when $p = 2$.

Proposition 2 [2, 6] *Let $p = 2$ and $m \geq 2$ be an integer. Set $\Omega^m = \Omega \times \cdots \times \Omega$ (m times) and $\Delta = \{(\xi_1, \dots, \xi_m) \in \Omega^m \mid \xi_i = \xi_j$ for some $i \neq j\}$. Define the Hamiltonian function*

$$\mathscr{F}(\xi_1, \dots, \xi_m) = \sum_{i=1}^{m} \left(R(\xi_i) - \frac{1}{32\pi^2} \log V(\xi_i) \right) - \sum_{\substack{i \neq j \\ 1 \leq i, j \leq m}} G(\xi_i, \xi_j)$$

on $\Omega^m \setminus \Delta$. If $\mathscr{F}$ has a nondegenerate critical point (Baraket-Dammak-Ouni-Pacard [2], *$V \equiv 1$ case), or, a "minimax value in an appropriate subset" (Clapp-Muñoz-Musso* [6]), *that is, if $(a_1, \dots, a_m) \in \Omega^m \setminus \Delta$ satisfies*

$$\frac{1}{2} \nabla R(a_i) - \sum_{j=1, j \neq i}^{m} \nabla_x G(a_i, a_j) - \frac{1}{64\pi^2} \nabla \log V(a_i) = \mathbf{0}$$

for $i = 1, 2, \dots, m$ and some additional conditions, then there exists a solution sequence $\{u_\rho\}$ which blows up exactly on $\mathscr{S} = \{a_1, \dots, a_m\}$.

For the precise meaning that $\mathscr{F}$ has a "minimax value in an appropriate subset", we refer to [6]. By this proposition, we know that if Ω has the cohomology group $H^d(\Omega) \neq 0$ for some $d \in \mathbb{N}$, or, if Ω is an m-dumbbell shaped domain (roughly, a simply-connected domain made by m balls those connected to each other by thin tubes), then there exist m-points blowing up solutions for any $m \geq 2$ [6].

In this paper, on the contrary, we prove the nonexistence of multi-bubble solutions to (1) on *convex* domains, under an additional assumption on the coefficient function V.

Theorem 1 *Assume $\Omega \subset \mathbb{R}^{2p}$ be a bounded convex domain. Let $\{\rho_n\}$ be a sequence of positive numbers with $\rho_n = O(1)$ as $n \to \infty$, and let $\{u_{\rho_n}\}$ be a solution sequence to* (1) *for $\rho = \rho_n$ satisfying $\|u_{\rho_n}\|_{L^\infty(\Omega)} \to +\infty$ as $n \to \infty$. Assume $\inf_\Omega V > 0$ and $R - \frac{1}{\alpha_0(p)} \log V$ is a strictly convex function on Ω. Then there exists a point $a \in \Omega$ such that, for the full sequence, we have*

$$\begin{aligned} &\rho_n \to 2\alpha_0(p), \\ &u_{\rho_n} \to 2\alpha_0(p) G(\cdot, a) \quad \text{in } C^{2p}_{\text{loc}}(\overline{\Omega} \setminus \{a\}), \\ &\rho_n \frac{V(x) e^{u_{\rho_n}}}{\int_\Omega V(x) e^{u_{\rho_n}}\, dx} \rightharpoonup 2\alpha_0(p) \delta_a \quad \text{in the sense of measures} \end{aligned}$$

as $n \to \infty$.

In Theorem 1, we can claim also that $a \in \Omega$ is the unique minimum point of the strictly convex function $R - \frac{1}{\alpha_0(p)} \log V$.

We remark here that, for the 2nd order case, the Robin function of $-\Delta$ with the Dirichlet boundary condition on a bounded convex domain Ω in $\mathbb{R}^N$ is strictly convex on Ω. This fact was first proved by Caffarelli and Friedman [4] when $N = 2$, and later extended to $N \geq 3$ by Cardaliaguet and Tahraoui [5]. By combining this fact and a kind of Pohozaev type identity for the Green function of $-\Delta$ under the Dirichlet boundary condition, Grossi and Takahashi [11] proved that blowing-up solutions with multiple blow up points do not exist on convex domains for various semilinear problems with blowing-up or concentration phenomena. In this paper, first we extend the above mentioned Pohozaev type identity to the Green function of $(-\Delta)^p$ under the Navier boundary condition, see Proposition 3 in Sect. 2. Once the identity is established, we can follow the argument in [11] without difficulty. However when $p \geq 2$, it is not known whether the Robin function of $(-\Delta)^p$ under the Navier boundary condition is convex or not on convex domains in $\mathbb{R}^{2p}$. Therefore at this stage, we cannot drop the assumption on V and we do not know whether Theorem 1 is true or not when V is a constant.

This paper is organized as follows. In Sect. 2, we prove a new Pohozaev type identity for the Green function, Proposition 3, which is crucial to our argument. For this identity, we do not need the assumption of the convexity of Ω. In Sect. 3, we prove Theorem 1 by using the key identity in Sect. 2 and the characterization of blow up points (2).

2 New Pohozaev Identity for the Green Function

In this section, we prove an integral identity for the Green function of $(-\Delta)^p$ with the Navier boundary condition, which is a key for the proof of Theorem 1. The corresponding identity when $p = 1$ was former proved in [11].

Proposition 3 *Let $\Omega \subset \mathbb{R}^N$ $(N \geq 2p)$ be a smooth bounded domain. For any $P \in \mathbb{R}^N$ and $a, b \in \Omega$, $a \neq b$, it holds*

$$\sum_{k=1}^{p} \int_{\partial\Omega} (x - P) \cdot \nu(x) \left(\frac{\partial (-\Delta)^{p-k} G_a}{\partial \nu_x} \right) \left(\frac{\partial (-\Delta)^{k-1} G_b}{\partial \nu_x} \right) ds_x$$
$$= (2p - N) G(a, b) + (P - a) \cdot \nabla_x G(a, b) + (P - b) \cdot \nabla_x G(b, a),$$

where $G_a(x) = G(x, a)$, $G_b(x) = G(x, b)$ and $\nu(x)$ is the unit outer normal at $x \in \partial\Omega$.

Proof We follow the argument used in [11], which originates from [3]. In order to introduce the idea clearly, first we show a formal computation. Let us denote $G_a(x) = G(x, a)$, $G_b(x) = G(x, b)$ and define $w(x) = (x - P) \cdot \nabla G_a(x)$. Since $\Delta^j((x - P) \cdot \nabla) = 2j\Delta^j + ((x - P) \cdot \nabla \Delta^j)$ for $j \in \{0\} \cup \mathbb{N}$, we have

$$\begin{cases} (-\Delta)^p w(x) = (x-P)\cdot\nabla\delta_a(x) + 2p\delta_a(x), \\ (-\Delta)^p G_b(x) = \delta_b(x), \end{cases}$$

where δ_a, δ_b are the Dirac delta functions supported on a, b respectively. Multiplying $G_b(x)$, $w(x)$ respectively to the above equations, and subtracting, we obtain

$$\begin{aligned} &\int_\Omega \left\{\left((-\Delta)^p w(x)\right)G_b(x) - \left((-\Delta)^p G_b(x)\right)w(x)\right\} dx \\ &\quad = \int_\Omega \left\{(x-P)\cdot\nabla\delta_a(x)G_b(x) + 2p\delta_a(x)G_b(x) - \delta_b(x)w(x)\right\} dx. \end{aligned} \tag{3}$$

By an iterated use of Green's second formula, we see

$$\begin{aligned} \text{LHS of (3)} &= (-1)^p \sum_{k=1}^p \int_{\partial\Omega} \left(\frac{\partial \Delta^{p-k} w}{\partial\nu} \Delta^{k-1} G_b - \frac{\partial \Delta^{k-1} G_b}{\partial\nu} \Delta^{p-k} w \right) ds_x \\ &= (-1)^{p+1} \sum_{k=1}^p \int_{\partial\Omega} \left((x-P)\cdot\nabla\Delta^{p-k}G_a\right)\left(\frac{\partial\Delta^{k-1}G_b}{\partial\nu}\right) ds_x \\ &= \sum_{k=1}^p \int_{\partial\Omega} (x-P)\cdot\nu(x) \left(\frac{\partial(-\Delta)^{p-k}G_a}{\partial\nu_x}\right)\left(\frac{\partial(-\Delta)^{k-1}G_b}{\partial\nu_x}\right) ds_x, \end{aligned}$$

here we have used $\Delta^{k-1}G_b = 0$ and $\Delta^{p-k}w = (x-P)\cdot\nabla\Delta^{p-k}G_a$ on $\partial\Omega$.

On the other hand,

$$\begin{aligned} \text{RHS of (3)} &= 2pG_b(a) - w(b) + \int_\Omega (x-P)\cdot\nabla\delta_a(x)G_b(x)\,dx \\ &= 2pG_b(a) - w(b) + \sum_{i=1}^N \int_\Omega (x_i - P_i)\frac{\partial\delta_a}{\partial x_i} G_b(x)\,dx \\ &= 2pG_b(a) - w(b) - \sum_{i=1}^N \int_\Omega \frac{\partial}{\partial x_i}\left\{(x_i-P_i)G_b(x)\right\}\delta_a(x)\,dx \\ &= 2pG_b(a) - w(b) - \sum_{i=1}^N \frac{\partial}{\partial x_i}\left\{(x_i-P_i)G_b(x)\right\}\Big|_{x=a} \\ &= (2p-N)G(a,b) + (P-a)\cdot\nabla_x G(a,b) + (P-b)\cdot\nabla_x G(b,a). \end{aligned}$$

Thus we obtain the conclusion.

To make this argument rigorously, we use standard approximations. Define $\delta_{a,\rho}(x) = \frac{1}{|B_\rho|}\chi_{B_\rho(a)}(x)$ where $\chi_{B_\rho(a)}$ is the characteristic function of the ball $B_\rho(a)$ with radius $\rho > 0$ and center $a \in \Omega$. Denote $\delta^\varepsilon_{a,\rho}(x) = j_\varepsilon * \delta_{a,\rho}(x)$ where $j(x) \ge 0$, $\operatorname{supp} j \subset B_1(0)$, $\int_{\mathbb{R}^N} j(x)\,dx = 1$ and $j_\varepsilon(x) = \varepsilon^{-N} j(\frac{x-a}{\varepsilon})$. For a point $a \in \Omega$ and for $\rho > 0$ and $\varepsilon > 0$ sufficiently small such that $B_{\rho+\varepsilon}(a) \subset \Omega$, $\delta^\varepsilon_{a,\rho}$ is well-defined and a smooth function on Ω. Let $u^\varepsilon_{a,\rho}$ denote the unique solution of the problem

$$\begin{cases} (-\Delta)^p u^\varepsilon_{a,\rho} = \delta^\varepsilon_{a,\rho} & \text{in } \Omega, \\ (-\Delta)^j u^\varepsilon_{a,\rho} = 0 & \text{on } \partial\Omega \ (j = 0, 1, \dots, p-1). \end{cases}$$

Define $\delta^{\varepsilon}_{b,\rho}$, $u^{\varepsilon}_{b,\rho}$ in the same way. Since $\delta^{\varepsilon}_{a,\rho} \to \delta_{a,\rho}$ as $\varepsilon \to 0$ in $L^q(\Omega)$ for any $1 \le q < \infty$, we have $u^{\varepsilon}_{a,\rho} \to u_{a,\rho}$ in $W^{2p,q}(\Omega)$ as $\varepsilon \to 0$, where $u_{a,\rho}$ is the unique solution of

$$\begin{cases} (-\Delta)^p u_{a,\rho} = \delta_{a,\rho} & \text{in } \Omega, \\ (-\Delta)^j u_{a,\rho} = 0 & \text{on } \partial\Omega \ (j = 0, 1, \dots, p-1). \end{cases}$$

Since $\delta_{a,\rho} \to \delta_a$ as $\rho \to 0$, we have

$$\lim_{\rho\to 0}\lim_{\varepsilon\to 0} u^{\varepsilon}_{a,\rho} = G(\cdot, a)$$

in $C^k_{\mathrm{loc}}(\overline{\Omega}\setminus\{a\})$ for any $k \le 2p-1$, and the same holds for $u^{\varepsilon}_{b,\rho}$.

Define $w(x) = (x-P)\cdot\nabla u^{\varepsilon}_{a,\rho}(x)$. A simple calculation shows that w satisfies

$$(-\Delta)^p w = (x-P)\cdot\nabla_x \delta^{\varepsilon}_{a,\rho} + 2p\delta^{\varepsilon}_{a,\rho}. \tag{4}$$

Multiply $u^{\varepsilon}_{b,\rho}$ to (4), w to the equation $-\Delta u^{\varepsilon}_{b,\rho} = \delta^{\varepsilon}_{b,\rho}$, subtracting, and integrating on Ω, we have

$$\begin{aligned} &\int_\Omega \{((-\Delta)^p u^{\varepsilon}_{b,\rho})w - ((-\Delta)^p w)u^{\varepsilon}_{b,\rho}\}\,dx \\ &\quad = \int_\Omega \big[2p\delta^{\varepsilon}_{a,\rho}(x)u^{\varepsilon}_{b,\rho}(x) + (x-P)\cdot\nabla_x\delta^{\varepsilon}_{a,\rho}(x)u^{\varepsilon}_{b,\rho}(x) - \delta^{\varepsilon}_{b,\rho}(x)w(x)\big]\,dx. \end{aligned} \tag{5}$$

The LHS of (5) is

$$\begin{aligned} &(-1)^p\sum_{k=1}^{p}\int_{\partial\Omega}\left(\frac{\partial\Delta^{p-k}w}{\partial\nu}\Delta^{k-1}u^{\varepsilon}_{b,\rho} - \frac{\partial\Delta^{k-1}u^{\varepsilon}_{b,\rho}}{\partial\nu}\Delta^{p-k}w\right)ds_x \\ &\quad = (-1)^{p+1}\sum_{k=1}^{p}\int_{\partial\Omega}\big((x-P)\cdot\nabla\Delta^{p-k}u^{\varepsilon}_{a,\rho}\big)\left(\frac{\partial\Delta^{k-1}u^{\varepsilon}_{b,\rho}}{\partial\nu}\right)ds_x \\ &\quad = \sum_{k=1}^{p}\int_{\partial\Omega}(x-P)\cdot\nu(x)\left(\frac{\partial(-\Delta)^{p-k}u^{\varepsilon}_{a,\rho}}{\partial\nu_x}\right)\left(\frac{\partial(-\Delta)^{k-1}u^{\varepsilon}_{b,\rho}}{\partial\nu_x}\right)ds_x \\ &\quad \to \sum_{k=1}^{p}\int_{\partial\Omega}(x-P)\cdot\nu(x)\left(\frac{\partial(-\Delta)^{p-k}G_a}{\partial\nu_x}\right)\left(\frac{\partial(-\Delta)^{k-1}G_b}{\partial\nu_x}\right)ds_x \end{aligned}$$

as $\varepsilon \to 0$ and then $\rho \to 0$.

The RHS of (5) is

$$\begin{aligned} &2p\int_\Omega \delta^{\varepsilon}_{a,\rho}(x)u^{\varepsilon}_{b,\rho}(x)\,dx \\ &\quad + \int_\Omega\sum_{i=1}^{N}(x_i - P_i)\left(\frac{\partial\delta^{\varepsilon}_{a,\rho}}{\partial x_i}(x)\right)u^{\varepsilon}_{b,\rho}(x)\,dx - \int_\Omega \delta^{\varepsilon}_{b,\rho}(x)w(x)\,dx. \end{aligned}$$

Now, integrating by parts, we have

$$\begin{aligned}
&\sum_{i=1}^{N} \int_{\Omega} (x_i - P_i) \left(\frac{\partial \delta_{a,\rho}^{\varepsilon}(x)}{\partial x_i} \right) u_{b,\rho}^{\varepsilon}(x)\, dx \\
&\quad = - \sum_{i=1}^{N} \int_{\Omega} \frac{\partial}{\partial x_i} \{ (x_i - P_i) u_{b,\rho}^{\varepsilon}(x) \} \delta_{a,\rho}^{\varepsilon}(x)\, dx \\
&\quad = -N \int_{\Omega} \delta_{a,\rho}^{\varepsilon}(x) u_{b,\rho}^{\varepsilon}(x)\, dx - \int_{\Omega} (x - P) \cdot \nabla u_{b,\rho}^{\varepsilon}(x) \delta_{a,\rho}^{\varepsilon}(x)\, dx,
\end{aligned}$$

thus

$$\begin{aligned}
&\text{RHS of (5)} \\
&= (2p - N) \int_{\Omega} \delta_{a,\rho}^{\varepsilon}(x) u_{b,\rho}^{\varepsilon}(x)\, dx \\
&\quad - \int_{\Omega} (x - P) \cdot \nabla u_{b,\rho}^{\varepsilon}(x) \delta_{a,\rho}^{\varepsilon}(x)\, dx - \int_{\Omega} (x - P) \cdot \nabla u_{a,\rho}^{\varepsilon}(x) \delta_{b,\rho}^{\varepsilon}(x)\, dx \\
&\to (2p - N) G(a, b) \\
&\quad - \int_{\Omega} (x - P) \cdot \nabla_x G(x, b) \delta_a(x)\, dx - \int_{\Omega} (x - P) \cdot \nabla_x G(x, a)(x) \delta_b(x)\, dx \\
&= (2p - N) G(a, b) + (P - a) \cdot \nabla_x G(a, b) + (P - b) \cdot \nabla_x G(b, a)
\end{aligned}$$

as $\varepsilon \to 0$ and then $\rho \to 0$. This proves Proposition 3. □

3 Proof of Theorem 1

In this section, we prove Theorem 1 along the same line in [11].

Step 1. We argue by contradiction and assume that there exists an m-points set $\mathscr{S} = \{a_1, \ldots, a_m\} \subset \Omega$ $(m \geq 2)$ satisfying (2). Set $K(x) = \frac{1}{2} R(x) - \frac{1}{2\alpha_0(p)} \log V(x)$.
$P \in \Omega$ is chosen later. Multiplying $P - a_i$ to (2) and summing up, we have

$$\begin{aligned}
&\sum_{i=1}^{m} (P - a_i) \cdot \nabla K(a_i) \\
&\quad = \sum_{i=1}^{m} \sum_{j=1, j \neq i}^{m} (P - a_i) \cdot \nabla_x G(a_i, a_j) \\
&\quad = \sum_{1 \leq j < k \leq m} \{ (P - a_j) \cdot \nabla_x G(a_j, a_k) + (P - a_k) \cdot \nabla_x G(a_k, a_j) \}.
\end{aligned} \tag{6}$$

Step 2. By Proposition 3, we obtain

$$\begin{aligned}
&(P - a_j) \cdot \nabla_x G(a_j, a_k) + (P - a_k) \cdot \nabla_x G(a_k, a_j) \\
&= \sum_{l=1}^{p} \int_{\partial\Omega} (x - P) \cdot \nu(x) \left(\frac{\partial (-\Delta)^{p-l} G(x, a_j)}{\partial \nu_x} \right) \left(\frac{\partial (-\Delta)^{l-1} G(x, a_k)}{\partial \nu_x} \right) ds_x.
\end{aligned}$$

By the convexity of Ω, we have $(x-P)\cdot\nu(x)>0$ on $\partial\Omega$. Also by Hopf lemma, we obtain $\frac{\partial(-\Delta)^{p-l}G(x,a_j)}{\partial\nu_x}<0$, $\frac{\partial(-\Delta)^{l-1}G(x,a_k)}{\partial\nu_x}<0$ for $x\in\partial\Omega$. Thus we see that the right hand side of (6) is positive, and get

$$\sum_{i=1}^{m}(a_i-P)\cdot\nabla K(a_i)<0. \tag{7}$$

Step 3. By assumption, $K(x)=\frac{1}{2}R(x)-\frac{1}{2\alpha_0(p)}\log V(x)$ is strictly convex. Thus, all level sets of K is strictly star-shaped with respect to its unique minimum point $P\in\Omega$. Choose P as the minimum point. Then

$$(a-P)\cdot\nabla K(a)\geq 0,\quad \forall a\in\Omega\setminus\{P\}. \tag{8}$$

In particular,

$$\sum_{i=1}^{m}(a_i-P)\cdot\nabla K(a_i)\geq 0.$$

Now, (7) and (8) leads to an obvious contradiction. Thus we have $m=1$ and the rest of proof is easily done by Proposition 1.

Acknowledgements This work was partly supported by JSPS Grant-in-Aid for Scientific Research (KAKENHI) (B), No. 23340038.

References

1. Adams, D.R.: A sharp inequality of J. Moser for higher order derivatives. Ann. Math. **128**(2), 385–398 (1988)
2. Baraket, S., Dammak, M., Ouni, T., Pacard, F.: Singular limits for a 4-dimensional semilinear elliptic problem with exponential nonlinearity. Ann. Inst. Henri Poincaré, Anal. Non Linéaire **24**, 875–895 (2007)
3. Brezis, H., Peletier, L.A.: Asymptotics for elliptic equations involving critical growth. In: Partial Differential Equations and Calculus of Variations. Progress. Nonlinear Differential Equations Appl., vol. 1, pp. 149–192. Birkhäuser Boston, Boston (1989)
4. Caffarelli, L.A., Friedman, A.: Convexity of solutions of semilinear elliptic equations. Duke Math. J. **52**(2), 431–456 (1985)
5. Cardaliaguet, P., Tahraoui, R.: On the strict concavity of the harmonic radius in dimension $N\geq 3$. J. Math. Pures Appl. **81**(9), 223–240 (2002)
6. Clapp, M., Muñoz, C., Musso, M.: Singular limits for the bi-Laplacian operator with exponential nonlinearity in $\mathbb{R}^4$. Ann. Inst. Henri Poincaré, Anal. Non Linéaire **25**, 1015–1041 (2008)
7. Djadli, Z., Malchiodi, A.: Existence of conformal metrics with constant Q-curvature. Ann. Math. **168**(3), 813–858 (2008)
8. Djadli, Z., Hebey, E., Ledoux, M.: Paneitz-type operators and applications. Duke Math. J. **104**(1), 129–169 (2000)
9. Druet, O., Robert, F.: Bubbling phenomena for fourth-order four-dimensional PDEs with exponential growth. Proc. Am. Math. Soc. **134**, 897–908 (2006)
10. Esposito, P., Robert, F.: Mountain pass critical points for Paneitz-Branson operators. Calc. Var. Partial Differ. Equ. **15**(4), 493–517 (2002)

11. Grossi, M., Takahashi, F.: Nonexistence of multi-bubble solutions to some elliptic equations on convex domains. J. Funct. Anal. **259**, 904–917 (2010)
12. Hebey, E., Robert, F.: Coercivity and Struwe's compactness for Paneitz-type operators with constant coefficients. Calc. Var. Partial Differ. Equ. **13**(4), 497–517 (2001)
13. Lin, C.-S., Wei, J.-C.: Locating the peaks of solutions via the maximum principle II: a local version of the method of moving planes. Commun. Pure Appl. Math. **56**, 784–809 (2003)
14. Lin, C.-S., Wei, J.-C.: Sharp estimates for bubbling solutions of a fourth order mean field equation. Ann. Sc. Norm. Super. Pisa Cl. Sci. (5) **6**(4), 599–630 (2007)
15. Malchiodi, A.: Compactness of solutions to some geometric fourth-order equations. J. Reine Angew. Math. **594**, 137–174 (2006)
16. Martinazzi, L., Petrache, M.: Asymptotics and quantization for a mean-field equation of higher order. Commun. Partial Differ. Equ. **35**, 443–464 (2010)
17. Robert, F., Wei, J.-C.: Asymptotic behavior of a fourth order mean field equation under the Dirichlet boundary condition. Indiana Univ. Math. J. **57**, 2039–2060 (2008)
18. Wei, J.-C.: Asymptotic behavior of a nonlinear fourth order eigenvalue problem. Commun. Partial Differ. Equ. **21**(9–10), 1451–1467 (1996)

MIX
Papier aus verantwortungsvollen Quellen
Paper from responsible sources
FSC® C105338

If you have any concerns about our products,
you can contact us on
ProductSafety@springernature.com

In case Publisher is established outside the EU,
the EU authorized representative is:
Springer Nature Customer Service Center GmbH
Europaplatz 3, 69115 Heidelberg, Germany

Printed by Libri Plureos GmbH
in Hamburg, Germany